London's Global Office Economy

London's Global Office Economy: From Clerical Factory to Digital Hub is a timely and comprehensive study of the office from the very beginnings of the workplace to its post-pandemic future. The book takes the reader on a journey through five ages of the office, encompassing sixteenth-century coffee houses and markets, eighteenth-century clerical factories, the corporate offices emerging in the nineteenth, to the digital and network offices of the twentieth and twenty-first centuries.

While offices might appear ubiquitous, their evolution and role in the modern economy are among the least explained aspects of city development. One-third of the workforce uses an office; and yet the buildings themselves – their history, design, construction, management and occupation – have received only piecemeal explanation, mainly in specialist texts. This book examines everything from paper clips and typewriters, to design and construction, to workstyles and urban planning to explain the evolution of the 'office economy'.

Using London as a backdrop, Rob Harris provides built environment practitioners, academics, students and the general reader with a fascinating, illuminating and comprehensive perspective on the office. Readers will find rich material linking fields that are normally treated in isolation, in a story that weaves together the pressures exerting change on the businesses that occupy office space with the motives and activities of those who plan, supply and manage it.

Our unfolding understanding of offices, the changes through which they have passed, the nature of office work itself and its continuing evolution is a fascinating story and should appeal to anyone with an interest in contemporary society and its relationship with work.

Rob Harris is a consultant and analyst in the commercial real estate sector, where he has spent over three decades advising developers, investors, occupiers and public sector bodies. His interests and experience range from advising occupiers on their use of space to the urban policies that help shape future cities: from the 'workstation to the city region'.

Rob started work at design practice DEGW in the early 1980s, where he contributed to innovative work on new developments including Broadgate in the City of London and Stockley Park, Heathrow. He worked at surveyors DTZ and Gerald Eve in a research capacity, and he was director of research at Stanhope Properties plc in the 1990s. He established Ramidus Consulting Limited in 2003 as a specialist, independent built environment research and advisory business.

Rob has a wealth of research experience that has involved projects throughout the property process, including design, development, management, investment and occupation. He has been involved in establishing and running a number of industry groups, including CoreNet UK, Federation of Corporate Real Estate, Society of Property Researchers and Workplace Consulting Organisation. Rob presents widely on a range of property market issues, and he has published industry reports recently for the British Council for Offices, Corporation of London, Greater London Authority, Investment Property Forum and the Royal Institution of Chartered Surveyors.

London's Global Office Economy

From Clerical Factory to Digital Hub

Rob Harris

Routledge
Taylor & Francis Group

LONDON AND NEW YORK

First published 2021
by Routledge
2 Park Square, Milton Park, Abingdon, Oxon OX14 4RN

and by Routledge
52 Vanderbilt Avenue, New York, NY 10017

Routledge is an imprint of the Taylor & Francis Group, an informa business

British Library Cataloguing-in-Publication Data
A catalogue record for this book is available from the British Library

Library of Congress Cataloging-in-Publication Data
Names: Harris, Rob (Consultant), author.
Title: London's global office economy : from clerical factory to
    digital hub / Rob Harris.
Description: Abingdon, Oxon ; New York, NY : Routledge, 2021. |
    Includes bibliographical references and index.
Identifiers: LCCN 2020046645 (print) | LCCN 2020046646 (ebook)
Subjects: LCSH: White collar workers—England—London—
    History. | London (England)—Commerce—History. | Office
    practice—England—London—History.
Classification: LCC HD8039.M4 G699 2021 (print) | LCC HD8039.M4
    (ebook) | DDC 331.7/9209421—dc23
LC record available at https://lccn.loc.gov/2020046645
LC ebook record available at https://lccn.loc.gov/2020046646

ISBN: 978-0-367-65529-7 (hbk)
ISBN: 978-0-367-64672-1 (pbk)
ISBN: 978-1-003-12996-7 (ebk)

Typeset in Bembo
by Apex CoVantage, LLC

Contents

Foreword
London's Global Office Economy
From Clerical Factory to Digital Hub

This is much more than a stimulating book on London's office economy. Though focussed on London it provides examples that are globally relevant for all to reflect on as we confront the changing nature of real estate, work and livelihoods.

In his exploration of evolutions in the London office economy the author skilfully weaves a historical narrative describing the changing roles and aspirations of the key players in the development of the modern office including developers, planners, users and owners.

The story begins with counting houses and exchanges, then located in a domestic setting embedded into the day-to-day life of the city. Office work as a distinctive component of the economy and the office building in which it took place emerged as office work became a definer of economic success and office buildings shaped the spatial structure of the City. The office economy grew and changed from the *clerical factory* (1830-1920), to the *corporate office* (1920-1970s) to the *digital office* of today (1980 onwards)

The desk top computer (IBM PC 1981) brought office users and Information and Communication Technology (ICT), in the process becoming an icon of the digital era. By 1990 technology had moved from desktop to laptop and palm with ubiquitous wireless connectivity. The commercialisation of the World Wide Web (Wifi 1997) and internet browsers paved the way for Apple's smart phone (2007), integrating information and communication, to become the icon of the *networked economy*.

The recession of 1989-92 triggered a reappraisal of the office economy. DEGW's move to new offices in redundant warehouses at King's Cross (1991) became an exemplar of *New Ways of Working* and a catalyst for urban transformation. The working environment was designed to support the way the practice worked across multiple locations in different countries. It reflected what we had learnt in the 1980s and how many global clients were restructuring their businesses driven by increased business pressures. In such tech enabled organisations as many as 50% of staff were mobile workers supported by laptops and cordless phones working in an array of settings and across multiple time-zones.

In the early 21st Century understanding grew amongst progressive global companies, developers and their professional advisers of the demand for building and management systems that responded to occupiers' continuously changing requirements. New environments were needed with digital technology creating parallel environments of physical and virtual space that challenged conventional institutional structures and business models.

To meet the demands of the digital era would require radical systemic thinking to change approaches in the following ways.

- Reappraise perceptions of flexibility and adaptability merging functional, organisational, financial and spatial demands into a holistic response.
- Create a built environment sector that places the interests of demand ahead of the supply process, providing occupiers with service and flexibility.
- Recognise financial value is only one of several values be considered. Long term value being less in the *tangible*, bricks and mortar and more in the *intangible* qualities of expression through design, experience of place and customisation of service.
- Reflect a paradoxical world where expectations are no longer binary as this or that. In a diverse world design must accommodate occupiers in space that is both accessible and secure, private and communal and independent yet collaborative.

Post the Millennium two events changed societal perceptions fundamentally. The Indian Ocean Tsunami (December 2004) raised public awareness of the power of nature, climate change and the future of the planet. Then the World Financial Crisis (2007–08) brought into question the market economy, based on unsustainable consumption and growth financed by credit. Confronting these two issues requires sets of values which cross political boundaries. They are also influenced by the growing realisation of the power of Artificial Intelligence (AI) that is transforming office work. Andy Haldane, the Bank of England Chief Economist posits a vision of repetitive administrative jobs being eroded and new skills and roles emerging for: **Heads** lateral thinkers prepared to take leaps of imagination; **Hands** makers undertaking bespoke design and production, **Hearts** providing empathy and building relationships.

Covid–19 has indirectly become a global catalyst in recognising the shortcomings and opportunities in our current approaches. The lockdown of society and the economy has provided a giant experiment in home working, online shopping, local deliveries, and the cashless society. From this ad hoc experiment there is much for the office economy to learn.

Traditionally studies of the office economy have focussed on the workplace as the individual setting of the worker in a stereotype commercial office shell. Less interest has been shown in exploring the role of "office" in moulding the form and economy of urban settlements. The city has spread far beyond the original central city to create polycentric networked metropolitan regions.

Looking ahead what might the future relationship be between, work, location and livelihoods?

As large parts of office work become commoditised and undertaken by Artificial Intelligence, so improving productivity and wealth generation will become focussed on creativity, innovation and quality of relationships. High performing talented individuals working individually or in high intensity teams, with productivity being measured by outcome not hours will become the norm. The role of the office will evolve further to become a place of: **Exchange** between colleagues, partners, and public; **Development** of employee, company and product; **Learning** from colleagues, consumers and competitors through shared experience, discourse and reflection; **Identity** as a symbol of the culture of the organisation and a place to dwell.

The last five years have raised government and community awareness of the disruptive power of the global giants of the networked economy. The World Wide Web launched as an open, shared service free to all has become dominated by the few such as Facebook, with over 2.7 billion monthly users worldwide; Google, with over 4 billion users); and Netflix, today the world's largest media service provider and production company.

Google with its proposed ground-scraper at King's Cross Central could show the way forward as an exemplar for city centre offices in the *Networked Office Economy*. Reflecting the proposals for integrated flexibility developed in the last two chapters of the book, the complex might become a neighbourhood within the larger district, inviting to its neighbours at the ground floor, and hosting embryo and infant businesses, training and development, as well as meeting, workshop and simulation space, in addition to the core requirements of the company.

Predicting the future, as Rob Harris has shown, emerges from hindsight, an understanding of the past, and sightings from the present. Are we at a point where the office is the city and the city is the office?

John Worthington MBE
Co-founder DEGW.
Inquisitive Urbanist.
Past Director Academy of Urbanism and Commissioner
Independent Transport Commission

Acknowledgements

This book not only brings together a wide range of secondary sources, it also reflects heavily the author's three decades of direct experience in the subject matter. In this respect, I wish to acknowledge an enormous debt to some inspirational people along the way: Dr Francis Duffy and John Worthington as founding partners of DEGW; and Sir Stuart Lipton in his role as chief executive of Stanhope Properties.

A number of people have provided invaluable support during the preparation of this book. I would like to express particular and deep thanks to the following: Amy Auscherman, Michael Bedford, Erik Brown, Mark Catchlove, Ian Cundell, Robert Houston, Chris Kane, Peter McLennan, Nigel Oseland, Jake Sales, and Antony Slumbers.

The author owes particular thanks to Kerry for so much patience during the writing, and for her insightful critiques of draft copy.

1 Introduction

1.1 Aims and perspectives

This book aims to describe the role of offices in cities, their contribution to the economy, the people who build and own them, the organisations that occupy and manage them, and the nature of the space itself as a place of work. It is largely a historical account, reaching back to the origins of the office in fourteenth-century Italy. These earliest buildings retained many features of markets and were meeting places for merchants to buy and sell goods.

The earliest recognisably modern 'offices' in sixteenth-century London resembled houses; indeed, they were called coffee houses and counting houses. These buildings provided a focus for increasingly sophisticated methods of trading that relied less on commodities and metals and more on promissory notes and bonds. Knowledge was emerging as an ingredient of trade alongside the goods themselves.

As the economy industrialised and world trade expanded in the nineteenth century, so specialist office buildings emerged, spread and grew in size. These were 'clerical factories' – full of process, noise, heavy equipment and monotonous work. This was the age of the clerk. It was the era that drew upon a whole new class of worker, and that included women, whose labour had been previously confined largely to factories and domestic service.

Towards the end of the nineteenth century, the American economy vertically integrated, creating much larger organisations whose function was 'managerial' rather than 'productive', resulting in the iconic skyscraper. Europe followed suit in the twentieth century. This was the 'corporate' office, with 'Company Man' – the symbol of armies of similarly dressed workers, undertaking largely repetitive, processed-based work in highly divisionalised, layered companies. The later twentieth century saw the technological revolution and the arrival of the digital office, crammed full of technology supporting densely organised workforces.

The book aims to give a full account of this history, offering the reader an illuminating, interesting and comprehensive perspective on the office – a hugely important influence on modern life. The book closes with some speculation over the future of the office, suggesting that it is moving into its 'fifth age', the network office.

The secret office

One of the compelling aspects of the history of the office is, in fact, just how little is known of it in the collective memory. Offices are such a commonplace feature of modern life – almost one-third of all workers go to work in one each day – and yet, comparatively little is known about their present or past. When urban planning awoke to the scale of office activity in the 1960s, it deemed that offices were a problem to be controlled rather than a critical engine of growth. For urban geographers, until the 1970s, the CBD, or downtown, was generally just a black hole in the city centre where very little apparently changed. Academic studies in economics and geography largely ignored the role of the office at least until the 1980s.

The same neglect of the influence of office activity is also to be found in history texts. Even in London, in many senses the birthplace of the modern office economy, offices are not seen to be a distinctive part of its evolution. Scanning the contents lists and indexes of most histories of London reveals barely any particular treatment of office buildings or the activities therein. Kynaston's four-volume history *The City of London* has no index reference to offices; neither does Ackroyd's *London: The Biography*, Gray's *History of London*, Jenkins's *Short History of London*, Porter's *London: A Social History* or Sheppard's *London: A History*.

Social histories are packed full of factories, farms and mines, but very few offices. Even Pevsner's magisterial *History of Building Types*, the first such survey ever, and which encompassed global architecture, offered no treatment of office buildings as a distinctive building form. Instead, separate chapters are devoted to 'Exchanges and Banks' and 'Warehouses and Office Buildings'.

The mainstream print and broadcast media often alight on 'office stories'. These might involve reporting on innovations in 'green buildings'; the question of economic productivity; 'sick building' syndrome; working from home trends and others. Various sources publish 'best' and 'worst' stories, such as 'the best place to work' or 'the worst building designed'. Most recently, of course, there has been the Covid-19 coverage which has yielded countless stories on a supposed existential crisis for the office. But these occasional forays into the world of the office are rarely given contextual treatment.

This book seeks to redress this neglect of office activity by shedding a beam of light across its history to explain how the office has evolved over time; its role today, and its possible future role. The book seeks to blend technical discussion with historic anecdote to tell the story of the office, to provide the reader with a broad and deep perspective on the emergence of the office as a function of the economy and its impact on all our lives.

An activity-based perspective

The underlying perspective of this book is to view the office economy in terms of the organisations that occupy office space, rather than those who supply it. In tackling the subject in this way, emphasis is placed upon an *activity-based* approach. This contrasts with, for example, investment, development, design or policy perspectives.

It is implicit in an activity-based approach that buildings are not inert and passive but an integral element in the performance of the organisations and individuals that occupy them. Whether the activities perform an economic function, such as with offices, or whether they perform a more social function, such as schools, they provide the settings that can enable activity, or hinder it. They can have a positive effect on the people working within them, or a negative one. They can respond to change, or they can become obsolete. In these ways and more, buildings are the skeletons onto which all the muscular, neurological and circulatory systems of organisations are hung. Described thus, the demand for buildings is derived from the production, organisation and delivery of goods and services. As those goods and services change, so the environments in which they are produced need to change.

Occasionally, change to activities is so profound that it becomes impossible to adapt the existing buildings to the new work practices. This happened during the Big Bang in the City of London in 1986, when newly formed finance houses demanded large, intensively serviced buildings with dealing floors the size of football pitches. The same has occurred in logistics in recent years: here demand has driven the appearance of vast distribution centres, some the size of small farms, where robotic machinery whirs and clunks in total darkness without the assistance of human input. In both cases, it was the nature of change within the businesses, or activities, that drove a demand for a new kind of space to accommodate different work processes.

The original inspiration for an activity-based perspective on real estate, and on offices in particular, was provided by John Rannells in his 1956 masterpiece *The Core of the City*. He placed the spotlight on an activity-based approach to understanding buildings, arguing that the renewal of the built environment should be driven by the changing demands of its users:

> It is necessary to take in somewhat more than relates directly to land use, since changes in demand for space or location may come about as secondary results of business decisions or consumer preferences which, in themselves, have no concern with questions of land use.[1]

Rannells also emphasised the fact that the nature of activities within buildings undergoes a perpetual process of change, as new assortments of occupiers "are always being formed by the continually changing ways of doing business which are characteristic of commercial enterprise".[2] In other words, there is a direct relationship between the changing nature of work and the settings needed to support that work, leading to a constant process of obsolescence and replacement.

London as a backdrop

The story told in this book takes place largely with London as a backdrop. It is recognised that a more international perspective could have been taken, but the story as told is so rich and full of interesting, thought-provoking material that the temptation to take in even more has been resisted. London contains one of the largest, most sophisticated office economies in the world; speculative, purpose-built offices first appeared in London, and during the later nineteenth century, London became one of the first 'global cities' of the modern era. London's heritage as the context to a history of the office economy is unparalleled.

In using London as a backdrop, the story told is grounded in real events with real people rather than in the realm of theory, helping to convey the richness of a fascinating story. Emerging in the nineteenth century as a global trading city, London has since continually reinvented itself, thereby providing a fabulous laboratory for the study of the changing nature of real estate, the office and the ever-evolving nature of the activities that it accommodates.

1.2 What is the office economy?

Some etymology

The following chapter describes the evolution of the office from earliest times to the present. But before doing so, this chapter seeks to define the office economy. We begin with some etymology: first of 'office' and secondly of 'economy'.

The modern word 'office' is rooted in an Old French contraction of two Latin words, *opus* (or work) and *facere* (to do), resulting in *opificium*, or 'work-doing', later to become *officium*. This translated into Middle English as *office*. From the early fourteenth century, the term was used to refer to ecclesiastical and civic roles; coming to mean a building or a room before the fifteenth century. The term has been used interchangeably to mean a role and a place ever since. The root of 'economy' goes back even further.

Xenophon was a prolific Greek historian and philosopher whose single-most relevant work here was *Oikonomikos*. According to Backhouse, this term is best translated as estate manager or estate management. Literally, 'oikos' is Greek for 'household' (by extension, an estate). The book itself was a treatise on managing an agricultural estate.[3] It is a curious fact that, during much of the twentieth century, 'estate management' was something in which a university degree could be obtained before embarking upon a career in surveying.

So, the office economy in the title of this book has both Latin and Greek roots, one emphasising the doing of work and the other the wider setting within which the work is done. This is a helpful way of summarising the office economy. However, in defining the office economy in modern times, we need more precision. The office economy can be defined in terms of an amalgam of four key features: economic progress, sectoral activity, jobs and floorspace.

Economic progress, sectors, jobs and floorspace

In terms of economic 'progress', economists tend to categorise economic development in terms of a number of stages. This is not fail safe, and it is open to criticism in terms of making assumptions about what 'progress' actually means. However, in broad terms, across millennia, agrarian economies generally gave way to industrial (or manufacturing) economies, which in the twentieth century yielded to post-industrial economies (along with enormous temporal variation around the world).

The most recent period has attracted a number of appellations, but its key attribute is a dominant role for 'service' activities (both personal and business-to-business services), and this is related to a rapidly expanding number of workers engaged in 'knowledge work'. The knowledge economy is broadly synonymous with the 'office economy' in the title of this book because the majority of knowledge workers work in offices. However, knowledge workers also work in laboratories, health institutes, teaching facilities and so on.

Economies comprise a range of 'sectors', including finance, manufacturing, retailing, transport and wholesaling. Some of these have offices as ancillary uses (for example, offices in warehouses or in manufacturing plants); while others occupy offices exclusively. The office economy is thus a subset of the service economy, of producer services and of the knowledge economy and is uniquely characterised by its occupation of commercial office buildings. In this respect, the 'core' of the office economy includes accountants, advertising agencies, insurance agencies, lawyers, public relations firms and real estate firms. These firms, or sectors, are characterised by their predominantly business-to-business nature and by the fact that they occupy *only* office buildings.

But, of course, the office economy is much broader than these sectors. Banks are also central to the office economy. Investment banks qualify for the reasons given previously; but retail banks also qualify, even though they have a portfolio of high street outlets for business-to-consumer activity. The sheer scale of their headquarters and administration activity qualifies them for inclusion.

The technology sector also qualifies as part of the office economy. Hardware firms are part of the office economy (despite their manufacturing plants); software firms occupy mostly offices, as do film, video and games firms, although they will variously also have studios and light industrial space. In the media sector, publishers are part of the office economy, despite their printing plants.

Some manufacturers also qualify to the office economy. Take, for example, consumer products conglomerate Unilever, manufacturer of numerous food, drink and household goods brands. It is obviously a large manufacturer, but it also has a large office footprint, where it houses the usual blend of management, administration, finance, marketing, sales and other office-based functions.

The examples given here are private sector organisations, yet the public sector is a very large consumer of office space. Central and local government and the civil service, higher education, the health sector, police and transport are among the largest office occupiers. The government's Central Civil Estate occupies nearly 16 million sq ft (1.5 million sq m) in London.[4]

In the UK, jobs data are collected according to the Standard Industrial Classification (2007) which includes five sectors which can be said to account for the great majority of office jobs: Information and Communications; Financial and Insurance; Professional, Scientific and Technical and Real Estate; Administrative and Support Services, and Public Administration.

There will of course be office jobs in other sectors that take place within office environments – manufacturing and retail headquarters, offices of utilities companies, offices within universities and hospitals. Nevertheless, the five employment categories capture the vast majority of office jobs with a reasonable degree of accuracy and clarity (Figure 1.1). The number of office economy jobs has risen from 3.8 million in 1978, to 6.6 million at the turn of the century, to 9.6 million in 2020.[5] This equates to a 250% growth in office jobs over four decades.

In 1991, the number of jobs in primary and manufacturing sectors (maker jobs) was exceeded by jobs in the office economy for the first time. This was a defining moment in the structure of the economy, representing the final step in the transition from a manufacturing economy to a service-based economy. In simple terms, after 1992, more people were going to an office each day than were going to a factory. And in the nearly three decades that have followed, maker jobs have shrunk by 44%, from 5.7 million to 3.2 million; while office economy jobs have doubled, from nearly 4.8 million to 9.3 million.

It appears then, that the office economy accounts for very nearly 10 million jobs in the UK. This is likely to be an underestimate, given the presence of office jobs in economic sectors not included within the five sectors described earlier. In round terms, this is one-third of the entire workforce (c32 million jobs). In London, the office economy accounts for around 2.5 million jobs (c42% of the total). The office economy in these terms is a fundamentally important component of the national and London economies.

Perhaps one of the simplest and most direct measures of the office economy is the amount of space that it occupies. Again, government data provide a helpful guide. Figure 1.2 shows the growth of office floorspace in England and Wales over four and a half decades from 1974 to 2019. The amount of office space has more than doubled over this period, from c440 million sq ft to around 946.3 million sq ft (40.9–87.9 million sq m). London accounts for 282.8 million sq ft (26.3 million sq m), or almost 30% of the national total.

The vigorous growth in floorspace during the 1990s, following deregulation of financial services and the rapid spread of digital companies, is particularly noticeable. The slight dip over the most recent quintennial is unexplained in official data but is likely related to the extension of Permitted Development Rights in 2013, allowing offices to be converted to residential uses without planning consent. But it does raise the intriguing question of whether the office market has reached its maximal phase of growth.

To most people, around 950 million sq ft is a fairly abstract number. So, to provide a sense of scale, we need to find a comparator. Most people are familiar

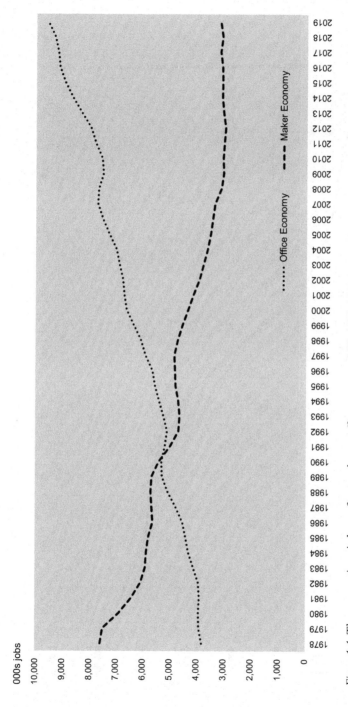

Figure 1.1 The economic switchover from maker to office economy, UK, 1978–2020

Source: ONS (2019).

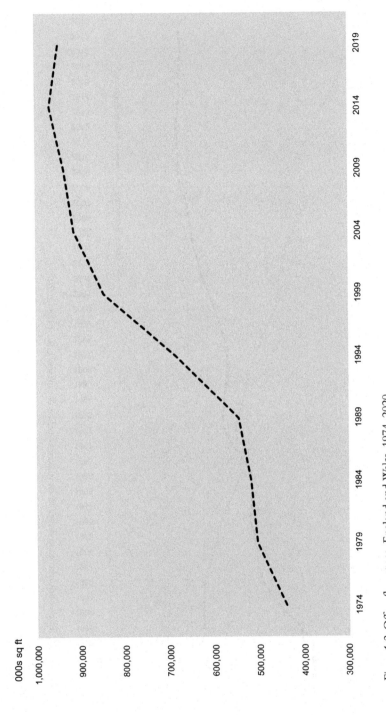

Figure 1.2 Office floorspace, England and Wales, 1974–2020

Source: Valuation Office Agency.[6]

with a soccer pitch, which typically measures 76,900 sq ft (7,140 sq m), or 1.76 acres (0.714 ha). On this basis, the c950 million sq ft (c88 million sq m) of offices is equivalent to around 12,000 soccer pitches.

1.3 The office economy: place and activity

Following the markets and counting houses, which were prototype offices, the earliest offices of the modern office economy were those occupied by the banks and insurance companies in the early 1800s. These were then joined in the twentieth century by brokers, accountants, lawyers, real estate advisors and advertising agencies. The growth of central and local government expanded the office sector greatly. Later still came management consultants, commercial architects, technology and communications firms and engineers. Also, from the 1980s, the digital and creative industries exploded onto the scene.

Until very recently, 'productive industries' formed the beating heart of the economy; office activity was seen to be an ancillary function; something that happened as a spin-off from manufacturing and other commerce, rather than as an economic motor in its own right. Indeed, our understanding of the role of offices in the economy, in society and in urban development was virtually non-existent until the 1950s and not properly explored until the 1970s. As Peter Daniels, one of the most incisive and prolific writers on the geography of office and service activities, opined in the introduction to a chapter on service industries in a book on industrial geography, it is perhaps surprising

> to discover a contribution on service industries in a text devoted to progress in industrial geography. For some reason the latter is invariably equated with manufacturing and most of the textbooks in economic geography during the last 15 years do little to dispel this notion.[7]

To some extent, economic and geography texts have been a little more helpful in terms of putting some context to the term 'office economy'. As the economy evolved in the later nineteenth century, office activities separated from production and other commercial activities at two levels. First, within individual firms, the activities of administration became more distinct and required a more tailored physical environment. As Cowan *et al* described:

> The financial and administrative arrangements necessary for rapid industrialisation are formidable, and all the apparatus of distribution and marketing requires a great deal of co-ordination and control.[8]

Secondly, within the city, office activities became increasingly concentrated within the centre leading, in US terms, to downtown or, more academically, the central business district (CBD). And, as Mumford so elegantly explained:

> no great corporate enterprise with a world-wide network of agents, correspondents, market outlets, factories and investors could exist without

relying upon the services of an army of patient clerkly routineers in the metropolis: stenographers, filing clerks and book-keepers, office managers, sales managers and their varied assistants right up to the fifth vice-president whose name or OK sets the final seal of responsibility on an action.[9]

These buildings were symbolic of the rise of corporatism and corporations, sprawling international organisations that sought to gather together thousands of workers in single-headquarter buildings for command-and-control management. These vast, unchanging and predictable 'corporate islands' planned ahead with a comparatively high degree of certainty, providing jobs for life.

But then this model was shaken to its core by the emergence of the 'knowledge economy', the 'knowledge worker' and the digital workplace. The world of office work – and the economy generally – changed, and are changing, rapidly and profoundly. The ubiquitous impact of technology, and its ability to both destroy and create jobs, lies at the core of the changes; while economic, social and environmental forces are also at work. The rise of 'flexible space', and real estate as commodity, reflect changes in wider society for on-demand products and services.

The recent changes have transformed the office from an inert backdrop to work and an expensive overhead into a *business driver*, or a strategic corporate resource; to be managed and deployed like other corporate resources – people, capital and equipment. Today's office workplace is seeking to integrate technology, place and people in a more seamless manner, providing workers with concierge-type services, providing for wellbeing and productivity and being managed to support corporate agility and enable change. It now provides a hub and connector, conveying the corporate brand and values, and providing a palette of settings for interaction, collaboration and innovation. In short, the office is moving from a 'castle' model to a 'condominium' model. And the story of the transformation is a fascinating one.

So, the office is both a place and an activity. While we have the quantitative measures described earlier, our activity-based approach describes the office in more qualitative terms. In these terms, the office economy firms include those which have a blend of a majority of the following features:

- knowledge-based work;
- technology-enabled, 'agile and connected', workstyles;
- networks of relationships replacing the 'corporate island';
- large and small firms undertaking fundamentally the same work;
- occupation of purpose-built office accommodation;
- office work being physically separated from any manufacturing or other activity;
- tendency to concentrate spatially, normally but not exclusively, in downtown, and
- mostly business-to-business activities.

In these terms, the office economy is more than a building type, more than an employment type and more than a locational characteristic. As with our Greek and Latin roots, it is a blend of place and activities.

1.4 This book

The audience

In his analysis of the Enlightenment, the great naturalist Edward O Wilson observed that the main branches of learning emerged in their present form – natural sciences, social sciences and the humanities – out of a unified Enlightenment vision that searched for an ordered, intelligible universe. However, as scientific knowledge expanded, reductionism, its key method, pushed thinking in the opposite direction. Consequently, scientists have become so professionally focused that it is "not surprising to find physicists who do not know what a gene is, and biologists who guess that string theory has something to do with violins".[10]

Similarly, within the real estate industry, there are many specialists whose interests and expertise are minutely focused. By contrast, this books seeks a synthesis. The book's preparation has involved drawing upon a wealth of ideas, information and data from numerous and diverse sources. This has been a conscious attempt to write a text that can both inform *and* entertain. Furthermore, it has been an objective to narrate an interesting story while at the same time providing readers with the ability and means to follow up particular lines of enquiry or interest.

For these reasons, the book is aimed at both informed and lay readers. For those informed about the dynamics of the built environment, whether as practitioners or academics, the book will provide a deeper understanding of a neglected part of the built environment. Such readers will find much material linking fields that are normally treated in isolation, in a story that weaves together the pressures exerting change on those who occupy office space with the activities and motives of those who supply and manage it. The book also signposts sources for more specialised writings.

For those readers who are not directly involved in practice or study in the built environment, the book provides an insight into what is now a hugely important aspect of our economy and also explains the history, workings and future of the built spaces. The story of our unfolding understanding of offices, the changes through which they have passed, and the nature of office work itself and its continuing evolution is a fascinating story, and should appeal to anyone with an interest in social and economic history.

A note from the author: experience, interest and objectivity

The author has been involved in the London office market, in a research and advisory role, for over three decades, with the vast majority of that time spent unravelling the dynamics of the office economy. This has been a tremendous privilege because the period coincided with one of the most innovative in the history of the office economy, allowing the author to experience first-hand many of the changes taking place and indeed, seeking in a small way to influence some of them.

The author started work in London, in 1984, with Frank Duffy and John Worthington at design practice DEGW, where he immediately began research relating to the early stages of what became the Broadgate development at Liverpool Street station in the City of London. The time was the run-up to Big Bang, and much of the work involved research to understand the building implications of financial deregulation in the City. His later work involved sector-specific studies in financial services, professional services and media firms. His doctoral thesis was based on the empirical work undertaken at Broadgate (supervised by Nigel Thrift and examined by Pete Daniels, both of whom appear in this story).

At the same time as the Big Bang, the UK planning system was being shaken up with a new Use Classes Order which led to the explosive growth of the business park sector. The author's involvement entailed research into the premises needs of 'high-tech' firms and specifically how these could be accommodated at Stockley Park, Heathrow.

The clients for the Broadgate, Stockley and other work comprised a number of discrete firms, but the linking individual patron was Sir Stuart Lipton, of Stanhope Properties Plc, who perhaps did more than any other developer to promote the importance of understanding customer requirements in the real estate sector.

Following spells in surveying firms, the author joined Stanhope Properties in 1990 to run its research programme, which continued to challenge traditional approaches in the development and construction industry. In 2003, he established Ramidus Consulting Limited, as a specialist, independent-built environment research and advisory business. Ramidus works closely with investors, developers, advisors and policy makers to provide insights into property market dynamics, economic and investment trends, and policy and planning matters. Throughout he has been focused on London and on the office economy.

Given the author's background, it is inevitable that the later years of this historic account are written through his lens of personal experience of real events, and in places somewhat influenced by them. While every attempt has been made to maintain objectivity, it is also hoped that the direct experience gives the narrative an added appeal.

Book outline

The most obvious way to write a history of the office economy would have been to pen a chronological account, working gradually from the fourteenth century and Enlightenment period, through the Industrial Revolution and into the twentieth century, through to the current digital era. However, there are a number of strong themes that benefit from being told as a series of histories within an overall framework, and this is the approach that has been taken.

Recording the office economy. Recording traces the history of the office from its very early beginnings, through the emergence of a recognisably modern office economy in the nineteenth century, and to the rise of the 'corporate' office in the twentieth century. Chapter 2 describes the growing maturity of the office economy, and its coming of age with rise of knowledge work and the digital workplace, which provided the catalyst to a rebirth in London's fortunes in the late twentieth century.

Explaining the office economy. It is a curious fact of history already noted that even as office buildings came to dominate city centres in the nineteenth century, there was only the crudest understanding of the internal dynamics of the office economy in geography, economics and urban planning. For decades, academic analysis of cities barely recognised offices as a distinct facet of urban structure, and when they did, the work was largely descriptive rather than analytical. Chapter 3 traces the office economy through classical economics, spatial explanations and the emergence of an activity-based approach, and the chapter sees the office economy finally recognised as a distinctive facet of urban growth.

Planning the office economy. As described in Chapter 4, urban planning has had an uneasy relationship with the office economy. Indeed, once office activity was recognised as a major force in the economy in the early part of the twentieth century, it was politicised and given particular treatment in economic and planning policies by central, regional and local government. Throughout much of the post-war period until the late 1980s, office activity was seen as a problem, a negative trend, and was subjected to policies aimed at limiting its growth.

Building the office economy. Chapter 5 follows the earliest developers in the seventeenth century, through the grandeur that was the Victorian period, through to the Klondike that was post-war re-building. The development process is full of insights into how the office economy has evolved. The twentieth century has also been characterised by office property booms and crashes, and the chapter describes the 1950s and 1970s booms in detail, together with the lessons learned. We examine the emergence of the developer entrepreneur and the institutions, and the paradox whereby the real customer of the development process is the industry itself rather than those who lease and occupy its buildings.

Chapter 6 begins with the Big Bang boom of the late 1980s and travels through the 1990s crash – perhaps the worst on record. The chapter asks whether the boom-to-bust model is inevitable or avoidable. The second part of the chapter takes a panoramic view of the spatial re-organisation of the office economy in London over the past three decades.

Mediating the office economy. Developers and institutions are principals in building the office economy, and they are surrounded by a moderately

14   *Introduction*

sized army of 'advisors', or professions. These come in various, and increasing numbers of, forms. Chapter 7 focuses on the roles of three in particular: architect, builder and surveyor. What is apparent from the early organisation of these professions is that the key drivers were more social and cultural than functional. The early professions were searching not only for functional identify but also social position and respect. This is critically important because today's built environment activities continue to suffer from some of these socio-cultural origins and constraints, manifested in a highly fragmented supply industry.

The chapter then explores two further aspects of mediation: first, market research, and secondly, the flexible space market. The market research described here is exemplified by the aforementioned collaboration between DEGW and Stanhope Properties, with a focus on work in the City of London. Secondly, the flexible space market has been around since the late 1980s but has evolved rapidly since the Global Financial Crisis of 2008. It is now challenging traditional forms of contract in real estate and is likely to become a greater force in the years ahead.

Working in the office economy. In Chapter 8 we trace the evolution of the office workplace by examining the nature of work and the organisations that occupy office space. We look at the work that they undertake, the workstyles that they adopt and the consequent change in demand for physical space. We examine the evolution of workplace technology, from the typewriter to Twitter. The chapter traces the evolution from corporate workplace to digital workplace and describes the main drivers of change on the office economy in the twenty-first century.

Managing the office economy. The office building has evolved from very simple to enormously complex structures with supporting systems and services. As this evolution has taken place, so the role of management has become critical. Chapter 9 sets out the evolution of the real estate management profession, and particularly its transition over the past three decades from a routine and reactive janitorial function into an integrated workplace resource management role. Over this time, the office has evolved from being a static backdrop to office activities, into a corporate resource which must be professionally managed in order to ensure that it has a direct and positive impact on the productivity and wellbeing of workers. This has led to real estate becoming a critical corporate management function.

Divining the office economy. Peter Drucker made the incisive observation that it is "the nature of knowledge that it changes fast and that today's certainties always become tomorrow's absurdities".[11] In this context, it is with some nervousness that Chapter 10 looks to the future of the office economy. It is probably no exaggeration to state that the office economy faces a more uncertain future than at any time in its modern history. Political, economic, social and technological changes are driving enormous change to disrupt many old assumptions. The final part of the chapter speculates that the next, fifth age, of the office economy might be characterised by the 'network office'.

Possible worlds

Figure 1.3 provides a 'map' of the journey on which this book takes the reader, the five ages of the office economy: the coffee houses and counting houses of the seventeenth century; to the 'clerical factories' of the nineteenth century; to the 'corporate offices' and then the 'digital offices' of the twentieth century, and speculating towards the end on the emerging 'network office'. The map shows some of the events, the innovations and the people that we meet along the way.

The changes that are running through the office economy today represent a period of extreme turbulence. Yesterday's truths, assumptions and norms are becoming tomorrow's archaic practices. And this means that old attitudes and perceptions must change. Where once a building was seen as a castle, it must now be seen as a condominium. It exists to meet a short-term need, and everything around its design, delivery and management must be focused on that need.

The remarkable fact is that the end of the journey shown here has more in common with the early stages of the story than with the stages between. As the twentieth-century geneticist JBS Haldane observed in the final paragraph of his essay *Possible Worlds* (1927):

> I have no doubt that in reality the future will be vastly more surprising than anything I can imagine. Now my own suspicion is that the Universe is not only queerer than we suppose, but queerer than we *can* suppose.

However, we do need to suppose, or at least postulate. We cannot predict the future nature of work and workplaces with certainty; but we can examine the implications of contemporary changes and postulate a possible future.

Notes

1 J. Rannells (1956) *The Core of the City,* Institute of Urban Land Use and Housing Studies, Columbia University Press, New York
2 *Ibid,* p42
3 R.E. Backhouse (2002) *Economics,* Penguin Books, London
4 Cabinet Office (2019) *The State of the Estate in 2018–19,* HM Government, London
5 Office for National Statistics (2019) *JOBS02 Workforce Jobs by Industry* (seasonally adjusted) https://www.ons.gov.uk/employmentandlabourmarket/
6 Valuation Office Agency (2020) *Business Floorspace Tables,* VOA, Table FS3.0
7 P.W. Daniels (1985) Service Industries: Some New Directions. In: M. Pacione (Editor), *Progress in Industrial Geography,* Croom Helm, London, p111
8 P. Cowan, D. Fine, J. Ireland, C. Jordan, D. Mercer, & A. Sears (1969) *The Office: A Facet of Urban Growth,* Heinemann Educational Books Ltd, London, p28
9 L. Mumford (1940) *The Culture of Cities* (2nd edition), Secker and Warburg, London, p227
10 E.O. Wilson (1998) *Consilience: The Unity of Knowledge,* Little, Brown & Company, London
11 P. Drucker (1992) The New Society of Organizations, *Harvard Business Review,* September–October, pp95–104

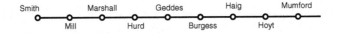

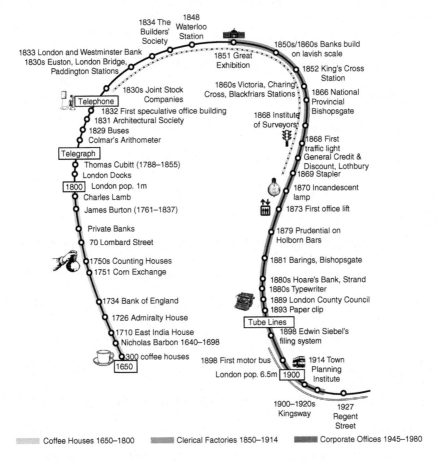

Figure 1.3 The five ages of the office economy

Source: © Ramidus Consulting Ltd.

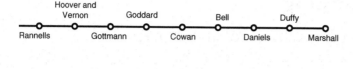

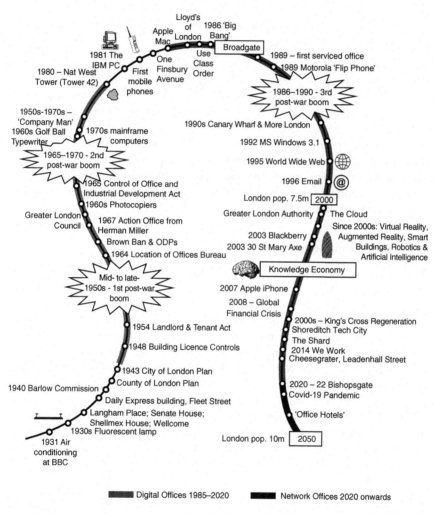

Figure 1.3 Continued

2   Recording

Emerging white-collar factories

This chapter traces the history of the office from its beginnings, through the emergence of a recognisably modern office economy in the late nineteenth century, and to the rise of the 'corporate' office in the twentieth century. The chapter describes the growing maturity of the office economy and its coming of age with the seismic events of 'Big Bang' in 1986 and the ensuing technological revolution which provided the catalyst to a rebirth in London's fortunes and its built environment into the twenty-first century.

2.1 Office ancient history

As much as five millennia ago, the people of ancient Egypt were recording transactions, inventories and accounts and preparing documents. It was the scribes of this time who gave modern people such insight into the workings of that civilisation. The people who learned to read and write were not office workers in any recognisably modern sense of the term; but they were widely employed to record everything from court proceedings, legal matters and medical procedures to stores. Most of what we know today of ancient Egypt is due to the efforts of scribes, armed with their wooden palette, brushes, reed pens and papyrus, Egypt's prototype of paper.

For a more recent step towards office activity as it is understood today, we have to fast forward to Renaissance Europe, at least the fourteenth century. At this time exchanges and banks were emerging. For example, in Bologna there was the Loggia dei Mercanti (dating from 1382) and in Barcelona, the Taula de Canvi (1383).

By the fifteenth century, Italian merchants and bankers were practicing recognisably modern office techniques. In 1494, Italian Franciscan monk and mathematician Luca Pacioli (c1445–1517) published his *Summa de Arithmetica, Geometria, Proportioni et Proportionalita*, the first text to codify double entry bookkeeping; known as the 'Venice system', thereby laying the foundations of the modern art of accountancy.

The book introduced the world to plus and minus symbols. Pacioli did not fit the modern construct of an accountant: he was a 'Renaissance man' with a

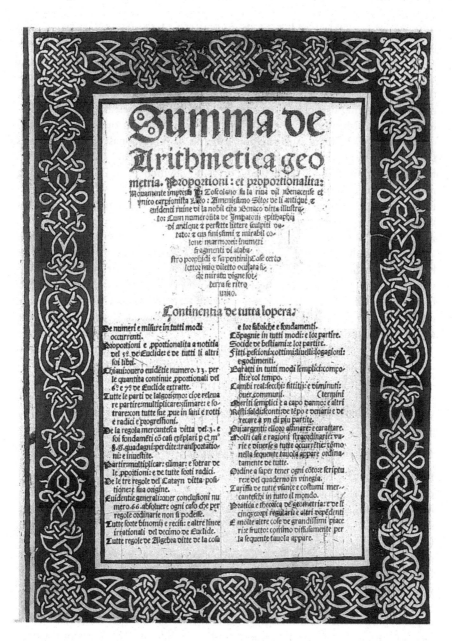

Figure 2.1 Pacioli's Summa de Arithmetica, Geometria

passion for the arts, architecture and astronomy; he socialised with Leonardo da Vinci.

Pacioli was in fact building on his country's role in the development of a banking system. Renaissance Italy was at the forefront of international trade and is often credited with originating banking as we know it today. Affluent trading nodes such as Florence, Genoa and Venice were key cities in the fourteenth century when banking was dominated by the Bardi and Peruzzi families, busily establishing branches elsewhere in Europe. The oldest bank still in existence is Banca Monte dei Paschi di Siena, which has been operating continuously since 1472. But it is to the political and banking dynasty of the Medicis that the origin of office in its modern sense might be traced. Giovanni Medici famously founded the Medici Bank in 1397, which established a counting room, or office, to oversee the affairs of the bank, which sought to introduce a financial system based on bills of exchange. The key development here was that trade, or one aspect of it, had begun to move away from the open market environment of commodities exchange to the private environment of the office.

Coincidentally perhaps, it was also in Italy that one of the earliest office buildings was built. Today's Uffizi Gallery in Florence was originally commissioned in 1560 by the extremely wealthy Cosimo de Medici, to accommodate the offices ('uffizi') of the Florentine magistrates, as well as Cosimo's growing collection of art. He commissioned painter and architect Vasari to design the offices – perhaps the first ever commission to create purpose-built offices. The 13 offices of the Magistrates in charge of overseeing Florentine production and trade were located on the ground floor of the building, while the first floor housed the administrative offices and workshops of the Grand Duchy, which were dedicated to the manufacture of precious objects. Originally, the building was topped by an open loggia, or roofed gallery.

Banks and exchanges

In some senses, Italy stole the march on the development of the office economy. Bills of exchange, accounting, legal structures and banking, even an office building, were all in place by the sixteenth century. Secular public banks also spread: Banco della Piazza di Rialto (Venice, 1587), Banco di Santo Spirito (Rome, 1591) and Banco di Santo Ambrogio (Milan, 1593).[1] But at this time, of course, what we know today as the nation state of Italy was in fact a collection of city states. The unification of modern Italy under one flag and one government, in a social and economic reform known as *il Risorgimento*, would not occur until 1861. Before this, Italy lacked the national ambition for global expansion and trade that was growing elsewhere in Europe, notably in Great Britain, Holland and Spain. It was almost inevitable that the locus of financial power should shift from southern to northern Europe.

The Italian influence was not lost altogether. In the fourteenth century, Bruges became a centre of commerce and trade, strongly influenced by Italian traders. By the beginning of the fifteenth century, Bruges had a fully functioning money market, reflecting the exchange rates of the leading European commercial centres, including Barcelona, London, Paris and Venice. The merchants and bankers met in a square known as the Place de la Bourse – hence the origin of the modern French bourse and the German borse.

While Bruges played an instrumental role in the development of stock exchanges, its influence had waned by the end of the fifteenth century. In its place, Antwerp assumed a growing role in European trade, providing accommodation for merchants and bankers. Indeed, the Antwerp beurs, or bourse, was the world's first purpose-built financial and commodity exchange (the building pre-dated stock trading). Built in 1531, the new Antwerp bourse was a voluminous, rectangular space with covered galleries on all four sides. It became the focal point for European trade: "novel financial instruments evolved in Antwerp to finance the growing volume of international trade in that city over the course of the sixteenth century".[2] This included the foreign bill of exchange.

Antwerp was unable to consolidate its position as a leading commercial centre, but it had established a template for other exchanges, including London, which opened its version in 1565. The building bore a striking resemblance to the Antwerp exchange, which was not surprising given that it was founded by land mercer Sir Thomas Gresham, who had been a royal agent in Antwerp. He described it as a "comely bourse for merchants to assemble upon".[3] It was

Figure 2.2 The Antwerp Bourse

renamed the Royal Exchange following a visit by Queen Elizabeth I in 1571; however, the original building was destroyed by the Fire of London in 1666. Further European stock exchanges were opened in Rotterdam (1599), Amsterdam (1609), Middelburg (1616), Hamburg (1616) Delft (1621) and Nuremburg (1621).[4]

These buildings were markets, or exchanges, more than bank offices as we understand the term today. But they were forebears of the modern office, early prototypes where merchants and bankers congregated for trade, centralised away from the land and the factories but, significantly, adjacent to the key ports. Indeed, in the sixteenth century, Britain had no public bank "or sophisticated bankers, much less a stock of metallic moneys stored in a central secure place".[5]

Before very long, paper was beginning to replace coins and bullion for purposes of exchange. In 1609, and based in the city's town hall, the *Amsterdamsche Wisselbank* (Amsterdam Exchange Bank) pioneered cheques, direct debits and transfers between accounts, issuing paper money to simplify trading.[6] In short order Amsterdam became the world's pre-eminent financial centre through to the Industrial Revolution.

The Bank of England would not open until 1694. Nevertheless, London's office economy was evolving and innovating, especially with the growth of joint stock companies. In comparison with its continental neighbours, London "saw the chartering of a wide range of companies especially during the second half of the sixteenth century",[7] together with a market in company shares. This is how the central market of brokers buying and selling equities emerged, a market that "would grow exponentially in the last decade of the seventeenth century".[8]

The countries of northern Europe were outpacing Italy rapidly in terms of international trade. By the late sixteenth century, the largest commercial centres were Amsterdam, Hamburg and, of course, London.

> In the sixteenth century the foreign trade of England was already assuming a world-wide character . . . exports penetrated into nearly every part of the globe . . . imports comprised the products of Europe, Asia, Africa and America. A large proportion of her population was now dependent for its livelihood upon foreign markets. . . . Her commercial organisation was highly developed on the basis of companies, regulated and joint stock, while the working of the credit system and the foreign exchanges reproduced in its essentials the mechanism of modern business life.[9]

The relationships between the ports and the centres of finance was a symbiotic one. Securing credit and trading on the commodity exchanges, for example, were key functions where the finance and the trade needed proximity. The relationship between the City of London and its docks was a particularly strong example.

The origins of English trade with America and Asia date back no further than the beginning of the seventeenth century, but by the close of the century,

America and Asia were providing one-third of England's imports, and the re-export of their produce accounted for nearly a third of all English exports. One of the most influential companies of the century, the East India Company, was granted a Royal Charter in 1600 to trade in the Indian Ocean, India and the East Indies (more of which follows).

As international trade expanded, so London grew rapidly and became more cosmopolitan. Indeed, during the Tudor dynasty, London experienced unprecedented growth, as people flocked there from around the country and from Europe and beyond in search of work and opportunity. Between 1550 and 1600 the population almost tripled, from 75,000 to 220,000. It then rose to around 400,000 by mid-century, reaching some 575,000 by 1700. At this time, London became, for the first time, the largest city in western Europe, or indeed, the whole of the western world with perhaps the sole exception of Constantinople, which by around 1750 London had also overtaken.[10]

Alongside the success and wealth came poverty and squalor: overcrowding was acute, and for many, sanitation and housing conditions were appalling. In an attempt to alleviate the problems, Queen Elizabeth I passed an Act of Parliament, the Restrictions on Building Act, in 1592, which decreed that "No new buildings shall be erected within three miles of London or Westminster" – a sort of Tudor Green Belt. The same Act sought to tackle overcrowding by preventing dwellings being subdivided – four centuries before the Parker Morris Standards in housing.

But Elizabeth's actions were rear-guard and largely unsuccessful. Living conditions improved barely at all during most of the seventeenth century. Indeed, the plagues of 1625, 1636 and 1665, and the Great Fire in 1666 (that led John Evelyn famously to declare: "London was, but is no more") were all worsened by the sheer density of humanity and the poor building and unsanitary conditions. But despite plagues and fires, London continued to expand ever outwards and at great density, and land speculation began in earnest.

The West End was developed as the well-to-do sought an alternative to the full-blooded commercialism of the City. The intense level of building activity there prompted one anonymous writer to complain about the effects of land speculation:

> The desire of profit greatly increaseth buildings and so much the more, for that this great concourse of all sorts of people drawing near unto the City, every man seeketh out places, highways, lanes and covert corners to build upon. . . . These sort of covetous builders exact great rents, and daily do increase them in so much that a poor handscraftman is not able by his painful labours to pay the rent of a small tenement and feed his family.[11]

Plus ça change, plus c'est la même chose. But despite the increasing rents, growing numbers of people did afford to pay, and by mid-century the population of London had reached nearly half a million. However, despite the best endeavours of the new developers, much of the physical fabric of London remained medieval in appearance.

It was also during this period that London originally discovered the delights of coffee and coffee houses. The first such business was set up on St Michael's Alley, off Cornhill, by the appropriately exotic Pasqua Roseé in 1652. Roseé was a Greek who developed a taste for the drink while working as a servant in Turkey. Being an entrepreneurial sort of gent, he started to import coffee to London and sell to businessmen and others, with great success – selling over 600 dishes of coffee a day. Just as many coffee shops and coworking centres are today used as venues for business, so too was the case over three centuries past. Competition followed, and by 1663 there were 82 outlets in London; while by the turn of the century there were well in excess of 500.[12] It is an interesting fact that that at the same time, there were 207 inns, 447 taverns, 5,975 alehouses and 7,000 vendors of gin![13]

The eighteenth century witnessed boom periods following the Treaty of Utrecht in 1713 and the Peace of Paris in 1763, resulting in Mayfair, Hanover Square, Portman Square and Russell Square and its surroundings. By the end of the century, London had a population of over 900,000 – twice the size of Paris.

Indeed, it is an enduring mystery as to why Paris fell so far behind London at such an early stage of modern trading. One theory lays the blame at the door of a little-known Scottish economist, John Law (1671–1729). Quite a colourful character, Law was sentenced to death in 1694 for killing a love rival with a sword in a dual in Bloomsbury Square. But he absconded and escaped to Amsterdam, before going to France. Eventually, Law persuaded the Banque Générale of the merits of establishing a centralised bank, and in 1716 Law set up the Banque Générale Privée, which developed the use of paper money, as pioneered in England, Genoa, Holland, Sweden and Venice.

Law invested heavily in the Mississippi Company, causing wild speculation: shares rose to sixty times their original value. The bubble burst in 1720, and Law closed the bank and fled the country, leaving behind hundreds of ruined Parisian investors. The experience left the French with a deep suspicion of banking: "The shock kept France from new attempts until in 1776 the Caisse d'Escompte was established".[14] The Banque de France was not founded until 1800, long after the Bank of Amsterdam (1609) and the Bank of England (1694).

The Bank of England sought to raise loans for investment by the government, thereby creating a publicly-financed national debt, and a market for government and bank stock. This in turn saw the "emergence of modern financial capitalism" which "required eye-to-eye deal making", boosting the number of stockbrokers and stockjobbers operating out of London's burgeoning coffee houses.[15]

The first office buildings

It was in the eighteenth century that offices in the modern sense of the term – purpose-built, stand-alone, dedicated spaces where administrative and commercial functions are undertaken – began to appear, with London taking the lead. This was coincident with the rise in banking and insurance services to provide credit and cover for the expanding trade passing through the Port of London. And London's skyline was continuously changing, as it grew into one of the most vibrant and cosmopolitan cities in the world.

The UK's rapidly growing manufacturing base and increasingly global trading role was manifested during this time with burgeoning small office suites in private houses, suitably well-appointed to reflect the status and material success of their owners. The symbols of the city prior to the mid-eighteenth century remained the castle, the cathedral, the palace and the marketplace, and the place of the office remained very small indeed.[16] There were, however, a few notable exceptions.

In the mid-eighteenth century, the growing wealth of the London populace was reflected in the growth of private banks, and banking houses began to spring up. They were of a domestic scale; there were no purpose-built bank buildings at this point. Black describes the banking house at 70 Lombard Street, built in 1757, which had similarities with substantial Georgian town houses:

> The ground floor with strong classical features . . . but no advertising that this was, in fact, a banking house. Nonetheless, the contrast with the domestic simplicity of the upper floors had the useful effect of marking the division between place of business and place of residence within the same building.[17]

Pevsner observed that

> The headquarters of Georgian and even early Victorian banks were exactly like private houses, with the business rooms on the ground floor, left and right of the entrance, and the managers' living quarters above.[18]

Black notes that the tradition of private banker living above the shop continued well into the mid-Victorian decades. He also describes the interiors of typical banking houses:

> Everyday business, such as paying in bills or cashing, would be done at a long mahogany counter dividing the floor of the shop. More personal discussions and financial arrangements, especially those involving an aristocratic clientele, took place in the parlour, a private space usually modelled on a Georgian drawing room.[19]

The title of 'oldest office building', in a modern sense of the term, is a contested one, and one that is unlikely to be solved categorically. On most lists of contenders is *Admiralty House on Whitehall*, today known officially as the Ripley Building. This is the oldest of a complex of buildings including the Admiralty Buildings between Whitehall, Horse Guards and The Mall. The three-storey

U-shaped brick building was designed by Master Carpenter Thomas Ripley and completed in 1726. It is often cited as the first purpose-built office building in Great Britain, and was built to handle the administration of the Royal Navy. However, its only official history describes the building as comprising a ground floor of two halls and three state rooms; a first floor of a library, the First Lord's office and "the state bedroom of the First Lord, a dressing room and a large boudoir for his wife", and second and third floors of "innumerable bedrooms" for family, guests and servants.[20]

Perhaps a more likely contender for 'oldest modern' office sat on the site of what is now the Lloyd's of London building in Leadenhall Street: East India House, built of course by the East India Company. Perhaps its most famous worker was essayist and poet Charles Lamb (1775–1834). And it is the insight that both the company and this one worker provide that paints such a useful description of the office economy of the time.

Lamb's most famous works include *Essays of Elia* and *Tales from Shakespeare*, the latter a children's book co-authored with his sister, Mary. He enjoyed a sophisticated social life, rubbed shoulders with the likes of Coleridge and Wordsworth and was a mainstay of the literary scene in London. But Lamb's life also had a darker side.

Lamb suffered badly from a stutter, which was largely responsible for his leaving Christ's Hospital boarding school in Newgate Street at 14 years of age, instead of going to university. "He was always to regret not having gone to university . . . Instead, his 'university' would be his beloved London".[21]

In 1792, Charles's family suffered a blow when his father John's employment as a clerk to London lawyer Samuel Salt came to an abrupt end with the latter's demise. The family had been living in Salt's house rent free and was compelled to leave with no financial support. Charles, aged 17, was obliged to find employment. This he did by taking a lowly position as a clerk, joining the Accountant's Office of the British East India Company, at East India House.

At 20, Charles met his first love, Ann Simmons, but "her rejection of him . . . was such that it precipitated a fit of insanity" and periods of depression. A year later, in September 1796, his sister Mary "stabbed their mother to death and wounded their father by embedding a fork in his head". To save Mary from permanent incarceration at Islington Asylum, Charles undertook to care for her at home, which he did for the rest of his life.[22] At 21 years of age, the young Charles was a writer, carer – and clerk. He worked for the East India Company from 1792 until his retirement 33 years later.

As noted earlier, the East India Company had been founded in 1600 to oversee all trade with Asia. The company moved its London headquarters to Craven House in Leadenhall Street in 1648 on leasehold, and it eventually purchased the building in 1710. East India House was reconstructed in the 1720s using the site of Craven House and neighbouring properties (today the site of the Lloyd's insurance building), and the new building designed by Theodore Jacobsen was finished in 1729.

In 1796, the company purchased an additional plot of land, and work began to extend its building. The new design and project was started by Richard Jupp and completed by Henry Holland in 1799. The scale of the building was commensurate with the scale of East India Company's trading, which resulted in a huge demand for generating, processing, receiving, despatching and filing vast amounts of paper records. The large numbers of workers created a large and complex bureaucracy. The East India Company was, in many respects, the first global corporation. It was managing trade relationships thousands of miles distant that required several weeks of travel to reach, unlike today's instantaneous communications.

Like many modern global corporations, the East India Company created a headquarters to reflect its dominant trading position (Figure 2.3). Externally, the neo-classical architecture was extravagant, finished off with a dominating six-column portico. The large tympanum that sat atop depicted King George III defending the commerce of the East, with a statue of Britannia crowning the top, and statues of Asia and Europe to the east and west, respectively.

Internally, the building was extraordinary.

> Reflecting the great wealth of the East India Company, no expense was spared on the interior decorations. Palladian architectural features, including Corinthian pilasters and heavy moulding, continued throughout the

Figure 2.3 East India House, Leadenhall Street

interior. Lavish Georgian furnishing including ornately carved boardroom tables and velvet upholstered chairs for the Chairman and Vice-President, were commissioned.[23]

But unlike many modern corporations, working conditions behind the grandeur of the façades and great halls was rather different. Huw Bowen takes extracts from Lamb's correspondence to paint a picture of working conditions in one of the earliest and grandest of offices.[24] During his early years in the Accountants' Office, Lamb often worked hard and late, to the point that he arrived home every night "o'er wearied and quite faint". He found the "strain unbearable at times and he suffered from what today would be regarded as work-related stress. In 1815 he wrote to William Wordsworth:

> On Friday I was at office from 10 in the morning (two hours dinner except) to 11 at night – last night till 9; my business, and office business in general, has increased so.

On 20 March 1822, Lamb was near morose in his latest letter to his dear friend Wordsworth:

> I grow ominously tired of official confinement. Thirty years I have served the Philistines, and my neck is not subdued to the yoke. You don't know how wearisome it is to breathe the air of four pent walls, without relief, day after day, all the golden hours of the day between ten and four, without ease or interposition. *Taedet me harum quotidianarum formarum*, these pestilential clerk-faces always in one's dish. Oh for a few years between the grave and the desk! They are the same, save that at the latter you are outside the machine.[25]

The Latin element of this quote translates as: *I am tired of these everyday forms*. It is also clear that as well as the long hours, management-worker relations were frayed. For example, Lamb fumed at a top-down initiative to cut costs, including retraction of the annual £10 holiday allowance; turning Saturday from a half-day to a full working day; taking away holidays on Saints' Days; taking away the perk of sending private letters postage free from India House and, perhaps most resented of all, the introduction of a clocking in/out system.

The contrast between Lamb's two pre-occupations – literature and ledgers – could not have been starker. Little wonder he was so demoralised by the latter. Lamb likened his retirement in 1825 to "emancipation from slavery".

But the writings of Charles Lamb also tell that the office was evolving. They were growing larger and more complex; they were becoming 'machines'. Hours were long, and the work was tedious and repetitive. Hierarchy was rigidly observed, with those at the bottom intensively managed hired hands. The external image was very different to the internal reality: grand architecture and images stating global role and reach, while internally cutting petty perks to make marginal savings.

2.2 The birth of the office economy

As the tentacles of European trade stretched out across the globe, London's international role, and the country's transformation into the greatest trading bloc the world had ever seen, was exemplified by the development of the docks in east London. Although they had been expanding for several decades, it was during the nineteenth century that the docks' full commercial potential was realised with the opening of purpose-built, enclosed docks. The West India Company opened the first large, enclosed dock (including Canary Wharf) in 1802; followed in quick succession by the East India (1805), the London (1805) and the Surrey (1807); and then later by St Katherine's (1828), Royal Victoria (1855) and Royal Albert (1880). By mid-century, half of Britain's trade passed through the port of London. Of course, the rapid expansion of the docks was driven by Britain's growing global influence, as suggested by the various specialities of the docks: softwood at Surrey; grain at Millwall; wool, sugar and rubber at St Katherine's; ivory, spices, coffee and cocoa at London; rum and hardwood at West India dock.

London's growing population and workforce also required new infrastructure, and more sophisticated means of travel began to emerge. Just east of today's Edgware Tube Station, where the elevated Westway becomes Marylebone Road, there is a small, non-descript cul-de-sac called Shillibeer Place. The street name commemorates the man who brought what we now call buses to London's streets. Born close by in St Marylebone in 1797, George Shillibeer (1797–1866) spent his early career in the Royal Navy before joining coach-builder Hatchetts in Long Acre. He was then offered work in Paris in 1825 to

Figure 2.4 George Shillibeer's first omnibus

build a horse-drawn vehicle, significantly larger than a typical stagecoach. His vehicle, capable of carrying over 20 people, went into service in Paris in 1827.

On returning to London, Shillibeer set up business on Bury Street to build a new omnibus, later referred to as a 'bus'. His first vehicle went into service on 4 July 1829, drawn by three horses and with a capacity of 22 people, between Paddington and the Bank of England. So began London's bus history. Today there are nearly 9,000 buses, operating on 700 routes, with a combined 19,000 stops.

By the early nineteenth century, office buildings began to take on their non-physical role – that is, to reflect the status and power of their occupiers. More and more banks, insurance companies and related businesses were seeking offices with a prestigious address in the city. One such example was architect Charles Robert Cockerell's Westminster, Life and British Fire Office on the Strand (1831), reputedly the oldest insurance office in the world. Cockerell (1788–1863) was also responsible for other 'named' buildings, including London and Westminster Bank, Lothbury (1837), Sun Fire Office, Threadneedle Street (1841) – all in classical style. Cockerell also replaced the ageing Sir John Soane at the Bank of England in 1833 (see Chapter 6), overseeing the building of its branch offices in major English cities including Liverpool and Manchester.

Some bank buildings were also becoming more ostentatious. Black describes, for example, The City Bank, which assembled a site in Threadneedle Street and then appointed architect William Moseley (1799–1880) to design the new office. At a cost of £31,145 including purchase of leaseholds, building, fixtures and fittings, Moseley produced

> an imposing Italianate palazzo design, distinguished by its rounded corner entrance with a heavily rusticated ground floor and a deeply projecting cornice at roof level. . . . The palazzo style allowed richness of detail and a conscious expression of symbolic capital without the loss of propriety necessary to convey the safe conduct of banking business.
>
> Contemporary commentators also drew attention to the refined taste and craftsmanship displayed in the cabinetwork and ornamental fittings which, when combined with the feature dome and slender granite columns, gave a real richness of effect to the interior.

Black goes on to explain that in common with many other new joint-stock banking companies, The City Bank was

> involved in the production of new symbolic spaces and sought to create a distinct identity for a new and more public form of the money economy. Palatial style and grandeur offered a distinct break with the Georgian restraint of the City's private bankers, signalling to the new metropolitan middle class an attractive combination of both innovation and security in their money transactions.[26]

Black states that "It was indicative of the pace and scale of building activity in the City's central financial district that designs which would have amazed just 30 years earlier were becoming almost commonplace".

Speculative building

It was during the mid-nineteenth century that multi-occupied office buildings emerged in number. As noted earlier, up to this point the norm was for offices to be accommodated on the lower level of a building with living quarters above. Such buildings were built primarily for the occupier, but then 'speculative' development began to occur with buildings built for the open market (described more fully in Chapter 5). One designer of such buildings was Edward l'Anson (1812–1888), the second of three generations of architects spanning the nineteenth century.[27]

L'Anson was born in St Laurence Pountney Hill in the City of London and educated at the Merchant Taylors' School. Following a two-year tour of Europe, l'Anson set up a practice in the City and developed a reputation for designing commercial buildings. His first major building was the Royal Exchange Buildings, designed for Sir Francis Graham Moon in 1837. L'Anson was eventually president of both the Royal Institute of British Architects and the Surveyors' Institution. L'Anson delivered many papers to the RIBA, but in an 1864 paper, *Some Notice of Office Building in the City of London*, he made particular reference to purpose-built office buildings. Pevsner relates how he told his audience that within his recollection

> "merchants dwelt in the City over their counting houses and next to their warehouses", but that thirty years ago, ie, in the 1830s, "certain houses [were] let out in separate floors and used as offices". . . . "The first building which I remember to have been erected for that special purpose was a stack of office buildings in Clement's Lane at the end nearest to Lombard Street." This dated from about 1823 and was by an architect called Voysey.[28]

From this we can determine that the first speculative office building in London (and the world) was built in the City of London in 1823. The process found favour, and before long, speculative building had become the norm. Kynaston refers to the burgeoning specialist offices from mid-century.

> City men also luxuriated in purpose-built office blocks, king-size warehouses and new, gleaming railway stations. Such office buildings were not new, but from the 1850s they started to mushroom almost everywhere, many in the vicinity of the Stock Exchange in order to house stock-broking and jobbing firms. One, built in 1854, was at 5 Throgmorton Street. Owned by the two leading brokers James Capel and John Norbury, this housed not only their own firm but also several others. Another,

constructed by the mid-1860s, was at Warnford Court just off Throgmorton Street, where one of the early tenants was the new firm of de Zoete & Gorton.[29]

Funding arrangements; construction contracting and intermediation all evolved quickly, and the West End and the City began to take on their modern shape in bubbles of property development. Some of these bubbles created whole new districts and were distinctive not only in a temporal sense, but also in their architectural footprint. The emergence of Regent Street, with the architecture of John Nash and James Burton, and the patronage of George, the Prince Regent (later George IV), is perhaps one of the most distinctive. John Blashfield built Kensington Palace Gardens in the mid-1840s, and John Duncan and Jacob Connop transformed the area that became Ladbroke Grove.

London was emerging as Europe's first 'global city'. Nowhere was this more visible than at the Great Exhibition in 1851. Designed by Joseph Paxton, and built in the centre of Hyde Park, the Exhibition boasted 100,000 exhibits from the farthest reaches of the Empire, as well as examples of technological advancement much closer to home. It attracted six million visitors in its short five-month life. Moreover, London was booming. Between the Great Exhibition and the eve of the First World War, London's population doubled from 2.4 million to around 4.5 million. This growth in people saw the physical fabric of the city expand to cover a larger area and to intensify in the central area.

It is remarkable that this rapid physical growth of Victorian London took place without the slightest hint of a strategic plan and virtually no public planning: the development of London took place in a truly *laissez faire* manner, entirely the result of private capital and private enterprise. The overriding and continuing influence of the private sector is shown by the fact that the opening of Kingsway and Aldwych remained virtually the only major twentieth-century public improvement in central London until the Hyde Park Corner underpass was built in 1962.[30] London's first directly elected government, the London County Council, was not to appear until 1889. Even then, although the Council presided over the whole urban area, except for the City of London, it lacked the strategic powers to have any great impact on the *laissez faire* development of the city. It had no strategic planning powers, it had no social welfare responsibilities and its housing responsibilities were limited to slum clearance.

"Nowhere was the new spirit of laissez faire more devastatingly illustrated than in the field of transport", observed Simon Jenkins, and the coming of the railways from the mid-1830s "shattered both the living and working arrangements of hundreds of thousands of Londoners".[31] The 1830s saw the arrival of Euston, Fenchurch Street, London Bridge and Paddington, while Waterloo opened in 1848 and King's Cross in 1852. The 1860s saw a frenzy of station openings, including Victoria (1860), Charing Cross (1864), Blackfriars (1866),

Cannon Street (1866) and St Pancras (1868). The location of the stations, away
from the central area, was no accident:

> the unwillingness of the wealthier inner London estates to permit railways
> to run through their land, meant that by the mid-fifties London was circled
> with termini each situated at some distance from the others and few . . .
> coming near the main employment concentrations.

According to Jenkins, the areas around the new stations became "appallingly
congested", and in 1855 "Parliament was told by a Select Committee that
more direct lines of communication should be established . . . between the
several principal points of the Metropolis". This led directly to London's (and
the world's) first underground railway line, the Metropolitan Railway. The line
ran between Paddington and Farringdon and opened in January 1863. It used
steam locomotives to pull gas-lit wooden carriages and carried 38,000 passengers on the opening day.[32] The Metropolitan District Railway quickly followed
in 1868 (South Kensington to Westminster), as part of a plan for an underground "inner circle" connecting London's main-line stations.[33] The Metropolitan and District railways completed the Circle Line in 1884.[34] The first
tunnels were shallow, constructed using the 'cut-and-cover' method, before
yielding to deeper, circular profile tunnels (from which was derived the appel-
lation 'tube'). Today, the system includes 250 miles of track, with 11 lines
linking 270 stations, handling five million passengers daily[35] and 1.35 billion
passengers annually.[36]

It was to be another hundred years following the completion of the Circle
Line before many of the rail termini were to realise their full contribution to
London: by providing the land for a necklace of office mega schemes around the
central area: Broadgate (Liverpool Street), King's Cross, Marylebone (Regent's
Place), More London (London Bridge), Paddington Central and Victoria.

Financial capital of Europe

During the first half of the nineteenth century London was established as
the financial capital of Europe and the centre of world trade. Its growth was
shaped – physically and economically – by a generation of international financiers, many of whose names in banking live on into the twenty-first century.
One of the most influential of these was Nathan Rothschild, a young German
immigrant who, following a run on gold and an economic crisis in London
in 1825, saved the Bank of England by organising a £10 million shipment of
gold to shore up its reserves. Rothschild was followed by a number of other
immigrants whose names came to symbolise merchant banking in the UK and
whose work and reputations established the City as a truly global financial centre. Many came from Europe, including Hambro, Kleinwort, Lazard, Schroder
and Seligman, while others came from the USA. One such was Peabody, who

emigrated from Baltimore in 1831 to found the firm that was to grow and split into JP Morgan and Morgan Grenfell. Such individuals helped cement London's role as a truly global city.

Nevertheless, the banking sector was some way behind the insurance sector. This was due partly to the intransigence and raw power of the Bank of England, and partly to a Victorian liberal view that limited liability companies were somehow more profligate and that family-owned businesses were somehow 'purer'. As a result, the banking sector was composed initially of only private banks, which were private partnerships rather than joint stock companies. This placed severe constraints on their ability to grow. As promissory notes became transferable to a third party, and as paper money became more common with Bank of England notes, private banks began to formalise procedures.

In the 1830s, banks were permitted to become joint stock companies in London, for deposit only and not note issue (issuing banks were permitted only within a 65-mile radius of the City). But even then, their numbers were few. The first joint stock company bank to open in the capital was the London and Westminster Bank, following the legislation of 1833, opening for business on 10 March 1834, with a City office at 38 Throgmorton Street and a West End branch at 9 Waterloo Place.[37] Before long the bank had outgrown its City office and had

> assembled a substantial freehold plot in the heart of the City opposite the northeast corner of the Bank of England. The symbolism of the location, immediately across the road from its great rival, could hardly be missed.[38]

Black notes that the bank's new headquarters opened in Lothbury on 26 December 1838, at a cost of £37,000, including freehold, building, materials and architects' fees!

The registration and incorporation of companies, without specific legislation, was introduced by the Joint Stock Companies Act 1844. Initially, companies incorporated under this Act did not have limited liability, but it became common practice for companies to include a limited liability clause in their internal rules. The Joint Stock Companies Act 1856 provided for limited liability for all joint-stock companies and required, among other things, that they include the word "limited" in their company name.

The expansion and growing complexity of commercial activity had to be supported by increasingly sophisticated trading. By 1875, Britain was supplying not only between one-third and one-half of the world's trade in manufactures, "but also the facilities that made possible most of the rest of trade and investment".[39]

> City banks advanced large sums of short-term credit to importers and exporters and arranged the finance of long-term investments overseas. The Stock Exchange was the centre of the world's capital market; and London's insurance companies had no serious international competitors.[40]

By the 1870s London was the world's principal money market, and it further consolidated its position through the expansion of foreign issues. For example, by 1890, some 30% of the capital stock of American railways, and an even larger percentage of the Argentine and Indian railways, were held in Britain.

As noted by Powell, "Business meant money and numbers and these meant counting houses". The result was the emergence of novel building forms – modern offices: "Banks, insurance offices and allied business increasingly built new premises in the City . . . many of these manifestations of Victorian commercial spirit were imposing and costly, expressing power, prestige and dependability".[41] Kynaston mentions Mansion House Chambers, built in 1872 on the south side of Queen Victoria Street, which contained some 500 rooms, and "epitomised the dawning age of the office block".[42]

Despite the growth of speculative building, many office buildings from this period were delivered for their occupiers. For example, in 1877 the Royal Bank of Scotland replaced its first London branch of 1874 with a new building, at 3-5 Bishopsgate (Figure 2.5). The building was elegant with the lower two floors having Ionic columns, arches and spandrels, and much carved detail. A balustrade introduces the upper two storeys which have more carvings and figureheads. In 1879, the first stage of Prudential's iconic head office on Holborn was completed to architect Alfred Waterhouse's design. These new buildings were far more complex than their predecessors, and new skills were evolving to deal with increasingly complex physical infrastructure.

Figure 2.5 Royal Bank of Scotland, Bishopsgate

Source: © Stephen Richards and licensed for reuse under creative commons.org/licenses/by-sa/2.0/

Across the Channel, Paris was also beginning to emerge from its suspicion of banking, as the administrative institutions necessary to oversee the increasingly complex web of commercial activity matured and began to cluster, thereby becoming a physical presence as well as a commercial one. In Paris, for example, the financial district began to take shape, literally, between the Bourse and the Opera. Pelegrin-Genel cites the example of the architect Hector Horeau, who proposed in 1871 the construction of an office building entirely without living accommodation. Then in 1876, Henri Germain, the founding president of Credit Lyonnais, decided to house his Paris headquarters on the Boulevard des Italiens, "a building that was destined to become a model for banking headquarters of the future" and featuring a large central atrium, glazed skylight, mezzanine with reception rooms and a first floor reserved for senior management.[43] The development in Paris, however, was eclipsed by the growth and change taking place in London.

In the absence of any formal planning framework, the powerful forces of international trade were, in the latter part of the nineteenth century, driving the physical growth of the city. London responded to the growing volume of trade in products and commodities with increasingly complex trading arrangements, and it began to provide a sophisticated infrastructure for trade. This took the form of financial, insurance, accountancy, legal and other professional services. Very quickly, clerical and administrative activities expanded in the City to support the growing diversity of markets and trades in manufactures and services:

> It was the City's ability to supply such a comprehensive range of financial and trading services which was largely responsible for its emergence as a centre of international repute. No other city could offer such a 'package deal' of interrelated activities.[44]

Furthermore, as the office economy was growing in scale and complexity, particularly in the banking and insurance sectors, so the demands on 'paperwork' – invoices, records, lists, letters, ledgers, accounts – all grew. And alongside the paper was the proliferating office 'technology', including typewriters, telegraphs, telephones, printers and adders (Chapter 8). Then there was the growing invisible function of the markets – information – which it was increasingly important to be in proximity. Information on markets, prices, opportunities, sailings, etc. – often accessible in the counting houses and coffee houses. The need to concentrate, physically, grew ever stronger, and with this, the need to have a physical presence, or office, close to the 'centre of things'.

The rise of the clerks

Industrialisation and rapid expansion in trade led inevitably to an unprecedented demand for administrators to manage orders, sales, receipts, inventories, accounts and so on. In short, paperwork was on the march, and with it the need for systems and processes for storing information. The emerging importance of office activity in the national economy is seen directly in the growth of white-collar work during the latter half of the nineteenth century and into the twentieth (Figure 2.6).

The data show that the transformation that led to the dominance of the office economy in the twentieth century had been gradually gathering pace during the latter half of the nineteenth. In 1851, for example, there were just 95,000 clerks in Great Britain; a figure that had risen sixfold by the turn of the century.[45] The growth was coincident with the introduction of office automation in the form of typewriters, telegraphy and telephony.

The remarkable increase in numbers is reinforced by their growth as a proportion of all jobs from 1.2% to 7.8% over the period. The rapidly expanding role of women in the workforce is also clearly shown, rising from a mere 2,000 to 583,000.

The scale of change as one millennium closed and another started is also captured in the employment figures for the heart of the nascent office economy: the City of London (see Figure 2.7). Here we can see how the total working population (office and other) doubled between 1866 and just after the turn of the century – from 170,000 to 332,000. It went on to grow by a further 50%, to over 500,000 before the outbreak of the Second World War.[46]

The growing importance of commerce in the City during the latter half of the nineteenth century and the early years of the twentieth is also reflected in the relative decline in the residential population as commercial uses gradually displaced residential uses. Figure 2.7 shows how the population of the City fell by nearly 80% in the latter half of the nineteenth century. And as the City was shrinking, so London generally was expanding. Over the same

Year	No of clerks (000s)			Clerks as % of occupied population		
	Male	*Female*	*Total*	*Male*	*Female*	*Total*
1851	93	2	95	1.6	0.7	1.2
1861	123	4	127	1.9	0.1	1.3
1871	194	7	201	2.7	0.2	1.9
1881	294	14	318	3.8	0.4	2.9
1891	381	33	414	4.3	0.8	3.2
1901	530	83	613	5.2	2.0	4.3
1911	677	166	843	5.2	3.1	4.6
1921	792	502	1,294	6.7	9.9	7.5
1931	882	583	1,465	6.7	10.4	7.8

Figure 2.6 Number of clerks, 1851–1931, Great Britain

Source: Pickard (1949).[47]

Year	City of London	
	Working population	*Residential population*
1866	170,000	93,000
1871	200,000	75,000
1881	261,000	51,000
1891	301,000	38,000
1901	332,000	27,000
1911	364,000	20,000
1921	437,000	14,000
1931	482,000	11,000
1935	500,000	10,000

Figure 2.7 City of London working population, 1866–1935

Source: Dunning and Morgan (1971).[48]

period, London's total population tripled. Throughout the period up to the First World War, the City of London remained primarily a centre of commerce and trading rather than office activity per se.

It is, however, a curious fact that very little is known about how the nature of employment changed over this time. For example, in one of the most comprehensive studies of the development of the City of London, Dunning and Morgan state that "How many people were actually employed in offices before the [second] war is not known".[49] The authors also state that "the only published pre-1941 information on economic activity that we have been able to find is that of the distribution of factory employment in 1907".[50] This showed that 39,524 people were then employed in 1,058 factories and that 25,547 of these were in the paper and printing industry. An alternative source, however, suggested that in 1911, banking and finance accounted for 9% of jobs in the City, while manufacturing and commodity businesses accounted for 46%.[51]

Office activity as we understand it today was very limited: much of it was ancillary to commercial and trading enterprises, and much occupied buildings that had previously been in residential and light industrial uses. However, as we saw earlier, purpose-built offices were established.

Whatever the precise numbers, as the office economy grew, working conditions did not always take a commensurate step forward. In wondering how much the growth and change mattered "to the wretches actually working" in the City, Kynaston suggested that what abided

day after day, week after week, year after year – was not so much the larger environment as the actual grinding, routine: the voluminous ledgers, the salient account books, the endless, pernickety correspondence. There lies the true, inner, inscrutable history of the City, as draining on human vitality as the endless pleas of endless suitors recorded on the peau de chagrin in Balzac's famous story.[52]

The same author argued that for almost everyone

in the vast clerical army, daily life in the City remained manual and laborious. It is true that in 1873 the first typewriter was spotted in the square mile, but it would be many years before it became a standard feature of office furniture – not least because of an innate reluctance to employ women typists, eventual superceders [sic] of the traditional male copying clerks.[53]

The growing dominance of office activities was not a sudden phenomenon but a slow-burning process, with office activity, industry and trade, through the docks that remained close to the heart of the City, co-existing for a long period of time. As late as 1907, there remained 39,524 people employed in 1,058 factories in the City.[54] However, with the gathering pace of change in the main business activities of the central London area, manufacturing gradually yielded to the expansion of office activity, reflecting the more general restructuring of the national economy. The physical fabric of the City, and London generally, responded and evolved to accommodate the changing economic activity.

Large, centrally located, purpose-built office buildings became the norm. This was the period in which office activity became a distinct urban function:

The office function was clearly due to grow, and to grow quickly. And in addition it would need to separate off from other facets of urban activity, because of the growing complexity and intricacy of organisation needed to finance, administer and support the industrial revolution.[55]

The key point here is the hint at specialisation: *the growing complexity and intricacy* of activities separately to *finance, administer and support*. Indeed, herein lie the roots of the complex web of activities that form the office economy ecosystem today. The scale of activities was becoming such that single, often family-owned companies could no longer import/export, ship, wholesale/retail, finance and underwrite their activities. Specialist functions were emerging – first the insurers and bankers, then the shippers, the wholesalers and the retailers. But more importantly for the office economy, legal, accounting and related professions began their inexorable growth.

The later nineteenth century was a period of enormous change for the office economy. The City of London grew rapidly, and few buildings existing before 1850 survived beyond the turn of the century. It was also a period of innovation for the office building, one of the most important innovations being the lift, or elevator. According to Turvey, lifts were operating in London hotels from the 1850s, but the first office lifts came later:

> in 1873 a hydraulic lift was installed in the eighteen-sixty-seven office block of Palmerston Buildings between Old Broad Street and Bishopsgate at a cost of £750. It enabled the letting of offices on upper floors that had never before been tenanted.

Turvey cites further early lift installations in the Foreign Office in 1875; at Maw & Thompsons's Aldersgate Street offices in 1877 and at Leadenhall House in Leadenhall Street in 1879.[56]

It was into this firmament of change, from roughly the 1870s through to the zenith of Britain's colonial expansion during the Edwardian period, that the office economy as a distinct facet of urban life was firmly established. New building forms had evolved to accommodate its activities, and the property supply industry necessary to its continuing development was largely in place. The structure and distribution of office activity in central London today is largely the result of the city's expansion and evolving role during this period.

The specific characteristics of London's evolving office economy described earlier left a physical legacy, a ghostly and complex street pattern dating back centuries. By contrast, the physical fabric of many North American cities was

Figure 2.8 Foreign Office, London, 1866

being made for the very first time. The famous grid iron pattern of urban development in the USA was reflective of its later development and the great differences in scale.

As the nineteenth century drew to a close and the twentieth got underway, the huge American economy was thundering along at great speed and laying down a very different urban structure. The process involved the American economy switching structural models, from an entrepreneurial, owner-managed form characterised by many layers of production and distribution, to a model dominated by large, vertically integrated companies that gained significant economies of scale and stripped out the add-on costs of intermediaries, brokers and distributors. The new economic model, and the rise of 'corporatism', gave rise to a fundamentally new building form: the skyscraper, with its distinctive modern form of regular internal space and repetitive façades. The buildings soon dominated the skylines of Boston, Chicago, New York, Philadelphia and the other great American cities.

2.3 The skyscraper

Office buildings have often been built to impress. In London, the buildings of the major banks and insurance firms appearing towards the end of the nineteenth and during the early years of the twentieth century were often designed, outside and in, to convey authority, stability, continuity and success. Of course, this feature of architecture reached its apotheosis in North America with the skyscraper, that came to symbolise wealth and economic success as well as cultural and technological advancement.

Skyscrapers can express the self-confidence of a city, or even of a nation, and the young American nation embraced the tall office building fully. The skyscraper was celebrated for its central role in the daily national heartbeat, from the very beginnings of the office economy in the 1870s and 1880s. The differences with European cities were stark, although most cities shared the effects of greatly increased densities in their core areas as well as the growing dominance of the office economy.

Curiously, it was in the conflagration that engulfed Chicago in 1871 that the origins of tall office buildings are to be found. After the devastation of the fire, the demand for new buildings was intense, sending land prices to all-time highs: if only "some means of piling up accommodation in the centre were to be found, the profits of the owners of real estate would multiply".[57] The solution was found by the wonderfully named William LeBaron Jenney (1832–1907). Born in Fairhaven, Massachusetts, Jenney went to study engineering and architecture at the École Centrale des Arts et Manufactures in Paris in 1853, where he discovered iron construction techniques (Gustave Eiffel, of the eponymous tower, graduated from the same school in 1855). He graduated in 1856 and returned to the USA in 1861. Following a spell in the army, working as an engineer, Jenney moved in 1867 to Chicago where he opened an architectural practice to work on commercial buildings.

The First Leiter building was one of Jenney's first commercial buildings, completed in 1879. It was renovated and extended in 1888, and demolished in 1972. Jenney designed this building, located at Washington and Wells Streets, as a department store for Levi Z Leiter. Originally five stories high (later extended to seven), the building marked a significant milestone in architectural engineering by bringing together an iron skeletal frame; terra cotta fireproofing materials on all of its structural members, and vertical transportation in the form of elevators. It also utilised a new type of glass in its windows.

Jenney designed both the Ludington Building and the Manhattan Building in Chicago (both delivered in 1891), but his name and reputation are firmly attached to the Home Insurance Building there, built between 1884 and 1885. This 42 m (138 ft), ten-story building is widely recognised as the first skyscraper, and was certainly the first high rise building in which an iron skeleton of vertical columns and horizontal beams carried the whole load of the building, and in which the external walls were reduced to panel infillings between the members of the frame. Saving on expensive foundations owing to the reduced weight of the building, and "the commercial possibility of buildings with more than twelve storeys, owing to the saving of space and increase of light on the lower floors which went with the elimination of load-bearing masonry, provided the economic basis for the skyscraper".[58]

As new technologies including fire resistant materials, lifts, lighting, telephones and sanitation made great strides, so buildings became become larger and taller still. For example, the first passenger elevator was installed by Elisha Graves Otis in New York in 1857. The first elevators were hydraulic, and their use was limited by the physical laws of the pressure of the water column. An elevator could be lifted by hydraulic force to a height of eighteen or twenty stories, but not much more than that, especially in the nineteenth century. To be liberated from this ceiling, architecture needed the electric elevator, which made its appearance about 1887. From then on, the point at which people could be lifted mechanically was no longer restricted by physical laws, and the sky became the limit.[59]

The incandescent electric lamp was invented around 1870. Sanitation too was critical to the development of the modern office buildings. Alexander Cummings's 1875 patent for a toilet with a 'stink trap' that prevented sewer gases from backing up through drains was an important development. And, as we have already seen, steel frame construction was born, enabling tall buildings, free from the limitations of load-bearing concrete structures.

The impact of these innovations was so profound that it might be argued that the subsequent 100 years of office development were mere refinements on this early model. They were of course part of a wider process of change that was having a fundamental impact on the organisation of the economy and that brought with it massive social change. Indeed, skyscrapers came to symbolise the new corporatism that was sweeping through the reorganising American economy, and many became synonymous with the corporations that occupied them. In New York, for example, there were the Singer Building (sewing machines), the Metropolitan Life Building (insurance), the Woolworth Tower (retailing) and the Chrysler Building (cars).

In some areas, concentrations of skyscrapers jostled with each other as corporations sought to stand out from their competitors. New York's famous 'Newspaper Row' contained the Pulitzer Building (aka the World Building) that housed the *New York World*; the Tribune Building; the Times Building and the Potter Building (home to the *New York Press*).

The buildings' patrons were themselves often exuberant, large-than-life characters. Frank W Woolworth teamed up with architect Cass Gilbert to create the Woolworth Tower. Its opening ceremony, on 24 April 1913, created some history. President Woodrow Wilson pressed a button in the White House to activate 80,000 light bulbs in front of 900 assembled guests. One of them, the prominent clergyman Samuel Parkes Cadman, was said to have cried out "the cathedral of commerce!"

A few years later, Walter P Chrysler set his architect a very simple brief for his own New York tower: make it taller than the Eiffel Tower. William Van Alen's design for the Chrysler Building typified many buildings of the period with its ornate design, intricate patterns, mixing of materials and complex structural style (Figure 2.9). As time passed, the buildings became much simpler, more economic and utilitarian in their design, but their iconic status simply grew.

Europe's relationship with tall office buildings has been very different to that of the USA: it did not, and has not, embraced the tall office building with such enthusiasm. London waited until 1967 for Richard Seifert's 111 m (365 ft) Centre Point; until 1981 for his 183 m (600 ft) Nat West Tower (since renamed Tower 42), and until 1991 for I M Pei's 235 m (771 ft) One Canada Square in Canary Wharf. Mainland Europe also shunned tall buildings until the 257 m (843 ft) Messeturm was completed in Frankfurt in 1990. The proliferation of tall buildings in the Far East in recent times perhaps underscores their iconic status as these emergent economies compete to express their positions in the world economy. Figure 2.10 shows the rise of the office skyscraper with a selection of iconic buildings, culminating in the Burj Khalifa in Dubai.

The muted response to tall buildings in the UK and Europe is deeply rooted in economic and cultural factors, not to mention the physical constraints imposed by the complex and aged nature of its city centres. Free from such constraints, the USA embraced the new office economy, driving many of the technological advances that were to enable the modern office.

While the arrival of the skyscraper was necessitated by the reorganisation of business, it was enabled by new technologies, not the least of which included lifts, telephony, steel frame construction and new sanitation techniques. In a process not dissimilar to that which is changing the office environment today, the birth of the modern office in the late nineteenth century was enabled and driven by a number of technological innovations.

It is an interesting footnote of history that the quintessentially American term 'skyscraper' is neither American in origin nor originally a reference to tall buildings. Its etymological origins are traced back not to an office building at

Figure 2.9 The Chrysler Building, New York City

all, but to a horse; specifically the winner of the 1789 Epsom Derby. Slipping into everyday usage as a term for high-standing horses and for tall men, by the beginning of the nineteenth century the term also referred to tall hats and bonnets, and later to tall tales.

By the mid-nineteenth century the term was applied to the small triangular sail set above the skysail on sailing ships in order to maximise the effect of a light wind in calm latitudes. The first use of the term skyscraper in the context of high rise buildings was in 1883, in reference to the tall buildings emerging in Chicago.

Year	Building	City	Height, ft (m)
2010	Burj Khalifa	Dubai	2,717 (828)
2015	Shanghai Tower	Shanghai	2,073 (632)
2017	Ping An Finance Center	Shenzhen	1,965 (599)
2014	One World Trade Center	New York	1,776 (541)
2018	CITIC Tower	Beijing	1,731 (528)
2004	Taipei 101	Taipei	1,670 (509)
2019	Lakhta Center	St Petersburg	1,516 (462)
1998	Petronas Towers	Kuala Lumpur	1,483 (452)
1974	Willis Tower	Chicago	1,450 (442)
1973	World Trade Center	New York	1,375 (419)
2003	Two International Finance	Hong Kong	1,361 (415)
1997	CITIC Plaza	Guangzhou	1,283 (391)
1931	Empire State Building	New York	1,250 (381)
1992	Central Plaza	Hong Kong	1,227 (374)
1990	Bank of China	Hong Kong	1,204 (367)
1930	Chrysler Building	New York	1,046 (319)
1997	Commerz Bank Tower	Frankfurt	850 (259)
1990	Messeturm	Frankfurt	843 (257)
1991	1 Canada Square	London	771 (235)
1913	Woolworth Building	New York	761 (232)
1909	Metropolitan Life Insurance	New York	699 (213)
1984	AT&T Building	New York	646 (197)
1908	Singer Building	New York	620 (189)
1980	Nat West Tower	London	600 (183)
1985	Hong Kong & Shanghai Bank	Hong Kong	518 (158)
1958	Seagram Building	New York	500 (152)
1890	Pulitzer Building	New York	308 (94)
1902	Flatiron Building	New York	285 (87)
1885	Home Insurance Building	Chicago	171 (52)

Figure 2.10 The rise of the skyscraper: selected office towers, 1885–2019

There is possibly no better account of the evolution of the skyscraper than that contained within John Zukowsky's wonderful, richly illustrated, two-volume *Chicago Architecture*.[60] The books provide a fascinating account of the economic and technical changes that led to the skyscraper and, at the same time, provide a striking and graphic contrast with how the office economy within Europe was being housed in the latter years of the nineteenth century.

2.4 The office of the world

The birth of the modern office economy in the nineteenth century changed London dramatically. Kyanston cites a passage from *The Builder* magazine of 1868 which expresses the scale of change taking place at the time in the City:

> The City of London . . . is becoming more and more the office of the world. Stately buildings replace the ugly and cramped houses of the Georgian era, and these buildings are almost entirely parcelled out in offices.[61]

The scale of London's achievement was captured further by Michie:

> the City of London was the only commercial centre in the world that possessed the range and depth of personnel, experience, institutions and facilities to handle the increasingly large and complex network of international trade that developed during the period.[62]

London had modernised. The city had accepted its declining role as 'the world's port' and was now embracing its role as the office of the world. London was no longer *doing* the trade, it was *organising* the trade, *funding* the trade, *insuring* the trade. Telecommunications meant that goods could be directed rather than received and then re-exported. From an office in the City, merchants could direct their operations across the globe. As Michie points out, while Britain's share of the world's shipping fleet fell sharply, the City of London

> dominated the movement of international shipping . . . it was the only centre with all the information all the time, it was important to be represented there if cargoes were to be picked up.[63]

The massive economic expansion of the Victorian period had culminated in the establishment of the office economy as a distinctive feature of the urban landscape: by the early years of the twentieth century, "new office blocks became a familiar presence; new banks, company headquarters, insurance offices were built on a massive scale, with intense and dramatic architectural effects".[64]

By the turn of the twentieth century the office economy had gained a momentum of its own, and it was now driving a new period of growth in London. Despite the growing international crisis that would erupt in 1914, by 1910, around 30% of all quoted securities in the world were listed on the London Stock Exchange.[65] By 1914, 80% of all issues on the London market were for overseas borrowers.[66] Improvements in communications reinforced London's role as the leading centre in an increasingly integrated world securities market. And its growth as a commercial and international trading centre, together with its position as the hub of world communications, led inevitably to its becoming the world's foremost shipping market. By 1914 the British merchant fleet accounted for around half the world's tonnage, and one-third of the fleet was then engaged in trade that never touched Britain's shores at all.[67]

It is also clear that, as the nineteenth century drew to a close and the twentieth beckoned, London's role as a hub of international trade and commerce was being given further impetus by a growing domestic population, and one with more money to spend and invest. As ever, it seems, insurance was at the heart of growth. Black *et al* describe how organisations grew in complexity as their range of services and products expanded, and as their activities involved the general populace more and more. For example, regional offices spread among key cities to service the rapidly concentrating populations and

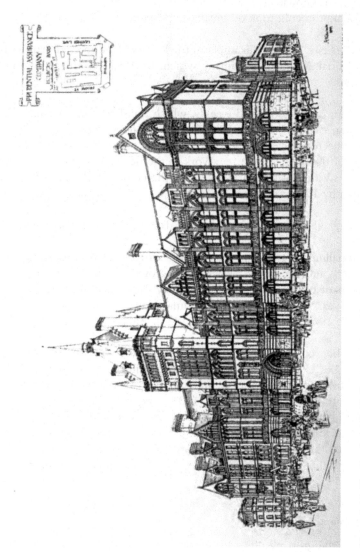

Figure 2.11 The Prudential's Holborn Bars by Alfred Waterhouse

industries. An early imperative was to introduce internal systems and processes to ensure consistency:

> As companies became larger and their workforces and operations more dispersed, knowledge management, including the deployment of organisational protocols, took on greater importance.[68]

The authors cite "a prime example of the large-scale, dispersed firm" as being the Prudential Assurance Company which, by 1891 controlled nine million life policies.

> Premiums were collected weekly, which translated into around 470 million transactions per year, each of which had to be recorded both locally, by the collector, and centrally. Collection work was undertaken by an army of over 11,000 field agents. . . . Work in the field was supported by another army: swathes of clerical workers, in the company's London head office, which by 1905 numbered 1,500.

As Black *et al* observe, "a miracle of data processing in an age without computers".

The appalling losses and huge cost of the First World War cast a long shadow over society and economy. Nevertheless, growth resumed, and the working population of the City of London grew by more than a quarter during the inter-war years. Among the new landmarks was a rebuilt Waterloo Station (opened in 1922); Wembley Stadium (1923), and a remodelled Regent Street (1927).

A number of corporate office buildings were also completed during this time. For example, the Midland Bank building on Poultry (1939), the biggest building in England at the time; the National Westminster Bank building on Threadneedle Street (1931); the Daily Telegraph building at 135–142 Fleet Street (1928); the Reuters & Press Association headquarters at 85 Fleet Street (1938) and Unilever House on the Thames at Blackfriars (1931). The distinctive Daily Express building at 121–128 Fleet Street (1932) was the first example of a curtain wall office building in Britain (Figure 2.12).

Getting around

Of course the growing office economy was drawing more and more people, and the growing public transport system was having a most profound effect on the long-term shape of London. In 1925 the first stretch of the Great West Road (the A4) was opened, between what is now Chiswick roundabout and Hounslow. This was to establish a major artery of growth westwards, leading to the development of other iconic buildings such as those of Beechams, Firestone and Gillette, among many others. The dynamism of this particular growth

Figure 2.12 Daily Express Building, Fleet Street
Source: © Ramidus Consulting Ltd.

corridor was to be felt throughout the twentieth century and was epitomised by the influx of the computer industry in the 1980s.

While it was in the centre of London that the impact of the burgeoning office economy was most obvious, not least with growing congestion, the public transport system was also expanding and connecting more places farther afield. Electrification and upgrading of the Underground together with the expanding road network were pushing London ever outwards, but also providing an ever greater system for connecting dormitory suburbs with the commercial centre, introducing mass commuting for the first time – each morning bringing thousands of residents into the centre and then in the evening sending then back to their suburban homes. Significant extensions were added to what are now known as the Bakerloo, Central, District, Northern and Piccadilly lines. Following the First World War, the Underground could not be described as a system: it lacked interconnectivity and coherence. This was to come through the influence of a single individual.

As Managing Director of London Underground in the 1920s and the first Chief Executive of London Transport, Frank Pick (1878–1941) arguably had one of the greatest influences on the look of modern London. He used his professional role to be a patron of excellent design, and he commissioned some of the most recognisable icons of London Underground's identity such as the distinctive red, blue and white roundel, the original Johnston typeface and the art deco architecture of many Underground stations designed by Charles Holden. Pick also commissioned striking advertising posters in a variety of styles, often working with famous artists of the day such as the surrealist Man Ray. He introduced free-standing seats bearing station names, station signs on platform walls and coin-operated ticket machines. Perhaps most famously of all, he commissioned Henry Beck to create the instantly recognisable diagrammatic map, based on an electrical circuit, in 1933. In short, Pick took a rambling and somewhat poorly managed organisation and, by the Second World War, had unified the system, bringing an identity, order, clarity and consistency to every corner of the increasingly complex transport system.

Also in the USA and Europe rapid developments in motor transport, particularly cars, trams and trains, acted as a spur to suburbanisation of populations and increased densities, and larger office buildings, in city centres. More dense, higher rise and more chaotic, American and European cities were becoming increasingly dominated by the economic activity that was evolving particularly at their centres. As their central areas intensified so cities spread rapidly outwards, encouraged by ever-denser networks of rail and tram transport, and the growing impact of the motor car. Cities began to take on a whole new level of complexity, with increasingly large and anonymous corporate offices at their centres. By the outbreak of the Second World War, the office economy had, in many respects come to define downtown.

Human filing cases

The growth of the office economy, and by extension the growth of the city, was not welcomed by everyone. Among many critics of seemingly uncontrolled urban growth, but perhaps one of its most severe, was American sociologist Lewis Mumford (1895–1990). He was forthright in his description of the office economy in North America. In a curious choice of words, Mumford observed that a "new trinity dominated the metropolitan scene: finance, insurance, advertising".[69] Mumford described, with a sense of parody, the mind-numbing work and layers of hierarchy:

> Plainly, no great corporate enterprise with a worldwide network of agents, correspondents, market outlets, factories, and investors could exist without relying upon the services of an army of . . . stenographers, filing clerks, and bookkeepers, office managers, sales managers . . . right up to the fifth vice president whose name or o.k. sets the final seal of responsibility upon an action.[70]

He went on describe the form of the office economy and the buildings that were springing up to house its processes:

> files, vaults, places for live storage and dead storage, parade grounds of documents, where the records of business were alphabetically kept, with an eye to the possibility of future exploitation, future reference, future lawsuits, future contracts. This age found its form in a new type of office building: a sort of human filing case, whose occupants spent their days in the circumspect care of paper: numbering, ticketing, assorting, routing, recording, manifolding, filing, to the end that the commodities and services thus controlled could be sold to the profit of the absentee owners of the corporation.[71]

The term 'human filing case' to describe office buildings, and the caustic 'circumspect care of paper' highlight Mumford's severe misgivings about the concentration of white-collar work in downtown.

Born in 1895, Mumford was one of the greatest observers of urbanism that ever lived, his long working life coinciding with the maximal period of growth in the office economy. He was also a great humanist. Leaving college early due to tuberculosis, Mumford spent enormous amounts of time studying, sketching and simply watching New York. What he saw was the dehumanising effect of modern urban life, its stultifying impact on the human soul and, ultimately, its denuding of human purpose.

Mumford regarded medieval cities as having been built on a human scale and balancing human needs, whereas to him the general advent of modernity signalled not emancipation and progress but a deterioration of community. This might be seen as just part of the long goodbye given to the mythical idyll of rural life that accompanied western industrialisation. A century before, William Wordsworth was mourning the passing of rural innocence and virtue in England in *The Prelude*: "*Cities Where the Human Heart Is Sick*". But Mumford was no iconoclast. He balanced his withering criticism with active participation in urban planning (in 1923 he co-founded the Regional Planning Association of America), and pursued a lifetime of writing on cities, architecture, environmentalism, technology and literature.

The Culture of Cities,[72] published on the eve of the Second World War, was one of Mumford's finest works. The book is a history of urbanism, beginning with the medieval town – which Mumford viewed fondly as akin to the garden city. His biographer Donald Miller described the book as "a study of the erosion of a balanced, decentralized civilization and its replacement by one with an oppressive metropolitan centralization of power, people and culture".[73] Mumford described the history of urbanism as a "fall from grace, a long plunge into chaos and moral confusion".[74] It is a lucid, passionate and immensely widely informed argument, finishing with a call for urban development to balance the needs of the institutional with the needs of the individual with a more integrated, more human urban landscape that performed a complex balance of roles:

> The city in its complete sense, then, is a geographic plexus, an economic organisation, an institutional process, a theater of social action, and an esthetic symbol of collective unity. On the one hand it is a physical frame for the common place domestic and economic activities; on the other, it is a consciously dramatic setting for the more sublimated urges of a human culture.[75]

The post-Second World War years were, however, to defy Mumford's prescription. The inter-war years had in fact been a period of transition: from the old nineteenth-century Britain with its pivotal role in world trade and commerce, not to mention its role as the world's creditor nation, to a post-war role that was very much more austere and economically challenged. Between the wars, London lost its pre-eminence; growth everywhere shrank dramatically and labour markets were hit hard. The physical, social and economic cost of the Second World War took many years to overcome. But it was also clear that London had changed in other ways, not least its office economy.

In the post-war years, London's office economy – as in Europe generally – grew rapidly in scale and complexity. Individually, organisations grew in scale: corporatism was the order of the day. Mergers and acquisitions grew rapidly as companies sought to dominate their respective sectors. It was as though one age had passed and another begun.

Twentieth-century growth

As we saw earlier, the ascendancy of the office economy was far from sudden. As Figure 2.13 shows, white-collar work had been growing absolutely and relative to manufacturing since the First World War. The table shows how the number of blue-collar jobs remained remarkably stable over a 50-year period.

The growth in total employment was almost entirely accounted for by the rise of white-collar work, which doubled in size as a proportion of the total workforce. What happened from the 1960s onwards was a more complicated

Year	White collar (millions)	Blue collar (millions)	Total (millions)	% white collar
1911	3.4	13.7	18.3	18.7
1921	4.1	13.9	19.3	21.2
1931	4.8	14.8	21.0	23.0
1941	–	–	–	–
1951	6.9	14.4	22.5	30.9
1961	8.5	14.0	23.6	35.5

Figure 2.13 Growth of white collar occupations, Great Britain, 1911–1961

Source: Daniels (1975).[76]

process, resulting in what has come to be referred to as the 'de-industrialisation' of Britain.

Of course, blue collar and white collar are very broad categories and tell us little about the nature of the work itself. Here we encounter the interminable problem in historical research of the lack of consistent, long-term historic data. Instead we have to rely on the piecemeal gathering of different but complementary data sets.

According to one estimate the number of office workers grew steadily between the 1950s and 1960s, with the number of people with an office occupation in the UK rising from 15% to 25% of the workforce between 1951 and 1971. Between 1951 and 1961 alone, while total employment in the UK rose by 7%, employment in office occupations rose by 40%.[77] The London Research Centre, who published their *Office Workers in London* report in 1986, provided one of the most accurate guides to office employment in London during the period concerning us here, the 1960s to the 1980s (Figure 2.14).

The first point to emerge from the data is that London's manufacturing jobs (including office workers) fell sharply. Office jobs in manufacturing rose from 27% of the total in 1961 to 38% in 1981. Meanwhile, the growing role of office workers in non-manufacturing industries is clear. This component rose from 34% in 1961 to 43% in 1981. The expansion of banking, finance and business services relative to manufacturing and the whole economy, on both of these indicators, provides clear evidence of their growing importance in the national economy.

These economic output figures are reflected in the growing concentration of employment in the service sectors (Figure 2.15). The relative strength of the services sector shown by these employment figures helps to explain the sustained demand for offices and their growing dominance of the city skyline.

The contribution of service activities to the economy in terms of, for example, output, balance of payments, investment and profit provide further context to the growth in demand for office space. Measured over time and related to demand factors such as employment, take-up, rental changes and construction

Activity	Employment	1961	1966	1971	1981	% change
Manufacturing	Office workers	385	381	354	246	(36)
	Other workers	1044	925	739	402	(61)
Non-manufacturing	Office workers	1019	1154	1173	1153	13
	Other workers	1938	1967	1818	1514	(22)

Figure 2.14 Growth of office workers, London, 1961–1981

Source: London Research Centre (1986).[78]

Service sector	Employment change, United Kingdom					
	1971–1976		1976–1981		1981–1984	
	000s	%	000s	%	000s	%
Distribution, hotels, etc	286.0	7.9	205.4	5.3	188.0	4.7
Transport and communications	(93.1)	(6.1)	(34.6)	(2.4)	(65.0)	(4.7)
Banking, finance, business, etc.	154.6	11.7	257.9	17.5	283.5	16.6
Other services	895.5	18.2	64.4	1.1	182.9	3.1

Figure 2.15 Employment change by service industry division, 1971–1984

Source: Howells and Green (1988).[79]

activity, these indicators provide a clear picture of the impact of economic trends on demand for office space.

Trends in the balance of trade and output figures from this time underline the growing importance of services to the national economy. For example, the balance of payments figures during the late 1970s and early 1980s showed a widening gap between visible and invisible earnings, or trade and services. A major component of invisible trade is financial and business services, and overseas earnings by UK financial institutions more than quadrupled between 1980 and 1987.

2.5 London reborn

In the early 1980s London was, in many respects, a successful city. It was a major centre of finance and insurance and an economic powerhouse; its political influence across the globe reflected its imperial past, and its cultural heritage attracted great numbers of visitors. London was already a global city.[80] Arguably it was the first, having wrapped its political, cultural and commercial tentacles around the globe in the latter part of the nineteenth century. It grew throughout the twentieth century, pulsing with the economic cycle, but forever broadening and deepening its international relationships.

However, London was at this time a very different place to the city we know today. The City of London was a much smaller, more insular place. Its most exciting building was the newly completed Nat West Tower, and the skyline was dominated by the 1970s rent slabs along London Wall. It was a world of fixed commissions, of jobbers and brokers, and of "Bank Walks". Bowler hats remained in vogue, and there were very few shops or hotels. The 1984 *City of London Draft Local Plan* designated much of the City a Conservation Area.

In the West End, today's prestigious shopping areas of Bond Street, Knightsbridge and Oxford Street were all successful, but around them the picture was rather different. Midtown was a ghost town; much of the South Bank was badly neglected (remember "cardboard city"?), and Victoria was populated by mandarins and the petro-chem industry.

Outside of the central area, the effects of 1970s economic decline were obvious. In the east, London's docks were closed, having succumbed to the technology of containerisation; and large parts of Outer London had taken a most severe economic battering. Following 150 years of continuous growth, London's population had been in decline for 30 years (Figure 2.16), mirrored by a fall in the number of jobs. Not only had London shared in the collapse of manufacturing, but the government had been actively encouraging companies to *leave* London (see Chapter 4).

In the mid-1980s, the fax machine and golf ball typewriter were leading-edge technologies. Personal computers were not widely available, and we had yet to experience the internet, email, mobile phones, social media, and digital media technology. The time pre-dated Canary Wharf, the Channel Tunnel, the Jubilee Line, the M25 and Terminal Five; and Big Bang was a cosmological theory, not a seminal economic event. London cuisine was, at best, mediocre; and coffee culture was a quaint French custom.

Then, something happened.

Over the following three decades, London and its office economy were transformed. The extraordinary growth impetus delivered by information and communications technology was a key factor, following the launch of the first IBM PC in 1981. The arrival of the 'digital' era had begun, with the impact of technology having a profound impact on every aspect of life. From the launch of the PC in 1981, through mobile phones, laptops, hand-held devices, and the internet, everyone's personal and business lives were radically changed through the ability to collect, receive, transform, analyse and communicate information.

But perhaps the single most catalytic event in the transformation was 'Big Bang', on 27 October 1986, the deregulation of financial services. The deregulation in London, which followed a similar event in New York in 1975, brought about a wholesale reorganisation of business activities in the City. The event itself followed the 1983 Parkinson-Goodison Accord, between the Thatcher Government and the London Stock Exchange, to settle a restrictive practices dispute between the Office of Fair Trading and the London Stock Exchange. The restrictive practices included the LSE's fixed minimum commissions; the single capacity rule that separated the activities of brokers and jobbers; the

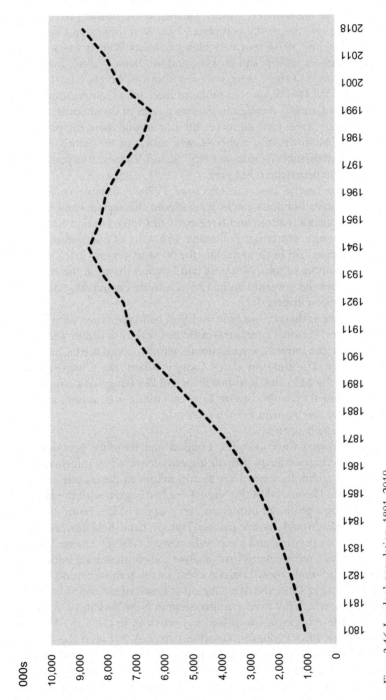

Figure 2.16 London's population, 1801–2019

Source: Sheppard and ONS.[81]

requirement that neither could be part of a broadly based financial group, and the LSE's exclusion of overseas firms from Stock Exchange membership. The Accord also presaged the introduction of electronic settlement systems.

The proposed changes ushered in an era of global financial trading in which markets became globally integrated. Investment banks from America, Europe and Asia poured into London, accompanied by the tech companies to provide their computers. In the run-up to Big Bang, international investment houses and merchant banks invested heavily in broking and jobbing firms. The feeding frenzy saw old City names absorbed into rapidly expanding financial services conglomerates. Examples included: Hoare Govett (Security Pacific); James Capel (Hong Kong & Shanghai); Quilter Goodison (Bruxells Lambert); Rowe & Pitman (SG Warburg) and Wood Mackenzie (Hill Samuel).

The combination of reorganising and expanding businesses, together with rapidly evolving technology for international dealing systems, led directly to a sharply increased demand for premises. Not only did the larger conglomerates wish to bring their merged companies together under one roof, but they also needed more highly specified buildings to cope with expanding technological demands.

Developments such as Broadgate and Canary Wharf were a direct response to these enormous changes in the office economy. The real estate responses are described more fully in Chapter 6, but in terms of the office economy, Big Bang unleashed a transformation that would result in London becoming one of a handful of truly global cities. Some of the achievements are described as follows.

The City of London became the largest money market in the world, its interbank market the largest in the world and its foreign exchange market the largest in the world. Other indicators are:

- *International banking.* There are over 250 branches and subsidiaries of foreign banks in the UK (mostly in the City), more than in any other country worldwide; and around half of European investment banking activity is conducted here.
- *Insurance.* London is the world's largest international insurance and reinsurance market and now controls more than £60 billion of annual premiums (compared to Bermuda at £25 billion, Switzerland at £19 billion and Singapore at £4 billion), employs 48,000 people and generates over 8% of London's GDP.[82]
- *Derivatives.* London is the biggest market in the world for interest rate derivatives traded over the counter with 43% of global turnover in 2016 (or four times as much euro-denominated trading as France and Germany combined).
- *Maritime.* London is the leading centre in the supply of professional and business services to the global maritime community. Shipbrokers in London account for 50% of the tanker and 30–40% of the dry bulk chartering business; and are involved in the sale and purchase of over half the world's tonnage.

The UK hedge fund industry is the largest in Europe, with UK-based funds managing around $472 billion in 2017, accounting for 72% of assets under management by Europe-based managers.[83] The UK's market is the second largest globally, after the USA. Over half (53%) of Europe's hedge fund managers are based in the UK, and the country is home to 39% of the region's institutional investors. Switzerland, Sweden and France all have between $34 billion and $43 billion, while the rest of Europe has $27 billion and Germany $11 billion. The remaining countries shown all have single-digit billions.

In terms of services exports, London exported around £100 billion worth of services in 2015. In contrast to the export of goods, this represented 45.8% of the UK's total of £219.2 billion. That was up from 42.6% in 2014 (though less than the 47.5% share recorded in 2011) and was more than twice as much as any other UK region. Financial services represented around 28% of service exports, with professional, scientific and technical at 17%, travel at 17% and information and communication at 16%.

Research group ZYen produces a Global Financial Centres Index (GFCI), based on 137 instrumental factors, grouped into "five key areas of competitiveness", namely: Business Environment, Human Capital, Infrastructure, Financial Sector Development and Reputation (Figure 2.17). In 2009 it argued that London and New York "remain the only truly global financial centres". The report alluded to one of London's enduring qualities in referring to "a genuine flight to safety".[84] A more recent edition of the GFCI (September 2018) ranks London second overall (behind New York); ranking first or second in four of the key areas of competitiveness, with a third position for infrastructure.[85]

Many measures of London's global city status focus on financial services, but London's office economy is much more than that. It is an economic agglomeration of enormous depth. Business services, creative and media services, life sciences, medicine and technology are all part of the city's extraordinarily rich tapestry of economic strengths. A diverse ecology is more resilient to external threats than a narrow one, and London is very diverse.

PWC's *Cities of Opportunities* 7 report ranks London number one, ahead of (in descending order) Singapore, Toronto, Paris, Amsterdam and

Rank	Business environment	Human capital	Infrastructure	Financial sector development	Reputational and general
1	London	Hong Kong	Hong Kong	New York	New York
2	New York	London	New York	London	London
3	Hong Kong	New York	London	Hong Kong	Hong Kong
4	Singapore	Singapore	Singapore	Singapore	Singapore
5	Chicago	Tokyo	Shanghai	Shanghai	Chicago

Figure 2.17 The Global Financial Centres Index, 2018

Source: ZYen (2018).[86]

New York.[87] The research assesses the competitiveness of cities across ten broad indicators (and 59 component indicators). London is rated as the leading city in three indicators: 'intellectual capital and innovation', 'city gateway' and 'economic clout'. And it is rated second in three factors: 'technology readiness', 'transportation and infrastructure' and 'ease of doing business'.

What these broad component indicators show, though, is that London's competitiveness is built upon London's standing as a major financial and economic centre, its central position in the world, the ability to grow and build upon technological improvement and its ability to attract a highly skilled workforce, from both within the UK and outside.

To illustrate this point, London's contribution to global creative and technology industries is enormous. London boasts a global cluster of designers, advertisers, film production specialists, games programmers and animators, artists, musicians and writers. As a result, London's creative industry is the city's second largest sector, worth $32 billion annually and generating 16% of the city's annual GVA. Almost 400,000 people work in London's creative industries, of which 77,000 are directly employed in film, video and broadcasting.[88]

Similarly, London's technology industries have been growing rapidly, again giving London critical advantage in a global context. For example, London & Partners reported that, in the second half of 2016, London received six times more investment from US tech firm backers than any other European city.[89] Tech investment by US venture capitalists totalled £325 million across 33 deals between 23 June and 15 September 2016. This compared to just £49 million in Dublin and £28 million in Paris, second and third highest, respectively. This London picture was reinforced by later, national, data from London & Partners which showed that the UK's tech sector attracted £6.7 billion of investment in 2016 – more than any other European country, and more than one-third of the European total.[90]

Within the tech sector generally, the financial technology (Fintech) sub-sector lies at the very heart of the Brexit discussion. In March 2017, EY published research on the UK Fintech market.[91] The report estimated that the sector represented c£6.6 billion in revenue in 2015, attracting c£524 million of investment and employing c61,000, or 5% of the total financial services workforce (Figure 2.18). The report noted that "more people work in UK FinTech than in New York FinTech, or in the combined FinTech workforce of Singapore, Hong Kong and Australia".[92] It should

Hub	Market Size (£ million)	Investment (£ million)	Employees
United Kingdom	6,600	524	61,000
New York	5,600	1,400	57,000
California	4,700	3,600	74,000
Germany	1,800	388	13,000
Australia	700	198	10,000
Hong Kong	600	46	8,000
Singapore	600	44	7,000

Figure 2.18 Global FinTech hubs compared, 2016

Source: EY (2017).[93]

be noted that, in official statistics, some Fintech jobs will be accounted for in Information and Communication activities rather than Financial Services.

The report ranks the UK Fintech ecosystem the largest in the world, ahead of California (2nd), New York (3rd), Singapore (4th), Germany (5th), Australia (6th) and Hong Kong (7th). The rankings are based on 'ecosystem attributes', including availability of talent; capital; policy and demand. But it also makes the point that, despite "the UK's leading position as a global capital for FinTech, its long-term position is not assured, as other regions accelerate policy initiatives, specialist regions emerge, and China emerges as a FinTech juggernaut".

Of course, as is illustrated in other parts of this book, London is much more than a place of commerce. Its cultural backdrop, including the entertainment industry, its heritage and its open and cosmopolitan outlook make London an international magnet.

- The population has risen dramatically, by one million since 1990, and is forecast to grow by another two million over the next two decades, reaching ten million people.
- London is a hugely cosmopolitan city, home to over 200 nationalities, speaking over 300 languages.
- London boasts access to some of the best graduate talent in the world, with more top world-ranking universities than any other city.
- The architecture is among the best in the world: since the mid-1980s the number of iconic buildings and structures has mushroomed.
- Foreign workers arrive in great numbers: 40% of the workforce was born overseas, and London now rates as France's sixth largest city.
- London is the most visited city in the world, with overnight visitors reaching 31.5 million in 2015, made up of 12.9 million domestic and 18.6 million overseas visitors.
- Theatre land continues to flourish: the West End took record box office sales of £705 million for 15 million attendees in 2017.[94]
- Residential prices defy economic gravity, as overseas buyers snap up luxurious developments.

The transformation of London's office economy following the 'tech revolution' that began in the 1980s was extraordinary. It was both more fundamental and more far reaching than the period of rapid change and innovation that had occurred a century earlier. London stormed ahead of its continental rivals in Paris, Frankfurt and Berlin. It consolidated its role as one of a handful of Global Cities – alongside New York, Singapore and Tokyo. It took the new technologies of global trade and re-built its office accommodation around them: London's commercial architecture and interior design from the late 1980s onwards was a world away from that of the 1960s and 1970s.

Perhaps even more importantly, London was in a prime position to respond to the emerging technologies of the late 1990s and 2000s – laptops, email, internet, social media and so on – developing competitive advantage in the software industries, video and film and countless other applications of digital technology. London is now a leading global city for tech investment and growth.

Over the past three decades, London's office economy has been reborn. The question obviously remains as to how strong its global position will remain, and then there is the question of further disruption that might be expected, and what the implications for the office economy might be.

Notes

1 N.A. Pevsner (1976) *A History of Building Types,* Thames & Hudson, London, p199
2 A.M. & L. Neal (2011) Amsterdam and London as Financial Centres in the Eighteenth Century, *Financial History Review,* Vol 18, No 1, pp21–46
3 The Royal Exchange, www.theroyalexchange.co.uk/christmas/changing-faces/. Accessed 12th July 2020
4 Pevsner (1976) *Op cit,* p199
5 Carlos & Neal (2011) *Op cit,* p25
6 Wilson B (2020) *Metropolis: A History of Humankind's Greatest Invention* Jonathan Cape, London, p181.
7 *Ibid,* p26
8 *Ibid*
9 E. Lipson (1947) *The Economic History of England Vol II the Age of Mercantilism* (4th edition), Adam & Charles Black, London, p184
10 F. Sheppard (1999) *London: A History,* BCA Books, London
11 R. Gray (1978) *A History of London,* Hutchinson & Co, London, p151
12 M. Green (2019) The Surprising History of London's Fascinating (but Forgotten) Coffeehouses, *The Telegraph,* www.telegraph.co.uk/travel/destinations/europe/united-kingdom/england/london/articles/London-cafes-the-surprising-history-of-Londons-lost-coffeehouses/. Accessed 27 October 2019
13 Wilson B (2020) *Metropolis: A History of Humankind's Greatest Invention* Jonathan Cape, London, p181.
14 Pevsner (1976) *Op cit,* p202
15 Wilson B (2020) *Metropolis: A History of Humankind's Greatest Invention* Jonathan Cape, London, p192.
16 Cowan (1969) *Op cit*
17 I.S. Black (2000) Spaces of Capital: Bank Office Building in the City of London, 1830–1870, *Journal of Historical Geography,* Vol 26, No 3, pp351–375
18 Pevsner (1976) *Op cit,* p200
19 I.S. Black (2000) Spaces of Capital: Bank Office Building in the City of London, 1830–1870, *Journal of Historical Geography,* Vol 26, No 3, pp351–375.
20 Thomas JPL (1960) *Admiralty House, Whitehall* Country Life, London p9.
21 The Charles Lamb Society, www.charleslambsociety.com/c&m.html. Accessed 23 April 2020
22 *Ibid*
23 British Library, https://blogs.bl.uk/asian-and-african/2017/01/east-india-company-headquarters-on-leadenhall-street.html. Accessed 24 April 2020
24 H.V. Bowen (2008) *The Business of Empire: The East India Company and Imperial Britain, 1756–1833,* Cambridge University Press, Cambridge, pp148–149
25 T.N. Talfourd (1838) *The Works of Charles Lamb Volume 1,* Harper & Brothers, New York, p196

26 Black (2000) *Op cit,* p362
27 P. Jefferson-Smith (2020) *The L'Ansons – A Dynasty of London Architects & Surveyors,* Clapham Books, London
28 E. l'Anson (1864) Some Notice of Office Buildings in the City of London RIBA Transactions XV 1864–65, pp25–36 As described by *Pevsner* (1976) *Op cit,* p214
29 D. Kynaston (1994) *The City of London Volume 1 a World of Its Own 1815–1890,* Chatto & Windus, London, p246
30 Gray (1978) *Op cit,* pp295–296
31 S. Jenkins (1975) *Landlords to London: The Story of a Capital and its Growth,* Book Club Associates, London, p100
32 B. Simpson (2003) *A History of the Metropolitan Railway Vol 1,* Lamplight Publications, Witney
33 J.R. Day & J. Reed (2010) *The Story of London's Underground* (11th edition), Capital Transport, London
34 *Ibid*
35 Transport for London, https://tfl.gov.uk/corporate/about-tfl/what-we-do
36 Transport for London (2018) *Annual Report and Statement of Accounts 2017/18,* 25 July
37 Black (2000) *Op cit,* p358
38 *Ibid,* p359
39 J.H. Dunning & E.V. Morgan (1971) *An Economic Study of the City of London,* Economists Advisory Group, George Allen & Unwin, London, p33
40 *Ibid*
41 C. Powell (1998) *The British Building Industry Since 1800: An Economic History* (2nd edition), E&FN Spon, London p14
42 Kynaston (1994) *Op cit,* p288
43 E. Pelegrin-Genel (1996) *The Office,* Flammarion, Paris, p26
44 Dunning & Morgan (1971) *Op cit,* p34
45 O.G. Pickard (1949) Office Work and Education 1848–1948, *The Vocational Aspect of Education,* Vol 1, No 3, pp221–243
46 Dunning & Morgan (1971) *Op cit,* p34
47 *Ibid*
48 *Ibid*
49 *Ibid,* p39
50 *Ibid,* p35
51 H. Rose (1994) *London as an International Financial Centre: A Narrative History,* The City Research Project, Corporation of London, p6
52 Kynaston (1994) *Op cit,* p247
53 *Ibid,* p289
54 Dunning & Morgan (1971) *Op cit,* p38
55 Cowan (1969) *Op cit,* p28
56 R. Turvey (1993) London Lifts and Hydraulic Power, *Transactions of the Newcomen Society,* Vol 65, No 1, pp147–164
57 M. Bowley (1966) *The British Building Industry: Four Studies in Response and Resistance to Change,* Cambridge University Press, Cambridge, p9
58 *Ibid,* p10
59 J. Gottmann (1966) Why the Skyscraper? *Geographical Review,* Vol 56, No 2, pp190–212
60 J. Zukowsky (Editor) (1987) *Chicago Architecture Volume 1 1872–1922 Birth of a Metropolis,* Prestel; J. Zukowsky (Editor) (1993) *Chicago Architecture and Design Volume 2 1923–1993 Reconfiguration of an American Metropolis,* Prestel-Verlag. München
61 Kynaston (1994) *Op cit,* p288
62 R.C. Michie (1992) *The City of London: Continuity and Change, 1850–1990,* Macmillan, London, p63
63 *Ibid,* p39

64 P. Ackroyd (2000) *London: The Biography,* Vintage, London, p717
65 H. Rose (1994) *Op cit,* p21
66 *Ibid,* p23
67 *Ibid,* p22
68 A. Black, D. Muddiman, & H. Plant (2007) *The Early Information Society: Information Management in Britain Before the Computer,* Ashgate, Aldershot, pp136–137
69 Mumford (1940) *Op cit,* p228
70 *Ibid,* p227
71 *Ibid*
72 *Ibid*
73 D.L. Miller (1989) *Lewis Mumford: A Life,* Weidenfield & Nicolson, New York, p355
74 *Ibid*
75 Mumford (1940) *Op cit,* p480
76 Daniels (1975) *Op cit,* p31
77 J.B. Goddard (1975) *Office Location in Urban & Regional Development,* New York, Oxford University Press, p4
78 London Research Centre (1986) *Office Workers in London,* Statistical Series No 55 LRC, London, p11
79 Howells & Green (1988) *Op cit,* p12
80 See J. Friedman and G. Wolff (1982) World City Formation: An Agenda for Research and Action International, *Journal of Urban & Regional Research,* Vol 6, pp309–344
81 1801–1991: Registrar General's Reports; Office of Population and Surveys. Cited in Sheppard (1999) *Op cit,* p364; 2001 and 2011 ONS, Census of Population; 2011–2018 ONS, Population Estimates
82 LMG & BCG (2014) *London Matters: The Competitive Position of the London Insurance Market,* London Market Group, London, p6
83 Preqin (2017) *Hedge Funds in Europe*
84 ZYen (2009) *The Global Financial Centres Index 5,* Corporation of London, London p6
85 ZYen (2018) *The Global Financial Centres Index 24,* China Development Institute, London, p12
86 *Ibid,* p18
87 PWC (2016) *Cities of Opportunity 7,* PricewaterhouseCoopers, London, p8
88 London & Partners (2011) *London Creative Industries,* Press Release, p10
89 London & Partners Press release, *London Europe's Number One Business Hub Say Leading US Tech Execs,* www.londonandpartners.com/media-centre/press-releases/2016/20160919-london-europes-number-one-business-hub-say-leading-us-tech-execs. Accessed 21 September 2016
90 London & Partners (2017) *UK Tech Sector Leads European Investment in 2016,* Press Release, 12 January
91 EY (2016) *UK FinTech: On the Cutting Edge,* HM Treasury, London
92 *Ibid,* p7
93 *Ibid,* p15
94 Figures from Society of London Theatre, *Press Release,* 29 January 2013. Accessed 1 October 2018

3 Explaining

A facet of the city

It is a curious fact of history that even as office buildings came to dominate city centres in the nineteenth century, there remained only the crudest understanding of the internal dynamics of the office economy. For decades, academic analysis of cities barely recognised offices as a distinct facet of urban structure, and when they did, the work was largely descriptive rather than analytical.

This neglect of offices in academic endeavour was reflected in spatial policy (Chapter 4) and in the supply process (Chapter 5), as both struggled to keep up with the changing dynamics of the office economy. The situation has changed in more recent times as studies have begun to explain the role of office activities in the changing structure of cities. This chapter explains how our understanding of cities, and particularly the role of the office economy within them, has evolved. There are a number of ways in which this story could be told; the approach taken here is to trace the chronological development of academic work.

3.1 Classical economics

The question of how land values are determined, and how different land uses are organised spatially, is one that has intrigued economists for centuries. In a modern context, this question is particularly relevant to the clustering of office activities in city centres, where values and densities are typically highest. Why can offices afford to pay the high rents, but not factories? Many cities have seen their centrally located manufacturing bases replaced not only by offices but also by other, mainly service, uses that can afford the rents determined by the high land values. In many cases it is possible to trace a ripple effect as land uses less able to afford rising land values are forced ever outwards towards the edge of the city. There is clearly an economic process at work.

The first attempts to answer the land value question took place when the economy was still largely agrarian. In his *Wealth of Nations* (1776), Adam Smith argued that landlords held a monopoly position, and value was not related to their effort or outlay, but rather to how much they could extract from a tenant. He was also of the belief that a landlord's profit costs "neither labour nor care, but comes to them, as it were, of its own accord, and independent of any plan or project of their own". Smith suggested that the "ease and security of their situation" lead landowners to a life of "indolence".[1] The belief that landlords might suffer collectively from the curse of indolence, of course, has survived in public discourse to the present day.

Perhaps more importantly, it was Adam Smith who speculated on the division of labour within industrialism and its role in driving productivity. The specialisation and concentration of workers in their subtasks, he argued, leads to greater skills and productivity than would be achieved with the same number of workers each undertaking the original broad task. He famously cited pin-making to make his point:

> A workman not educated in this business . . . nor acquainted with the use of the machinery employed in it . . . could scarce, perhaps, with his utmost industry, make one pin in a day, and certainly could not make twenty.[2]

To achieve an output of 20, Smith observed that the pin-making process is divided into 18 distinct stages. He also recognised the role of machine capital in the manufacture process. In short, he was recognising the blending of human skills and technology, in the context of a wider organisation (the firm) to enhance productivity. This becomes highly relevant when we consider the 'invisible glue' that today holds knowledge economy firms together in tight business ecosystems, or business clusters, in our cities. For much of the office economy today, the degree of specialisation that separated people and sub-tasks in Smith's 'production' process; is now reflected, only to a greater degree, in 'knowledge intensive' businesses. 'Knowledge capital' has in some ways replaced machine capital and allowed infinitely more specialisation in roles that can be applied across widely differing economic activities. The interdependencies between these specialisations helps form the spatial glue.

But, of course, in the mid-eighteenth century "cities were relatively unimportant in the landscape and viewed as parasitic on the honest toil of agriculture", and the lack of concern "for urban land continued until the late-19th century".[3] In these largely agrarian times, cities were regarded with some suspicion, and even manufacturing activities were in their infancy in terms of being a significant land use. Consequently, when David Ricardo published *On the Principles of Political Economy and Taxation* (1817), one of the most famous treatise of its era, he was able to argue that the most fertile lands will be the first to be farmed, with increasingly less fertile land utilised as demand grows. In doing so, Ricardo was introducing the notion of differential value, or rent, based on utility. Of course, this argument had little to do with centrality (and the notion of the highest rents being in the centre of cities), simply to some concept of the 'quality' of land. Ricardo also had harsh words for landlords: mere parasites and members of the unproductive class.

Ricardo's thesis was quickly followed by Johann Heinrich von Thünen's *Der Isolierte Staat in Beziehung auf Landwirtschaft und Nationalekonomie (The Isolated State with Respect to Agricultural Economics*, 1826), in which he developed the question of locational rent more fully. He argued that agricultural land uses bid for the use of land, which is assigned to the highest bidder, and that the use to which a piece of land is put is a function of the cost of transport to market and the rent a farmer can afford to pay. The most distantly cultivated land yields no savings in transportation, which means that the land there will be rent free. Von Thünen had introduced the notion of a bid-rent curve, whereby values fell with distance from the market (or centre).

John Stuart Mill's *Principles of Political Economy* (1848) said little about the mechanics of land value, but he highlighted the importance of ownership. With the key components of production being labour, capital and natural agents, then

> the only person, besides the labourer and the capitalist, whose consent is necessary to production, and who can claim a share of the produce as the price of that consent, is the person who, by the arrangements of society, possesses exclusive power over some natural agent.[4]

Because land is the principal natural agent, landowners are "able to require rent for their land" because "it is a commodity which many want, and which no one can obtain but from them. If all the land of the country belonged to one person, he could fix the rent at his pleasure".

Not to be outdone by either Smith or Ricardo, Mill also had a few choice words for landlords.

> They grow richer as it were in their sleep, without working, risking or economising. What claim have they, on the general principle of social justice, to this accession of riches?[5]

'Landlord and tenant' is phraseology that survives to the present day, not least in the context of the office economy, through the Landlord & Tenant Act. This piece of legislation is examined more fully in Chapter 5, but given the negative comments of such eminent classical economists, it is worth pausing to explain the origin of the terminology which continues to carry enormous influence over the market structure.

During the medieval period, under English feudal law, all land belonged to the king, who granted landed title to members of the nobility, who, in turn, raised knights and soldiers. The 'lords of the land' protected villagers, or serfs, whose main role was to farm the land. Where their holdings were large, the lords leased land to aristocratic peers, who in turn, often sublet. Landholdings in medieval England were called 'fiefs'. In return for the protection, each serf committed a portion of his output (produce rather than money) to the landlord; this was his 'rent'. When feudal land grants were of indefinite duration, they were referred to as freeholds, while fixed-term grants were termed non-freehold, or leasehold.

Because the Church forbade usury, or lending at interest, landed lords started to buy and sell leases of land and, by around 1500, the *leasehold* had become a form of transaction. The practice of transferring land in the form of leaseholds forms the legal foundation for the commercial leases that we have today, albeit no longer with the feudal right to exploit the output of the serfs. We continue with the feudal terminology of 'landlord and tenant' which, as we shall see in Chapter 8, has interesting ramifications in the very different world of the knowledge economy rather than a feudal, or even industrial, economy.

Urban land values

While the classical land economists helped our understanding of the role of land ownership, none of them helped us understand land value in an industrialised economy, let alone an office economy. It was to be another four decades after JS Mill before the first formulation of a coherent theory of urban land values began to emerge. Alfred Marshall was born in 1842, the son of William Marshall, an early worker in the office economy as a clerk at the Bank of England. Alfred set off in 1875 on a grand tour of the USA, courtesy of the largesse of an uncle. Upon returning in 1876 and taking a position at St John's Cambridge, Marshall became engaged to Mary Paley, a descendent of the theologian and philosopher William Paley, and then a student at Newnham College.

The couple married in 1877 and moved to Bristol, where Alfred secured the positions of Foundation Principal and Professor of Political Economy at the recently created Bristol University College. Alfred became absorbed in the administrative duties of his role, and he asked Mary to take on much of his teaching work. Mary compiled and published her lectures as *The Economics of Industry* (1879), with Alfred as the disputed co-author.

Alfred's workload led to a period of ill health and his resignation from Bristol in 1880. He and Mary headed to Palermo, Sicily, for a period of convalescence, and while recovering, Marshall began writing what would eventually become his most famous work. Soon after their return to England, Alfred was appointed in 1883 to Balliol College, Oxford. In 1885, he was elected Professor of Political Economy at Cambridge University, where he would direct economics for the next two decades. The *Principles of Economics* was published in 1890.

Whilst not quite as long in preparation as Charles Darwin's *On the Origin of Species* (1859), it is probably no coincidence that the title page of *Principles* contained the famous epigram, "*Natura non facit saltum*", or "Nature does not make jumps".

Marshall argued that the land use that captures a particular site is the one "from which the most profitable results are anticipated". He argued that potential users of land make bids for various sites based on their respective location advantages, and the highest bidder captures the land in each case.[6] This concept of price bidding (shown in Figure 3.1) came to dominate urban land economics, resulting in "a voluminous literature" and contained "the essentials of present day theory".[7]

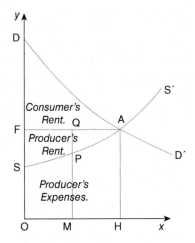

Figure 3.1 Alfred Marshall's classic supply–demand graphic
Source: Marshall (1890).

It also provided one of the earlier insights into the spatial organisation of the office economy.

The frenetic economic activity of the latter three decades of the nineteenth century which led to the birth of the modern office, and the consequent increases in building densities and inflated land prices at the core of most cities, continued to attract the attention of economists curious to explain the forces at work.

Most basically, the work suggested that land prices are determined by the interactions of the owners of real estate and those who wish to rent or buy buildings for transacting their business or as living space. This pricing mechanism reflects perceived levels and sources of demand. As demand for land intensifies, so less intense land uses and less profitable uses are displaced to fringe or transitional areas where the cost of centrality is less onerous. The classical explanations for these processes that cause activities to be tightly clustered around urban centres are traced back through inelegantly named theories such as agglomeration economics and bid rent theory. The explanations typically involve individual enterprises in decisions to maximise the benefits of a convergent public transport system; exploit access to labour, gain a competitive position and so on. In so doing they suggest an explanation for the tendency of central areas to rise vertically.

Perhaps the most significant work from these early years, and certainly the one that cast the longest shadow, was *Principles of City Land Values*, first published in 1903 by Richard Hurd.[8] One of the earliest proponents of urban modelling, Hurd suggested that a cursory glance revealed similarities among cities, and that closer analysis revealed that their structural characteristics responded to definite principles. The key principle that he identified, adopted and adapted from Ricardian theories of agricultural land value theory was that economic rent in cities was based on locational advantage:

> Since value depends on economic rent and rent on location, and location on convenience, and convenience on nearness, we may eliminate the intermediate steps and say that value depends on nearness.[9]

In other words, a desire for businesses to be close together forces land values upwards: straightforward supply and demand dynamics. Those who can afford the higher prices stay; those that cannot relocate. Ultimately, it was this 'price war' that services won over manufacturing and other lower-margin activities, and which led to the office economy physically concentrating in city centres.

Hurd developed this thinking, going on to propose a more general citywide model of structural form, in which he argued that cities grew outwards in a sectoral pattern:

> Cities originate at their most convenient point of contact with the outside world and grow in the lines of least resistance or greatest attraction . . . Growth in cities consists of movement away from the point of origin in all directions, except as topographically hindered . . . Central growth takes place both from the heart of the city and from each subcentre of attraction, and axial growth pushes into the outlying territory by means of railroad, turnpikes and street railroads.[10]

Hurd's was the first comprehensive model of urban structure, and he set the agenda for the next 40 years as urban geographers, economists and others strove to find and explain pattern in urban structure. Indeed, so seminal was his work considered by some that Paul Wendt was moved to comment half a century later: "No more comprehensive analysis of the dynamics of city growth has appeared since his work".[11] Praise indeed, but though comprehensive in its context, Hurd left much detail to be elaborated upon. Not least, like all his contemporaries, he failed to give any separate treatment of the office economy.

3.2 Spatial explanations

The classical economic interpretation of land values is helpful, but it is limiting. It says little about the forces driving the spatial arrangement of urban activities. As the twentieth century dawned and the nascent office economy consolidated its role, particularly in larger urban conurbations of the USA, and as the internal combustion engine wrought great change on cities, particularly by encouraging suburbanisation, so the analysis of urban morphology evolved and took a turn towards a more sociological approach that was sensitive to the differences between urban areas, or spatial patterns. In particular, there was the work of a group of scholars collectively referred to as the Chicago School, after the city in which they worked. During the 1920s and 1930s, these early theorists generated a number of models of urban morphology that are taught to undergraduates of geography to this day. They were ecologists and sociologists, and for them the urban system was a metaphor for an organism in which they were trying to understand emergent processes.

Perhaps the most famous of the models, known as the concentric zone theory, was put forward by Ernest Burgess in 1925.[12] The model is based on the notion that urban growth is a centripetal force, expanding outwards in a series of concentric zones, each with their own characteristics (rather like the ripple patter described earlier). The central area is surrounded by a zone in transition where high value uses are pushing outwards. And this zone yields to residential uses, and gradually to yet less intensive land uses.

Recognising the growing influence of the motor car, in 1932 Frederick Babcock proposed a modification to the model, suggesting that the concentric ring theory would be impacted by the sub-optimal distribution of route ways.[13] This would result in sites a given distance from the centre being either more or less accessible to the centre. This led to the axial theory, where development away from the centre grows outward further along major transport routes than in the interstices. A second model, proposed by Homer Hoyt in 1939,[14] was also strongly influenced by the road infrastructure. His proposal was that differences in land use near the centre would be perpetuated as the city expanded, thereby creating distinctive wedges of activity stretching from the centre, often along route ways. This gave rise to the sector theory of urban growth. In all three of these models, the close relationship between motorised transport and urban growth can be clearly detected.

The final spatial model was that proposed by Chauncy Harris and Edward Ullman. Theirs was the multiple nuclei theory,[15] a model that proposed a cellular structure for cities in which distinctive clusters of land use activities developed around specific points, or nuclei. The model recognised that most cities do not originate from a single central area but evolve, rather, by the gradual coalescence of a number of separate nodes into a single urban mass.

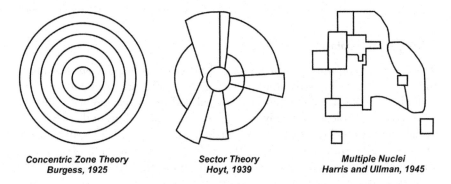

Concentric Zone Theory
Burgess, 1925

Sector Theory
Hoyt, 1939

Multiple Nuclei
Harris and Ullman, 1945

Figure 3.2 Spatial explanations of urban structure

A common characteristic of the models of urban morphology up to this point was their two-dimensional representation of the dynamics of urban change over time. Furthermore, they remained focused on residential and industrial activity, saying actually very little about the growing influence of the office economy. Nowhere was this truer than in the area that lay at the centre of each of the models: what had become known in the USA as downtown, or more formally, the central business district (CBD). The lack of any real understanding of the CBD up to this point possibly reflects the fact that it sat at the centre of each of the models, with all the change, development and growth taking place around, for example in the zones of transition. Ironically, the CBD was isolated by its own centrality, and not regarded as needing any particular explanation. Yet it was here that the office economy was emerging as the dominant land use.

The CBD represents the theoretical point of maximum accessibility in the urban system. It is the focus for the public transport system, both intra- and inter-urban systems. And it is home, generally, to those activities that require the highest levels of contact, whether business-to-consumer or business-to-business, and companies arrange themselves in order to minimise the friction of distance. As we have seen, the intensity of land use has historically led to an interpretation of urban morphology that focuses on competition, and price, for land. The classic bid-rent curve – so beloved of urban geographers and a set piece on all first degree courses for over 40 years – sits at the heart of this analysis.

The first major work to recognise that the CBD was not a homogeneous lump of economic activity but rather a complex interplay of relationships and activities feeding off each other and the benefits of centrality was that by Robert Haig.[16] This famous urban geographer observed that through a process of competition for space, the CBD becomes divided into functional sub-areas where strongly linked economic activities agglomerate. This complexity can be easily seen in central London. For many years London was a rich mosaic of 'sub-markets' of economic activity that clung together in their historic clusters before being nudged out by the higher-rent-paying white-collar activities. Of course, the City of London itself was a cluster, albeit one broken down into a number of concentrations of distinctive business activity, including banking, stock broking and jobbing, insurance, shipping, fur trades, publishing and commodities. To this day, the insurance sector remains tightly packed into the EC3 postal district in the south-east corner of the City. Such clusters were often associated with the historic foci of markets such as the Bank of England, the Lloyd's insurance market, the wharfs of the fur trade or the Commodity Exchange. But apart from the City, there were many other sub-markets within London's CBD.

In the West End, government-related activities grouped around Whitehall and the civil engineering practices clustered close by to facilitate lobbying when Acts of Parliament, for example, had to be passed before constructing canals and railways. Similarly, large oil companies were clustered in the Victoria area; while private banks and small headquarters were predominantly in Mayfair and

St James's – handy for the 'gentlemen's clubs'. Other examples of central area clustering included:

- Clerkenwell, jewellery;
- Fleet Street, printing;
- Harley Street and Wimpole Street, medicine;
- Inns of Court, solicitors and lawyers;
- Mayfair, property companies and surveyors;
- North of Oxford Street, clothing;
- Shoreditch, furniture and printing, and
- Victoria, government.

While the urban ecologists had begun to peel back the layers of activities within cities to reveal dynamism and process, the models they worked with continued to operate at the citywide level, rather than at the level of the activities therein. And they were predominantly descriptive rather than explanatory.

While the Second World War brought the physical growth of the office economy in London to a temporary halt, and post-war rationing and austerity measures delayed its resumption until the early 1950s, the period saw the ascendancy of the office economy in the national economy. However, the general approach to defining rather than explaining urban morphology continued into the post-war years. The work of Raymond Murphy and James Vance typified much of this effort.[17] They were primarily concerned with the desire to provide a uniform method for aerial definition that could replace the fixing of boundaries by haphazard and local procedures. They postulated, for example, that the CBD, as a spatial unit, could be defined through a series of measures that reflected its essential characteristics. In this way the physical attributes of the CBD were to be measured independently of a citywide analysis of urban structure. Their measurements included the definition of uses considered to be quintessentially CBD activities.

Echoes of this approach were to be found in the development of planning policy in central London many years later, where those activities not considered to have an essential requirement for a central location were to be encouraged to relocate (see Chapter 4). The physical parameters used by Murphy and Vance included the volume of space dedicated at any one time to various uses, and building height and intensity indices, which could be used to define more clearly the boundaries of the CBD. They provided no insight into the emerging influence of the office economy, and like other work from the time, there was no analysis of the internal dynamics of the patterns that they mapped.

3.3 Towards an activity-based approach

In the previous chapter we discussed Lewis Mumford's magnificent *The Culture of Cities*.[18] As implied by the title, Mumford sought to bring more of a cultural, or social, emphasis to urban analysis. For example, he argued that while cities had expanded physically, "the essential social nucleus" had been treated as an afterthought. He proposed that "social facts are primary, and the physical organization of a city, its industries and its markets, its lines of communication and traffic, must be subservient to its social needs".[19] He suggested that it was necessary to define the interrelationships between activities in order to lay down "the outlines of an integrated city"[20] which would create a more humanistic city.

One attempt to address this imbalance shifted the focus of study from spatial definition to a more explanatory approach with a less static, less definitional interpretation of urban morphology. Ralph Turvey[21] argued that cities are economic systems working towards, but never reaching, equilibrium:

> It is impossible to present a comparative static analysis which will explain the layout of towns and the pattern of buildings: the determining background conditions are insufficiently stationary in relation to the durability of buildings.[22]

Turvey was suggesting that a simplistic economic representation such as the rent-bid model had major shortcomings and that analysis had to be more behavioural:

> In other words, each town must be examined separately and historically. The features of London, for example, can be fully understood only by investigating its past: it is as it is because it was what it was.[23]

But perhaps the most significant work to emerge during the 1950s was John Rannells's study of central Philadelphia, *The Core of the City*.[24] This work is so important because of its emphasis on the role of uses, or activities, in shaping the morphology of the city.

In a reference to earlier work, and particularly the work of the urban land economists, Rannells highlighted the difficulties and problems inherent in distinguishing between land use factors associated with location, and those unrelated to location. Thus: "location is only one factor in the operation of any given activity".[25] More importantly, he made the critical distinction between broad land use changes associated, in classical terms, with utility maximisation as a factor of access, and the influence of operational decisions on the level and structure of final demand for space:

> It is necessary to take in somewhat more than relates directly to land use, since changes in demand for space or location may come about as secondary results of business decisions or consumer preferences which, in themselves, have no concern with questions of land use.[26]

The emphasis was abruptly shifted away from concepts of rational decision-making, based on the assumption of 'complete knowledge', into an area in which decisions that ultimately result in a land use manifestation are made for other, unrelated reasons. The decisions reached are, instead, related to the operational requirements of the companies themselves, and their linkages to one another. This is fundamental to our understanding of the evolution of the office economy, and particularly its concentration in CBDs.

Rannells was less concerned with defining any preconceived area than with identifying the interlinked systems of activity that were located in the central area. His work was not locational in the sense of spatial definition; rather, he used location-specific data to work out patterns of relationships. Rannells explained his approach:

> The underlying systems of activity, which result in varying patterns of urban land use, are organised around establishments, classified by kinds of business. These establishments, in turn, are the units for analysis of land use.[27]

These "definite locations" are chosen for a variety of reasons. Rannells recognised the availability of suitable accommodation, services and facilities, "as well as the associated activities of closely linked establishments or the market which their operations require".[28] However, wherever these requirements are to be met, access to other establishments is crucial because:

> members of every group benefit by nearness to establishments with which they have linkages and by the availability of services of all kinds which they require.[29]

Quite clearly, Rannells's work differed markedly from that which it followed. His was the first critical examination of the role of economic linkages in shaping urban structure. As he argued, the locational characteristics of activities are "visible manifestations of the many systems of activity into which the city's life is organised".[30] However, he took the analysis much further by examining the impact of the activities undertaken within buildings and the nature of the urban fabric. He argued that activities "carried on within an establishment determine, in general, the quantities and qualities of space used by itself for work, for display, for processing, for storage or for circulation".[31]

The most important contribution of Rannells's work was to shift the focus of study away from the spatial delimitation of activities and towards an examination of the characteristics of the activities that congregate in the urban area, and the interactions between them that explain the process of change over time. It was not, however, an analysis of the unique features of the office economy.

The Core of the City was very quickly followed by Edgar Hoover and Raymond Vernon's seminal work, *Anatomy of a Metropolis*. This work examined the changing distribution of people and jobs within the New York metropolitan region. The whole of the fourth chapter was devoted to "the white collar corps", one of the earliest attempts to separately articulate the *raison d'être* of the office economy. The chapter opens on a historical note, observing that for a century and a half,

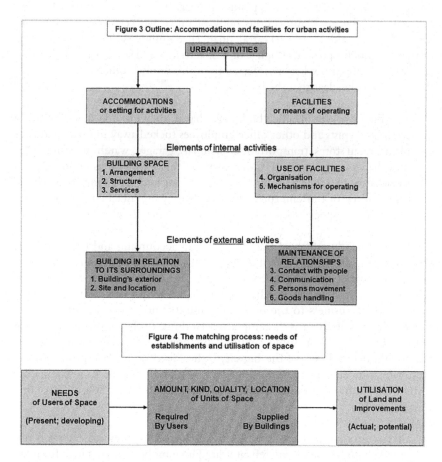

Figure 3.3 Key Diagrams from Rannells's *Core of the City*
Source: Reproduced from Rannells (1956).

Lower Manhattan has been associated with an interrelated mass of enterprises and institutions which collectively make up the financial community. Like its counterpart in London, the community – much of it – is cramped into a few narrow and cavernous streets, held together by seemingly potent forces.[32]

Hoover and Vernon's mention of potent forces is an early reference to the clustering behaviour of the office economy. They concede that such businesses do not need proximity because they can utilise "the telephone or teletype", but they are driven to do so because the professionals "all have their ears and minds attuned to a common background of facts and surmises against which commitments are made and avoided".[33] This is resonant of Hurd's (1903) observation mentioned earlier that "value depends on nearness". They go on to describe the scale of "the great amorphous body of office workers" in New York in 1956. Thus in addition to 311,700 financial community jobs, there were a further "895,000, spread through scores of activities and occupations". And, in a move to distinguish what are referred to in this book as office economy jobs, the authors make clear that these

> are office workers in office buildings; that is, they exclude the 685,000 or so clerks, typists and other office employees tucked away in manufacturing plants, retail stores, transportation terminals, garages, warehouses and other miscellaneous structures. For our purposes the distinction is important; for these latter 685,000, we must assume, are located according to the locational needs of the enterprises to which they are attached and not with regard to any special locational needs of the office itself.[34]

Noting that nearly a fifth of the jobs "are the accounting and legal services, the advertising and employment agencies, the architects and engineering services, and a variety of similar activities", the authors associate these jobs with what we refer to here as the office economy, as being firms "which look for much of their business to the office community and to one another". They suggest that while there is "a temptation to think of this clustering of office work as being based on some nebulous preference of executives for the fast tempo and bright lights of the big city", the "forces that have produced the clustering can be much more objectively defined". Then, in a fascinating insight, the authors suggest office professionals, or "the office elite", have one trait in common:

> they cannot say with any measure of certainty what their critical problems are likely to be during any given year. These problems may range from tax questions to television advertising, from Guatemala's economy to California's politics.[35]

Such complexity and uncertainty "largely determines the location of the accountants, lawyers, advertising agencies and similar specialists who contribute to the operations of the office elite", because like "the suppliers to the garment trade, they are influenced in their locational tendencies not only by the need to be near their market but also by the need to be near the services on which they in turn draw".[36] Thus:

> The location of each new increment of office activity was governed by the location of the pre-existing mass and by the ties of communication to the mass.[37]

Hoover and Vernon were getting much closer to a clearer understanding of the forces that pulled activities towards the CBD, thereby creating a cluster of activities dominated by the office economy. These are more commonly referred to today as agglomeration economics.

As the 1950s drew to a close and the 1960s began, while most work on urban morphology remained doggedly focused on spatial definition, there were the first stirrings of interest in the evolution and physical arrangement of the office economy. An early modification to Murphy and Vance's work was followed by Edgar Horwood and Ronald Boyce,[38] who separated the core of the CBD (the area where quintessentially CBD land uses were at their most intense) from the frame in which it was located.

Even with this elaboration (and with the notable exception of Rannells), the study of the structure of cities had not really moved on very far from pre-war approaches: it was still essentially Chicago school-style spatial description. As late as 1964, William Alonso's influential *Location and Land Use: Toward a General Theory of Land Rent* was based on the central tenet that cities could be modelled on a featureless plain principle, ignoring the effects of "hills, low land, beautiful views, social cachet, or pleasant breezes", which while important, "no way has been found to incorporate them".[39]

Mention should also be made of the contribution of French geographer Jean Gottmann. Gottmann (1915–1994) was born in the Ukraine under the Russian Empire, and his parents were murdered during the Russian Revolution in 1917, whereupon he was adopted by an aunt. They fled to Paris in 1921, where he experienced a cultured and cosmopolitan upbringing, eventually attending the Sorbonne. He then fled Nazi-occupied France, arriving in the USA in 1941, where he cemented his reputation as one of the greatest twentieth-century geographers. In 1956, following an approach from Robert Oppenheimer, director of the Twentieth Century Fund, Gottmann took the post of research director for metropolitan studies based in New York. In 1961 he published *Megalopolis*.

At over 800 pages, *Megalopolis* is literally a weighty tome, in which chapter 11 deals with "The White Collar Revolution". The chapter traces the growth of white-collar or tertiary occupations, and the process and the consequences of their rapid growth. For example, Gottmann describes the centripetal forces bringing about the clustering of large offices in downtowns, among which are "the need for good and reliable information" and "the need for a pool of labor". He describes the wide diversity of information being gathered, analysed, classified, filed and distributed, noting that the larger the organisation, "the longer and more variegated will be the chain of kinds of data that organization will need". He goes on:

> Every category of data in this long chain will be best handled by a specialist, or even by a profession, if the demand for that particular link is widespread enough.[40]

If this sounds like a nod to Adam Smith, it certainly was. Gottmann states that the "pooling of skills expands and sub-divides itself, according to a mechanism of the division of labor well described long ago by Adam Smith". He goes on to describe the increasing specialisation of work in the office economy, suggesting that "among the many different services needed, some are available within the office . . . while others are complementary to the office".[41] Here he is hinting at the invisible glue holding the concentration of firms together – in this case, in effect, supply chain relationships, or what later became known as agglomeration economics.

Gottmann also recognises another aspect of the glue holding firms together, notably paper.

> Office work also needs and produces large amounts of documents on paper. The volume of papers necessary to the financial operations transacted in Manhattan seems to be one of the reasons keeping close to one another the diverse establishments between which flows this great stream of documents. The huge volumes of mail that flow through Megalopolis are quite astonishing compared to the size of the mail carried over similar distances in any other part of the world.[42]

Gottmann refined his thinking on white-collar work in a later paper, "Why the Skyscraper?" He recognised that the first skyscraper in Chicago, and some of the early ones in New York, were built by insurance companies, firms "whose business is entirely bureaucratic. Their work is all on paper and entirely in transactions". He went on:

> The expression 'office industry' is a debatable one, but so far as it is in use let us stress its significance: to perform its function, an office industry does not need anything but space for executive offices, filing and clerical work, meetings, discussions, and the like.[43]

Gottmann took his conclusions a step further by suggesting that

> those sectors of employment which were, and still are, regarded by the classical economists as external activities because they did not seem to be important elements in the economic process (production through processing and distribution to consumption) – those external activities are now becoming major sectors of employment.[44]

Finally, he hinted at the nature of the matrix that binds the office economy together into the CBD.

> The daily work of decision making requires consultations with many specialists; the specialists cannot be expected to gather where there are only one or two customers, but rather where there are several thousand. Also, the specialists prefer to be near one another.[45]

Gottmann's 'office industry' is broadly what this book refers to as the office economy. It was understandable that he was cautious with the term in the mid-1960s, because even at this time, and despite the efforts of the authors cited here, there remained only the most rudimentary understanding of the office economy as a distinct facet of the modern urban economy.

Quantitative analysis

Turvey's point about historical contingency was taken up in a number of studies between the mid-1960s and early 1970s. This work took a more explanatory approach, although its emphasis was spatial rather than structural, and all placed the modern structure into the context of historic development. Davies undertook a major study of Cape Town.[46] He recognised that the core was itself heterogeneous with its own internal dynamics and changing structure and, like Horwood and Boyce, he defined a hard-core and surrounding frame. However, Davies himself conceded that: "currently there may well be diminishing returns in any delimitation studies per se".[47] Two of the primary drivers of urbanisation – access to goods and services, and contact – are ignored in spatial delineation. To describe urbanisation, as was common until very recently, simply in terms of being an expression of the competition for sites through the land pricing mechanism (e.g. through bid rent curves) ignores the character and motivation of those activities locating there. Of equal interest here is the fact that office activity was not seen as a distinctive activity within the fabric of the city.

In 1966 Harold Carter and Gwyn Rowley's published research focused on Cardiff.[48] Their approach was to explain the modern structure of the city in terms of a process of development, dating back to the medieval period, through the rapid expansion in the nineteenth century, taking into account physical features constraining the city's growth. The authors demonstrated the dynamic qualities of the CBD as a patchwork of specialised areas, a patchwork that had "emerged due to the needs created by the industrial hinterland".[49] David Ward took a very similar approach in his study of Boston[50] by recognising distinctive areas of specialised business activity, and tracing the development of the CBD from a nucleus in the nineteenth century. It was not his purpose to define areas, and, indeed "At no point . . . is the CBD as such defined nor is there any attempt at objective definition".[51] However, the study shed little new evidence on the influence of occupier characteristics on the development of sub-districts. Despite the fact that some studies had given an explicit office focus,[52] there was still no real explanation of the internal dynamics of the urban economy, nor of the complex relationships existing between the activities tightly clustered therein.

The quantitative revolution swept through urban geography and social sciences generally, from the mid-1960s to the mid-1970s, with the force of a tornado. Suddenly, 'scientific method' became *de rigueur*, and model building and multivariate analysis were used to explain urban form. The revolution had its critics, none more so than the pre-eminent David Harvey who highlighted the gap between the sophisticated theoretical frameworks popular at the time and our ability to say anything meaningful about real events.[53]

While Harvey was undoubtedly correct in a broad sense, and while the statistical regime ultimately yielded to more systemic approaches, some useful work was undertaken in pursuit of explaining the internal dynamics of cities generally and of the office sector in particular. One such area of investigation was that which examined the clustering of economic activities at the sub-local level and which measured the strength of information flows between and within organisations, and their influence on local level clustering. Bertil Thorngren,[54] Gunnar Törnqvist[55] and Brian Goddard[56] all published work based on contact between organisations and its relevance to the geography of office activity. For the first time, the internal dynamic of the office economy as a discrete sector of the urban landscape was being examined in detail. An early progenitor of such an approach was Donald Foley, who in 1956 published a paper in which he traced changes in the spatial distribution of administrative offices in San Francisco.[57]

The work of Brian Goddard[58] is of specific interest here not only because it typified much of the work taking place at the time, but also because his empirical work was carried out in central London. Goddard's early work[59] traced the changing location of publishing and advertising agencies' offices over an extended period, using successive editions of post office directories to note changes of address, new offices and office closures. He found considerable change in the distribution of activities and suggested that "the magnitude of the shifts suggest some fundamental changes in location factors".[60] Figure 3.4 shows changes in the advertising sector between 1955 and 1966.

Postal area	Deaths	Births	Change	Moves in	Moves out	Migration change	Total change
W1	41	91	50	26	22	4	54
SW1	16	17	1	5	14	(9)	(8)
WC1	20	32	12	8	13	(5)	7
WC2	32	27	(5)	12	19	(7)	(12)
EC1	7	8	1	5	2	3	4
EC2	7	5	(2)	2	3	(1)	(3)
EC3	2	1	(1)	–	3	(3)	(4)
EC4	33	25	(8)	10	11	(1)	(9)
SE1	3	1	(2)	1	–	1	(1)

Figure 3.4 Changes to the locational characteristics of the advertising sector, central London, 1955 to 1966

Source: Goddard (1967).

The changes show a texture to demand that we have not seen up to this point in other work. For the first time, there was evidence of dynamics within the office economy and evidence of spatial outcomes. Goddard's analysis showed that:

> The bulk of change between the two dates was not due to migration of firms between different parts of the centre but to the establishment of new firms in one area and the extinction of others elsewhere.[61]

It can be clearly seen from Goddard's work that a business sector's centre of gravity can be changed over time with the establishment of new firms. This has important lessons when examining the contemporary office market.

In later work, Goddard analysed taxi flows in central London in order to identify functional areas.[62] Whilst confirming the presence and importance of activity clusters, the study offered little explanation; however, his study of office linkages and location was more enlightening.[63] His empirical evidence of patterns of telephone and face-to-face meetings suggested strong evidence that the localised spatial groups of office activities were based on functional connections.

Although Goddard's work made a significant contribution to the general understanding of the distribution of sub-districts in central London and of the level of contact between individual establishments, it was limited in its description of the relevance of operational characteristics to concentration. The techniques used (among them multivariate analysis) lent themselves to aerial definition, rather than an explanation of the processes that operate to influence spatial concentrations of activities. With a few notable examples, no one had peeled back the layers of spatial form to reveal the internal dynamics of the office economy. As late as 1972, Brian Goodall's standard text, *The Economics of Urban Areas*,[64] gave no separate treatment to office activity. Goodall's preference was to stick closely to a narrow bid rent interpretation of urban land use because he believed that while spatial definition failed because urban areas never attained equilibrium, "the price mechanism always operates". He went on:

> Space and location are economic commodities subject to supply and demand forces. The urban real property market requires of each site or real property its highest and best use and this is largely determined by competition between potential users for available properties.[65]

Goodall's discussion of the CBD focused on retailing, and his reference to office activity showed only the most rudimentary understanding of the office economy:

> Besides retailing, the other major sub-core comprises offices. These offices are involved in executive and policy-making decisions in commerce,

industry, and government, and are not dependent on the general buying public but need a central location in order to obtain necessary supplies of labour.[66]

There appears to be no appreciation of the fact that office activity had already by this time become an integral part of economic change in its own right. Its physical centrality is instead associated with the classical economic argument of access to labour. Furthermore, the interdependencies that exist in the office economy are given the most fleeting of references:

> They are especially reliant upon complementarity. Although some services are internal to an office, typing for instance, many are external and require frequent face-to-face contact between members of individual offices.[67]

Much of the work between the 1960s and early 1970s was a development of earlier work rather than a radical step forward in our understanding of the office economy. Whereas the former had demonstrated the spatial structure of cities, the later work added statistical coherence to the distribution of activities. For example, the work of Goddard and Törnqvist provided some understanding of the flow of information between businesses and its impact on location. However, it was limited in two critical areas. First, with notable exceptions, it did not provide a focus on offices as a distinctive aspect of urban dynamics. Secondly, it did not provide an interpretation of the forces accounting for the processes of succession and change in either the supply of built space or in the occupiers themselves.

So it was that, as the 1970s began, and with the exception of a small number of important texts, there remained a primitive understanding of the office economy as a feature of the urban landscape in its own right. The office economy had been largely ignored even though offices had over the previous three-quarters of a century come to dominate many city centres both in terms of land use and employment. The following two quotes from eminent geographers in the mid-1970s sum up the situation well.

> our understanding of the office system is in its infancy compared with progress in our knowledge of the role of retail or industrial land uses, for example, in the function and structure of cities.[68]
>
> To a great degree our understanding of the operative processes and consequent usage patterns in the city centre is completely inadequate.[69]

Peter Daniels commented similarly: "Prior to the mid-1950s the interest of geographers in office functions and their location was limited or non-existent" and that "during the early-1960s, it was realised that office-based activities were making a significant contribution to the over-centralisation of economic activity in London and the South East".[70] He argued that, despite the obvious physical impact of office buildings on selected parts of the met-ropolitan areas, it is surprising that "insufficient attention has been given to their significance".

The fact that the general understanding of the office economy remained relatively so poor for such a long time probably reflects the fact that it was not until the 1970s that the uneven development of manufacturing and ser-vices came to be perceived as a policy issue. Up to this point, public policy, academic study, employment and the bulk of wealth generation were concen-trated in the production of goods. But, just as the impact of deindustrialisa-tion began to show through in regional data on unemployment, earnings and poverty, so trends in the balance of trade and output figures underlined the growing importance of services to the national economy. As we saw earlier, overseas earnings by UK financial institutions more than quadrupled in the 1980s.

What was needed was a systematic approach that sought to explain the rela-tionships between the activities, an approach that sought to relate these evolv-ing relationships to changes in the physical expression of urban form. Only with a more specific and activity-related approach that untangled the web of relationships that existed between the businesses in the office economy would a more general understanding of the office economy emerge. What was the glue that held it together so tightly? What was the driving force of ever-increasing densities?

3.4 The office economy as a facet of urban growth

These questions were at least partially answered in the late 1960s, when the role of offices as a distinct facet of urban growth was acknowledged by Peter Cowan and colleagues. Their work traced the office economy back to the late nineteenth century, describing how, as the Industrial Revolution reached its climax, "the role of the city changed utterly".[71] As the industrial processes evolved and changed, so too did their administration and governance. Office activities grew in importance and began to assume their own, distinct iden-tity in the urban environment. Again, Cowan and colleagues identified the link between changing economic relations and the rise of the office. Until the Industrial Revolution,

> the office function was generally an integral part of other facets of urban life. But it is a commonplace that, as an organisation or an organism grows in complexity, certain parts or functions become separate from others. The office function of communication and control is just such a case. When an organisation is fairly simple, as in a cottage industry or a small workshop, the office function is integral with the production process itself. But as the enterprise grows and increases in complexity . . . so the office function begins to separate off.[72]

This transformation from a two-dimensional style of aerial analysis to approaches that were more process or activity-based was described by Carter:

> Inevitably one is led towards a more realistic evaluation of the central area not being made up of a CBD . . . but of a number of closely associated areas or sub-cores, constantly subject to pressures and with . . . changing boundaries.[73]

The chief mystery is why it had taken so long to take an activity-based approach which recognised the pivotal role of the office economy in the CBD.

One explanation for the absence of office functions within analytical frameworks for urban analysis through to the latter half of the twentieth century lies in the question of classification.

> Whereas manufacturing, transportation, retail trade and wholesale trade are economic activities whose existence is easily recognised and catalogued, many aspects of office activity are more difficult to classify.[74]

In other words, office activity is somewhat intangible. Despite a few notable exceptions, including Hoover and Vernon's *Anatomy of a Metropolis* (1959), examining the changing distribution of people and jobs within the New York metropolitan area, offices were largely ignored. This would not have mattered so much if strategic policies governing the location of jobs, transport and housing were being made in the absence of a detailed understanding of the office economy. The significance was highlighted by Jean Gottmann:

> But . . . those sectors of employment which were, and still are, regarded by the classical economists as external because they did not seem to be important elements in the economic process (production through processing and distribution to consumption) – those external activities are now becoming major sources of employment. They were, of course, peripheral to the economic processes of old.[75]

Cowan *et al* observed that the practical consequences of ignoring or neglecting the office function "are equally serious, and we can see them all around us". They highlighted one particular example, in London,

> where policies and controls directed towards limiting the growth of the city failed because they omitted to control office employment. The Barlow Report (1950) set out to limit London's growth, using the control of industrial employment as its main mechanism.[76]

The lack of analysis of the office economy held back the search for, and explanation of, a rich, highly granular nature of change that helps to explain urban

morphology generally and the rise of the office economy in the CBD specifically. As Daniels argued:

> It may therefore be unwise to rely on conventional ideas about locational
> tendencies which largely reflect the influence of, for example accessibility
> or land rents. It is necessary to take closer account of organisational and
> structural change as well as the effects of technology on the future of estab-
> lished concentrations of service activity.[77]

Figure 3.5 illustrates the detail in Cowan's work, with his representation of the
various business clusters in the City of London just before the Second World
War. The map illustrates both office and other industrial clusters.

Cowan's work was followed by Peter Daniels's milestone book of 1975,
Office Location.[78] This was the first major work that examined not only
the locational characteristics of the office economy, but also its internal
dynamics. It attempted to explain the forces acting upon and within office
activities that give rise to their locational characteristics and their evolving
functionality.

The book was the first detailed, activity-based approach to untangle the
web of relationships within the office economy in order to explain its physical
expression. It helped describe the glue that held offices tightly together, and
it examined the nature of change that led to new demands being placed on
buildings. In just one example, Daniels examined trends in office mechanisation (what we now call IT) in North America in order to help explain trends in
the UK, demonstrating its impact on the development of office activity and, in

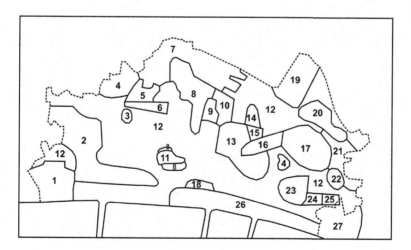

Figure 3.5 Pre-war business clusters in the City of London
Source: Reproduced from Cowan *et al* (1969).

particular, on the evolution of large office buildings firstly in the United States and, later, in the UK.

Daniels described static and dynamic elements as determining the location of office activities within urban areas. The static elements are those factors which make various parts of the urban area attractive to different types of office activity and account for clustering of interdependent office activities. The dynamic elements are the factors producing reorganisation and realignment of office location patterns.

> It is worth reiterating that one of the major obstacles to the study and analysis of office activities in Britain and elsewhere is the totally inadequate supply of relevant, comprehensive, and easily accessible statistics.[79]

He was drawing attention to the fact that, going into the final quarter of the twentieth century, there remained a paucity of data on what had by now become a dominant driver of the economy – office activity. In one highly prescient comment, Daniels highlighted the potential for a distributed model of work (Figure 3.6).

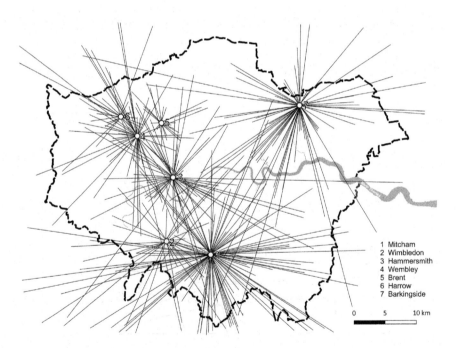

1 Mitcham
2 Wimbledon
3 Hammersmith
4 Wembley
5 Brent
6 Harrow
7 Barkingside

0   5   10 km

Figure 3.6 Journeys to work to decentralised offices

Source: Reproduced from Daniels (1975).[80]

Daniels suggested that transport nodes in outer London should be used "to provide the framework for locating offices at public transport nodes". He went on:

> By choosing public transport orientated locations it should be possible to counter-balance the disutilities of slower trip times and general inconvenience of public transport with the possibility of direct trips to the destination offices.[81]

In work such as that of Daniels, the emphasis in urban analysis during the 1960s and early 1970s on spatial definition was gradually yielding to one where cities were seen in terms of a much richer and more complex set of relationships and activities than had previously been recognised. Cities came to be seen as mosaics of activities in which change was not only prevalent but inevitable; and the office economy was coming to be recognised as a discrete facet of those mosaics. Interestingly, it was the perceived impact of office activities on congestion and transport infrastructure that provided the spur for closer scrutiny of the characteristics and location of office activities.

The final phase in our journey through the study of the office economy covers the period from the late 1970s through to the present time. This has been, above all else, the 'post-industrial age', or the 'information age', during which technology has had a profound impact on every aspect of life.

3.5 Post-industrial economics

The reason for the renewed impetus to understanding urban structure in the early 1980s was, of course, the escalating significance of developments in technology. In particular, there was the potential threat to the future of large urban concentrations of economic activity as the result of the anticipated liberating effects of technology on the locational imperatives of business. Amongst the earliest exponents of the influence of technology as a driver of structural change in urban morphology were Daniel Bell[82] and Brian Berry.[83] In 1973, Berry spelt out the concept of technology as a decentralising agent, suggesting that the new communications technologies would "concoct a solvent" to dissolve "the core-oriented city in both time and space, creating what some refer to as urban civilisation without cities".[84] There emerged a whole series of studies that had as their focus the impact of technology on the economy and urban morphology.[85]

Concepts of economic progress

Economist and statistician Colin Clark (1905–1989) published his *Conditions of Economic Progress* in 1940. At the time, Clark was a British émigré living in Australia, where he was appointed as Government Statistician,

Director of the Bureau of Industry and Financial Advisor to Queensland Treasury. One of Clark's many achievements was to pioneer the use of gross national product as the basis for studying national economies. In *Conditions*, however, he examined how economies evolve over time. In so doing, Clark subdivided the economy into three tiers: Primary, Secondary and Tertiary.[86] Broadly speaking, these represented agriculture, manufacturing and services, respectively.

Clark envisaged a progressive transition of economies from one level to the next. In this sense a service-dominated economy could be seen to be 'more advanced' than one based mainly on primary industries. In Clark's schema, services included education, finance, government, health, recreation, research, trade and transport. Tellingly, from a modern standpoint, there is no separate mention of professional services (legal and accounting) or, indeed, anything that we would recognise as office activity today, except for finance.

The notion of movement, or progression towards economic maturity, was reinforced a couple of decades later by American economist Walt Whitman Rostow (1916–2003). A somewhat controversial character, Rostow worked in the Office of Strategic Services during World War II; later as National Security Advisor to President Lyndon B Johnson, and as a foreign policy advisor to John F Kennedy. He was a vehement opponent of communism and a supporter of American involvement in Vietnam. But with regard to the story here, Rostow's academic legacy involved his work, *The Stages of Economic Growth: A Non-Communist Manifesto* (1960), which set out five progressive steps in economic development.[87]

First, the *traditional society* is almost wholly 'primary', characterised by subsistence agriculture. Next, the *preconditions for take-off* sees external demand for raw materials; the development of commercial agriculture; increasing spread of technology; changing social structures and shared economic interests. Thirdly, *take-off* sees growing urbanisation and industrialisation, alongside technological breakthroughs. Fourthly, the *drive to maturity* sees diversification of the industrial base; multiple industries expand and new ones appear. Manufacturing shifts from investment in capital goods towards consumer durables and domestic consumption. Finally the *age of mass-consumption* sees growing disposable income, consumerism and dense urbanisation.

In a later work, *Politics and the Stages of Growth* (1971), Rostow added a sixth stage, *beyond consumption* (the search for quality), when a high economic security and mass consumption is considered the norm; when consumer products diversify and services proliferate.[88] Whilst widely cited, Rostow offered little insight into how people would work at different stages. Even the sixth stage refers in the general to 'consumption' and 'services', with no specificity in terms of the economic activities behind them. However, Rostow's late addition to the five stages played into the view that the growth of service activity represented the emergence of a post-industrial economy, complete with office activity.

Goodwin offered a different perspective in the mid-1960s, seeking to discuss office activity in terms of "management centers" – cities in which "there is a concentration of headquarters offices of nationally important companies. It is a place apart from the production centers". He then isolated the workers themselves as a key dynamic of management centers, in so far as they are managers "who sell their managerial talents irrespective of the nature of the company that employs them. A management center is a reservoir of managerial talent available for hire".[89] In writing this, Goodwin is hinting at the key invisible force that maintains dense concentrations of economic activity, agglomeration economics (Chapter 10).

Goodwin argued that the classification of cities had given "little or no attention . . . to management per se or to management centers". Indeed,

> A recent bibliography of central place studies, which lists the significant studies of tertiary activities, does not contain among its more than five hundred entries a single reference explicitly to management centers.

By the mid-1970s, economic attention was turning to the growing role of the service economy, not least because of the policy consequences of the coincident decline in manufacturing activity. Gershuny (1978),[90] Gershuny and Miles (1983)[91] and Harris and Taylor (1978)[92] all produced excellent work. Perhaps the seminal work of the time was by Daniel Bell.

Service economy

American sociologist Daniel Bell (1919–2011) was the son of Jewish migrants. His father died when Daniel was just eight years old, and life was tough living in Manhattan's Lower East Side. Nevertheless, he graduated from Stuyvesant High School and took a degree from the City College of New York before becoming a journalist and then receiving a PhD from Columbia University in 1961.

Bell was a professor at Harvard University and is particularly remembered for his contributions to the study of post-industrial society, publishing *The End of Ideology* (1960), *The Cultural Contradictions of Capitalism* (1976) and the tome that concerns us here, *The Coming of Post-Industrial Society* (1973). In the last of these, Bell divided society into three parts: social, polity and culture. The polity "regulates the distribution of power and adjudicates the conflicting claims and demands of individuals and groups", while the "culture is the realm of expressive symbolism and meanings".[93] The social structure encompasses the economy, technology and the occupational system, and it is this with which the concept of a post-industrial society is concerned.

Bell clearly saw the significance of a declining role for manufacturing. He argued that the changeover to a post-industrial society is signified not only by the change in sector distribution – the places where people work – but in the pattern of occupations, the kind of work they do. He cited the rise of white-collar work as illustrative.[94] Bell suggested that the concept of a post-industrial society can be more easily understood in terms of "five dimensions, or components, of the term", as follows:

- economic sector: the change from a goods-producing to a service economy;
- occupational distribution: the pre-eminence of the professional and technical class;
- axial principle: the centrality of theoretical knowledge as the source of innovation and of policy formulation for the society;
- future orientation: the control of technology and technological assessment, and
- decision-making: the creation of a new intellectual technology.

Bell's work was part of a growing recognition of the long-term decline of manufacturing and the rise of the office economy. The mass redundancies in manufacturing starting during the 1970s have been fully recorded, not least by Blackaby in his *Deindustrialisation*. Blackaby systematised theoretical approaches to economic restructuring, particularly the rapidly reducing manufacturing capacity.[95]

Conversely, the rise of the service economy generally, and the office economy in particular, was much less fully documented. Both Clark and Rostow referred to services as a catch-all term for pretty much anything that was not manufacturing, almost as though services were the accidental by-product of a predominantly maker-economy. Bell made some progress by identifying a growing professional and technical class, but, as Howells and Green noted in the late 1980s:

> The problem is that the emphasis on manufacturing as the main locomotive force in the economy, which has persisted until very recently, has left us with a very limited understanding of the service sector in relation to economic development.[96]

For service sector, read office economy.

The authors point to a research tradition in which service activity was painted as something which passively follows population and manufacturing growth. But they suggested that it was "no longer an adequate conceptualisation of the place of the service sector in national economic development", and that evidence was increasingly suggestive that services were a "leading rather than lagging element in economic growth".

So what is service activity? A starting point in answering this question is official data from the Office for National Statistics, which produced five key facts about the services sector:

- the service sector generated 79% of UK Gross Domestic Product (GDP) in 2013;

- the percentage of jobs in the service sector rose from 33% in 1841 to 80% in 2011;
- 91% of London's economy is in services, higher than in all other areas of the UK;
- the UK economy is more reliant on services than any other G7 country, and
- the service sector has driven the economic recovery since the downturn in 2008.[97]

In other words, it is a fairly meaningless term. It cannot be possible to discuss as a single phenomenon something that makes up very nearly four-fifths of all economic endeavour. Yet, it is into this sprawling, amorphous collection of activities that the great majority of economic activity discussed in this book is normally allocated for the purposes of official statistics, mainstream media reporting, research and writing.

Producer services

The launch of the personal computer in 1981 led to a proliferation of books on the impact of 'information' in the economy. Jean Gottmann's *The Coming of the Transactional City*,[98] Thomas Noyelle and Thierry Stanback's *The Economic Transformation of American Cities*[99] and Mark Hepworth's *The Information City*[100] are a few of the more widely read. With the benefit of at least a decade's more experience, these later studies presented a more positive view than Bell and Berry of the impact of technology on urban structures. They were suggesting that the competitive positions of already densely developed urban areas would intensify, and that innovation and demand would encourage reinvention, with Jean Gottmann going as far as to suggest that "the centrality of the city is constantly being re-born".[101]

Mark Hepworth[102] examined the office economy in terms of the interrelated roles of information and the new information technology in the process of economic development. He pointed to large, multi-locational corporations as the dominant users of advanced telecommunications infrastructure, suggesting that their increasing use of private networks would have profound implications for the future of cities and urban systems in terms of uneven development. Not least, he foresaw the issue of what was later to become known as 'off-shoring':

> Arising from the distance-shrinking effects and capabilities of computer network technologies is an intensification of inter-urban competition for new markets and resources, particularly jobs.[103]

John Marshall suggested that the growing role of information in the economy led to "the development of an integrated office-based sector of employment", in which "the office block has replaced the factory as a symbol of modern urban development".[104] Daniels had earlier suggested that office space provided accommodation for at least 30–35% of the labour force in advanced economies.[105] At last, researchers were beginning to explain urban and economic development not simply in terms of its spatial outcomes, but in terms of the internal dynamics of the businesses that were the engine of change.

William Coffey argued forcefully in an echo from earlier times that up to this time, the study of services generally, and the office economy in particular, was plagued by its 'non-productive' label, and that "research on services was conceptually stagnant for several decades, being largely confined to the limited framework of central place studies".[106] Coffey suggested that it was only in the later 1980s and into the 1990s that producer services (the office economy) "were finally recognised as important economic phenomena in their own right – no longer simply 'the nonproductive remainder' of the economy".[107]

> The 'modern' era of research on services dates only from the early- to mid-1980s; it is only since this time [that researchers] have begun to analyse in a concerted and systematic way questions concerning the economic role and the locational dynamics of the full range of services activities, including producer services.

Coffey also made a direct link between the economic activities themselves and the buildings they occupied (offices), with the critical link being information technologies.

> Existing structures are often technologically insufficient to support the new information-related functions of firms, producing a demand for new office buildings capable of meeting modern technological and spatial conditions.[108]

Howells and Green distinguished "those activities which do not produce or modify physical objects . . . and purchases which are immaterial transient and produced mainly by people".[109] However, it remains a very wide definition, catching a huge number of disparate activities within its umbrella. So, to bring further focus, the authors proposed distinguishing producer and consumer services. The former are those that provide output that is consumed or used exclusively by other industries (what we call today business-to-business activity); while the latter include those that provide output going directly to consumers or households (such as retailing and leisure services).

The role of producer services was given extensive treatment in a series of papers in the late 1980s by various combinations of Peter Daniels, Andrew Leyshon and Nigel Thrift. For example, in their work on the City of London,[110] they explained key changes in the demand for commercial office space in the specific context of the restructuring of financial markets. Their work on the operational adaptation and spatial development of accountancy firms[111] yielded immense detail on the sector's internal dynamics and its resulting demand for office space. For the first time, the economic and spatial characteristics of urban areas were being examined from the perspective of the internal dynamics of the economic activities at their heart – the office economy. The work of Daniels and colleagues directly responded to John Rannells's call over three decades earlier for a more activity-based approach to understanding urban processes.

Increasing numbers of studies drew attention to the concentration of producer services in large metropolitan areas and empirically verified the 'glue' that held the companies so tightly together, in terms of the location of markets (clients and suppliers), highly skilled labour markets and the agglomeration benefits associated with minimising transaction costs.[112] Other studies showed how producer service companies created competitive advantage in urban areas including quality, creativity and innovation, spatial proximity, flexibility, timeliness of delivery, scope of services offered and adaptability to client needs – all a long way from the simpler explanations based on classical economics of earlier times.[113]

What all these studies showed was the evolution of the office economy and its changing spatial expression. They provided evidence of the 'glue' that held producer services together in tight proximity; they explained the internal dynamics of the businesses that ultimately gave rise to urban morphology, and they helped explain how change occurs. The decades of conceptually stagnant research on the office economy had been ended. At last there was a credible framework within which to analyse the office economy.

But even here we have difficulties in narrowing down to the office economy. Economic geographers such as John Marshall described producer services as including those services "which supply business and government organisations, rather than private individuals, whether in agriculture, mining, manufacturing, or service industries". As such, they are

> concerned with financial, legal, and general management, innovation, development, design, administration, personnel, production, technology, maintenance, transport, communication, wholesale distribution, advertising and selling, whether in primary, manufacturing, or service organisations.[114]

This definition is too wide to describe the office economy, because it includes services within primary and manufacturing industries. Marshall *et al*[115] went a stage further and broke producer services activities into three sub-groups (Figure 3.7).

Clearly the 'information processing services' most closely match the profile of the office economy, including, as they do, those activities that are typically concentrated in city centres and office buildings more generally and which, typically, have no final demand or individual consumer function, but which operate at a business-to-business level. This is helpful because it allows us to distinguish the largely office-based activities (information processing services) from non-office services, such as distribution and wholesaling. It does, however, make one large omission, which is central and local government.

Information processing services	Product/process, research and development
	Marketing, sales, advertising market research, photography, media
	Engineering (civil, mechanical, chemical, electrical, etc.) and architectural design
	Computer services, management, consultancy and administration
	Financial planning, accountancy, investment management, auditing
	Banking and other loan institutions
	Insurance
	Legal
	Training/education/personnel and industrial relations
	Purchasing
	Office services
	Property management/estate agency
Goods related services	Distribution and storage, wholesalers, waste disposal, transport management
	Installation, maintenance and repair of vehicles, communications and utilities
	Building and infrastructure maintenance
Personnel support services	Welfare services
	Cleaning, catering, security, safety
	Personal travel and accommodation

Figure 3.7 Different types of producer services

Source: Marshall *et al* (1987).

The typology is also dated, having been prepared during the mid-1980s, i.e. prior to the digital era.

Knowledge work

In 1992, management guru Peter Drucker predicted that the traditional factors of production – land, labour and capital – would become secondary to knowledge.[116] He was describing the ascendancy of the knowledge worker. Drucker was correctly interpreting the implications of the propensity within business, across all sectors, to collect, analyse, manipulate and extrapolate information and data. He was also watching the impact of computers and telecommunications technologies, and observing the symbiotic relationship between the outsourcing of non-core activities (e.g. advertising, marketing, payroll and recruitment) and the emergence of specialist firms to fulfil those functions.

As Drucker and others were quick to notice, new technologies based on knowledge and information production and dissemination, were increasingly driving change in the economies of many countries. Further, many of the technologies

> which emerged in the late 1950s, expanded with the proliferation of personal computers, and then surged dramatically with the widespread use of

email and the Internet – have considerable potential to re make the nature of work and the economy.[117]

It was during the early 1990s that the term 'producer services' yielded to the 'knowledge economy', typified by 'knowledge-intensive businesses', and investment in knowledge-based assets or 'intangibles' enabled by "powerful and cheap computers and the 'general purpose' information and communication technologies".[118] This, in turn, helped transform demand for offices; but the change in collective understanding was slow for some.

> The economic, industrial and social landscape is changing, often in ways that are poorly understood and documented. Many of our statistics and frameworks are still based on a world of accounting conventions where the only thing that really matters is machines and physical infrastructure.[119]

However, change in the economy was in fact quite rapid. Today, the knowledge economy accounts for over a fifth of UK output, and one in eight jobs. It has been responsible for nearly 40% of all economic growth and upwards of two million jobs in the UK since 1970.[120] Despite the growth of the knowledge economy, there remains the threat of further disruption to the nature of jobs, with the single greatest threat being the potential for artificial intelligence (AI) to automate swathes of routine, and not so routine jobs.

The concept of the knowledge economy recognises the fundamental shift from a focus on making goods and investing in physical assets towards an economy dominated instead by investment in intangible assets and activities. As Ian Brinkley argued, the knowledge economy is a story of how

> new general purpose technologies have combined with intellectual and knowledge assets – the 'intangibles' of research, design, development, creativity, education, science, brand equity and human capital – to transform our economy.[121]

He also noted that it is "a universal process", operating across all sectors, domestic and international, public and private, large and small firms.

> In doing so, the traditional boundaries between sectors such as manufacturing and services are disappearing and previously unnoticed industries – such as the creative sector – have emerged as major employers, generators of value-added, and exporters.

The critical point here is the universality of change, which has meant the rise of the office economy in more traditional sectors as well as emerging industries. The shift from investment in physical assets towards intangibles as the basis of economic activity is central to the knowledge economy story as "firms and organisations rely increasingly on the exploitation of knowledge to secure competitive advantage and better performance".

One of the earlier attempts to describe what the knowledge economy looked like in a practical sense was provided by Diane Coyle and Danny Quah.[122] Figure 3.8 is adapted from their work and is, perhaps, one of the more useful summaries of the characteristics of the knowledge economy. The authors refer to the 'old economy' and 'new economy', but this terminology is broadly equivalent to industrial and post-industrial, or manufacturing and knowledge economy.

The new economy is a somewhat vague term, and the table can be questioned in a number of areas. But it highlights some very significant trends within the economy that are leading to the re-organisation of jobs and the creation of new types of employment.

3.6 Man's art of trade

Jean Gottmann referred to the organisation of his beloved New York Megalopolis as "certainly one of the most remarkable achievements of Man's art of trade".[123] Quite a claim! Yet, given what we have explored in this chapter, and

Issues	*Old economy*	*New economy*
Economy-wide		
Markets	Stable	Dynamic
Scope of competition	National	Global
Organisational form	Hierarchical, bureaucratic	Networked
Structure	Manufacturing core	Services core
Sources of value	Raw materials, physical capital	Human and social
Business		
Organisation of production	Mass production	Flexible production
Key drivers of growth	Capital and labour	Innovation and knowledge
Key technology driver	Mechanisation	Digitisation
Competitive advantage	Lower costs	Innovation, quality, speed
Importance of research	Low to moderate	High
Relations to other firms	Go it alone	Alliances, outsourcing
Consumers		
Tastes	Stable	Changing rapidly
Skills	Job-specific	Broad and adaptable
Education needs	Craft or degree, one off	Lifelong learning
Workplace relations	Adversarial	Collaborative
Nature of employment	Stable	Risk and opportunity
Government		
Business relations	Impose regulations	Encourage growth
Regulation	Command and control	Market tools, flexibility
Government services	Nanny state	Enabling state

Figure 3.8 The old economy and the new economy

Source: Reproduced from Coyle and Quah (2002).

as a more general claim for the organisation of the office economy in larger cities generally, it is one that can be understood. We began in the agrarian economy of the eighteenth century and ended in the technology-driven knowledge economy of the twenty-first. Nearly 300 years of seeking to account for the spatial arrangement of cities, and more than a century to explain the office economy. The earliest work sought to explain land value (rent) in terms of the power that ownership invested in the lords of the land: they owned it; they could charge what they liked. In the nineteenth century, the concept of value became a reflection of what a buyer or renter was prepared to pay, and this in turn reflected their expectations for profit maximisation.

As the twentieth century got underway, explanation moved to describing two-dimensional patterns – the morphology of cities. Land use patterns were defined in terms of the ability of different uses to pay differential rents: the most profitable located in the centre, the notional point of maximum accessibility. The problem was that, in these maps, the central area was usually a black hole, almost literally. A central area would be shaded and labelled 'CBD', but there would be no description of what really went on there. There was no analytical understanding of the office economy as a driver or urbanisation in its own right.

We had to wait until the 1960s, and the quantitative revolution for any serious analysis of the structure of the CBD. As we have seen, John Rannells made a great step forward, explaining that locational characteristics of activities are visible manifestations of the complex systems of activities into which the economic life of cities is organised. But was anybody listening? In the late 1960s, there remained only the most rudimentary grasp of the office economy as a distinct facet of the modern urban economy. Brian Goodall felt able to comment that CBD offices were involved in executive and policy-making decisions, and just need a central location to gain access to labour. A hundred years after the first skyscrapers in New York and Chicago, and the best explanation on offer was labour supply.

But somebody was listening. Peter Cowan and colleagues brought us the office as *a facet of urban growth*. They set aside the notion of an homogenous CBD, looking at it instead as a number of closely associated areas or sub-cores, constantly subject to pressures and with changing boundaries. Then Pete Daniels provided the first major work that examined the internal dynamics of the CBD purely from the perspective of the office economy, untangling the web of relationships therein to explain its physical expression.

Nevertheless, bearing in mind that in the 1970s, in western Europe, 'deindustrialisation' and the rise of the service economy was well underway, it is perplexing that so little was known about the office economy. And, as late as 1988, John Marshall described the growing importance of office activities as leading to an integrated office-based sector of employment, with the office building replacing the factory as a symbol of modern urban development. Really? This is a remarkably naïve comment given the scale of the office economy in the late 1980s, and it reveals the paucity of academic understanding of the role of the office economy in the modern urban economy until quite recent times.

It should come as little surprise that the incredibly limited understanding of the office economy (excepting a few insightful writers) right through to the 1990s, was reflected in an equally naïve approach to the office economy in spatial planning.

Notes

1   A. Smith (1776) *The Wealth of Nations,* Penguin Books, p357
2   *Ibid,* pp109–110
3   W. Alonso (1964) *Location and Land Use: Toward a General Theory of Land Rent,* Harvard University Press, Cambridge, MA, p2
4   JS Mill (1848) Principles of Political Economy, with Some of their Applications to Social Philosophy Longman, Green, And Co, London, Book II, Chapter XVI, p255
5   JS Mill (1848) Principles of Political Economy, with Some of their Applications to Social Philosophy Longman, Green, And Co, London, Book V, Chapter II, p495
6   A. Marshall (1890) *Principles of Economics,* Macmillan, London, Cited in Alonso (1964) *Op cit,* p4
7   Alonso (1964) *Op cit,* p6
8   R.M. Hurd (1903) *Principles of City Land Values,* Arno Press, New York
9   *Ibid,* p77
10 *Ibid,* p14
11 P.F. Wendt (1956) *Real Estate Appraisal,* Holt, New York, p107
12 E.W. Burgess (1925) The Growth of the City: An Introduction to a Research Project. In: R.E. Park, E.W. Burgess, & R.D. McKenzie (Editors), *The City,* University of Chicago Press, Chicago
13 F.M. Babcock (1932) *The Valuation of Real Estate,* McGraw-Hill, New York
14 H. Hoyt (1939) *The Structure and Growth of Residential Neighbourhoods in American Cities,* Federal Housing Administration, Washington, DC
15 C.D. Harris & E.L. Ullman (1945) The Nature of Cities, *Annals of the American Academy of Political Science,* Vol 242, pp7–17
16 R.M. Haig (1926) Towards an Understanding of the Metropolis Parts 1 & 2, *Quarterly Journal of Economics,* Vol 40, pp179–208, 402–434
17 R.E. Murphy & J.E. Vance (1954a) Comparative Study of Nine Central Business Districts, *Economic Geography,* Vol 30, pp301–336; R.E. Murphy & J.E. Vance (1954b) Delimiting the CBD, *Economic Geography,* Vol 30, pp189–222
18 Mumford (1940) *Op cit*
19 *Ibid,* p482
20 *Ibid*
21 R. Turvey (1957) *The Economics of Real Property,* Allen & Unwin, London
22 *Ibid,* p47
23 *Ibid*
24 Rannells (1956) *Op cit*
25 *Ibid,* p5
26 *Ibid*
27 *Ibid,* p17
28 *Ibid,* p54
29 *Ibid,* p55
30 *Ibid,* p35
31 *Ibid,* p36

32 E.M. Hoover & R. Vernon (1959) *Anatomy of a Metropolis: The Changing Distribution of People and Jobs within the New York Metropolitan Region,* Harvard University Press, Cambridge, MA, p88

33 *Ibid*, p89

34 *Ibid*, p98

35 *Ibid*, p99

36 *Ibid*, pp103–104

37 *Ibid*, p106

38 E.M. Horwood & R.R. Boyce (1959) *Studies of the Central Business District and Urban Freeway, Development* University of Washington Press, Seattle

39 Alonso (1964) *Op cit,* p17

40 J. Gottmann (1961) *Megalopolis: The Urbanized Northeastern Seaboard of the United States,* The Twentieth Century Fund, New York, p577

41 *Ibid*, p586

42 *Ibid*, pp587–588

43 Gottmann (1966) *Op cit,* p199

44 *Ibid*, p205

45 *Ibid*, p207

46 D.H. Davies (1965) *Land Use in Central Cape Town: A Study of Urban Geography,* Longmans, Cape Town

47 *Ibid*, p87

48 H. Carter & G. Rowley (1966) The Morphology of the Central Business District of Cardiff, *Transactions of the Institute of British Geographers,* Vol 38, pp119–134

49 *Ibid*

50 Ward (1966) The Industrial Revolution and the Emergence of Boston's Central Business District, *Economic Geography,* Vol 42, pp152–171

51 H. Carter (1976) *The Study of Urban Geography* (2nd edition), Edward Arnold, London, p226

52 See for example: M.J. Bannon (1972) The Changing Centre of Gravity of Office Establishments within Central Dublin, 1940 and 1970, *Irish Geography,* Vol 6, pp480–484; J. Davey (1972) *The Office Industry in Wellington: A Study of Contact Patterns, Location and Employment,* Ministry of Works, Wellington

53 D. Harvey (1973) *Social Justice and the City,* Arnold, London

54 B. Thorngren (1970) How Do Contact Systems Affect Regional Development? *Environment & Planning A,* Vol 2, pp409–427

55 G. Törnqvist (1968) Flows of Information and the Location of Economic Activities, *Geografiska Annaler,* Vol 50, pp99–107

56 J.B. Goddard (1968) Multivariate Analysis of Office Location Patterns in the City Centre: A London Example, *Regional Studies,* Vol 2, pp69–85; J.B. Goddard (1970) Functional Regions within the City Centre: A Study by Factor Analysis of Taxi Flows in Central London, *Transactions of the Institute of British Geographer,* Vol 49, pp161–182

57 D.L. Foley (1956) Factors in the Location of Administrative Offices Papers and Proceedings, *Regional Science Association,* Vol 2, pp318–326

58 J.B. Goddard (1967) Changing Office Location Patterns within Central London, *Urban Studies,* Vol 4, pp276–285; Goddard (1968) *Op cit;* Goddard (1970) *Op cit;* J.B. Goddard (1973) Office Linkages and Location, *Progress in Planning,* Vol 1, No 2, pp111–232

59 Goddard (1967) *Op cit*

60 *Ibid*

61 *Ibid*

62 Goddard (1970) *Op cit*

63 Goddard (1973) *Op cit*

64 B. Goodall (1972) *The Economics of Urban Areas,* Pergamon Press, Oxford

65 *Ibid*, p95

66 *Ibid*, p98
67 *Ibid*
68 Daniels (1975) *Op cit*, p2
69 Carter (1976) *Op cit*, p245
70 Daniels (1985) *Op cit*, p116
71 Cowan *et al* (1969) *Op cit*, p26
72 *Ibid*, p28
73 Carter (1976) *Op cit*, p218
74 Cowan (1969) *Op cit*, p22
75 Gottmann (1966) *Op cit*
76 Cowan (1969) *Op cit*, p23
77 Daniels (1985) *Op cit*, p132
78 Daniels (1975) *Op cit*, p221
79 Cowan (1969) *Op cit*, p160
80 *Ibid*
81 *Ibid*, p211
82 D. Bell (1973) *The Coming of Post-industrial Society: A Venture in Social Forecasting*, Basic Books, New York
83 B.J.L. Berry (1973) *The Human Consequences of Urbanisation*, St Martins, New York
84 *Ibid*, p54
85 See for example: J.B. Goddard & R. Pye (1977) Telecommunications and Office Location, *Regional Studies*, Vol 11, pp19–30; B. Thorngren (1977) Silent Actors: Communication Networks for Development. In: IdeS Pool (Editor), *Social Impact of the Telephone*, MIT Press, Boston; R. Pye (1979) Office Location: The Role of Communications and Technology. In: P.W. Daniels (Editor), Spatial *Patterns of Office Growth and Location*, Wiley-Blackwell, London
86 C. Clark (1940) *The Conditions of Economic Progress*, Macmillan, London
87 W.W. Rostow (1960) *The Stages of Economic Growth: A Non-Communist Manifesto*, Cambridge University Press, Cambridge
88 W.W. Rostow (1971) *Politics and the Stages of Growth*, Cambridge University Press, Cambridge.
89 W. Goodwin (1965) The Management Center in the United States, *Geographical Review*, Vol 55, No, 1 pp1–16
90 J. Gershuny (1978) *After Industrial Society? The Emerging Self-service Economy*, Macmillan, London
91 J. Gershuny & I. Miles (1983) *The New Service Economy*, Frances Pinter, London
92 D.F. Harris & F.J. Taylor (1978) *The Service Sector: Its Changing Role as a Source of Employment*, Research Series 25, Centre for Environmental Studies Ltd, London
93 Bell (1973) *Op cit*
94 *Ibid*, p134
95 F.T. Blackaby (Editor) (1978) *Deindustrialisation*, Heinemann, London
96 J. Howells & A. Green (1988) *Technological Innovation, Structural Change and Location in UK Services*, Gower Publishing Company Limited, Aldershot, p2
97 ONS (2016) *Five Facts about . . . the UK Service Sector*, www.ons.gov.uk/economy/economicoutputandproductivity/output/articles/fivefactsabouttheukservicesector/2016-09-29. Accessed 10th June 2020
98 J. Gottmann (1983) *The Coming of the Transactional City*, University of Maryland Institute for Urban Studies Monograph Series No 2, College Park
99 T.M. Noyelle and T.J. Stanback (1984) *The Economic Transformation of American Cities*, Allanheld Osmun, Totowa, NJ
100 M. Hepworth (1987) The Information City, *Cities*, Vol 4, pp253–262
101 Gottmann (1983) *Op cit*, p15
102 M. Hepworth (1990) Planning for the Information City: The Challenge and Response, *Urban Studies*, Vol 27, No 4, pp537–558

103 *Ibid*
104 J.N. Marshall (1988) *Services and Uneven Development,* Oxford University Press, Oxford, p32
105 Daniels (1985) *Op cit*
106 W.J. Coffey (2000) The Geographies of Producer Services, *Urban Studies,* Vol 21, No 2, pp170–183
107 *Ibid*
108 *Ibid*
109 Howells & Green (1988) *Op cit,* p2
110 A. Leyshon, N.J. Thrift, & P.W. Daniels (1987) *The Urban and Regional Consequences of the Restructuring of World Financial Markets: The Case of the City of London,* Working Papers on Producer Services Series No 4, University of Bristol and University of Liverpool
111 A. Leyshon, P.W. Daniels, & N.J. Thrift (1987) *Large Accountancy Firms in the UK: Operational Adaptation and Spatial Development,* Working Papers on Producer Services Series No 2, Saint David's University College, Lampeter and University of Liverpool
112 Coffey (2000) *Op cit*
113 D.P. Lindahl & W.B. Beyers (1999) The Creation of Competitive Advantage by Producer Services Establishments, *Economic Geography,* Vol 75, pp1–20
114 Marshall (1988) *Op cit,* pp13–14
115 J.N. Marshall, P. Damesick, & P. Wood (1987) Understanding the Location and Role of Producer Services in the UK, *Environment & Planning A,* Vol 19, pp575–595
116 P.F. Drucker (1992) The New Society Organisations, *Harvard Business Review,* September–October, pp95–104
117 W.W. Powell & K. Snellman (2004) The Knowledge Economy, *Annual Review of Sociology,* Vol 30, pp99–220
118 I. Brinkley (2010) *Innovation, Creativity and Entrepreneurship in 2020,* The Work Foundation, London, p4
119 I. Brinkley (2008) *The Knowledge Economy: How Knowledge is Reshaping the Economic Life of Nations,* The Work Foundation, London, p10
120 A. Sissons (2011) *Britain's Quiet Success Story: Business Services and the Knowledge Economy,* The Work Foundation, London, p5
121 Brinkley (2008) *Op cit,* p9
122 D. Coyle & D. Quah (2002) *Getting the Measure of the New Economy,* The Work Foundation, London
123 Gottmann (1961) *Op cit,* p562

4 Planning

A tale of indifference and ineptitude

At first sight, a chapter about urban planning might seem an odd inclusion in a history of the office. So integral have they been to economic growth and change in twentieth-century cities that it might seem odd that offices have been treated in anything other than a benign manner by urban planners. Yet this is far from the case. Throughout much of the post-war period, office activity was seen as a negative trend; it was politicised and given particular treatment in economic and planning policies by central, regional and local government, particularly with policies aimed at limiting its growth.

4.1 Learning from a Victorian polymath

Early approaches to urban planning largely ignored the economic engine of change. Seminal works like, for example, Abercrombie's *Town and Country Planning* (1933) made virtually no reference to economic change. Rather, planning adopted a top-down approach in which it sought to impose social order to 'guide' cities towards 'just' outcomes rather than 'productive' outcomes. Inevitably, this led to a focus on control and restraint rather than on growth, and on spatial form rather than on economic structure. Early planners were intent on imposing their own ideas of optimal form.

> Top-down plans were designed to stop city growth and canalise it elsewhere, rather than accommodate it. Plans for garden cities . . . were implicitly based on the idea that city growth was evil in some way, and that what was required was some radical exorcising of the forces that had given rise to such forms.[1]

The antipathy towards office activity was due to a profound lack of understanding of the forces that were underpinning the inexorable rise of the office economy. Reflecting the more specific context of the property industry, it showed an obsessive concern with supply-side dynamics. Consequently, a growing concern with 'the office problem' during the 1960s and 1970s manifested itself in a myopic focus on limiting the growth of office floorspace, leading to reactive and restrictive planning policies that consistently failed to address the requirements of office occupiers. Right up to the eve of Big Bang in the City of London in 1986, there was an almost complete absence of policies based on the operational and occupational needs of the office economy.

Sir Patrick Geddes (1854–1932) was one of the founding fathers of modern urban planning. A great influence over Lewis Mumford (featured in Chapter 2), Geddes was one of the last great Victorian polymaths. He was a biologist (having been tutored under Charles Darwin's great friend and defender Thomas Henry Huxley at the London School of Mines), a botanist (holding the Chair of Botany at University College Dundee), sociologist (co-founding the Sociological Society in 1903) and urban planner. He was well travelled, having worked in India, Palestine and the USA, and was extremely well-connected (knowing Charles Darwin well). His ideas and studies encompassed history, the arts, science and philosophy. Like Mumford, he was a humanist, and greatly concerned over the enslavement of society on the altar of technological progress, and the implications for environmental degradation.

But one of the most important aspects to Geddes's broad knowledge and experience was the fact that he understood the interconnectedness of life: he was in essence an ecologist. His fascination for the organisation of social structures and their physical arrangement in space drew extensively from systems and philosophical thought. He saw society as a natural organism, one that required nurturing and caring for to ensure its long-term health, and one that depended upon the wellbeing of those other organisms sharing the same ecosystem for its own survival.

Geddes became Professor of Botany at University College Dundee in 1888, but the part-time nature of the post allowed him to spend much time in his favoured Edinburgh, at his Outlook Tower, complete with its camera obscura. He pursued his interests in applied sociology and his vision of cities as constantly evolving organic entities.

Geddes's seminal work, *Cities in Evolution* (1915), failed to have the impact that, from the current backdrop of history, one might have expected it to have. Almost as long in gestation as Darwin's own masterpiece, it seemed to lose some of the originality of Geddes's earlier thinking published in separate papers. Instead, it was a paper by one of Geddes's colleagues that was to have a longer-lasting impact. D'Arcy Wentworth Thompson was a master of morphological thinking, and his book, *Growth and Form* (1917), was "the most complete statement of physicalism that we had until the rebirth of morphology through the recent science of fractals, chaos and complexity".[2]

Nevertheless, Geddes pushed hard for cities and their planning to be seen in evolutionary terms, by arguing that they were the result of complex interactions, and in which numerous small decisions added up to a general picture of change, or adaptation. Batty and Marshall suggest that Geddes can

> be regarded as the first to imprint the analogy with evolution in our study of cities, and in his early years he developed many insights into the biology of cities which resonate strongly with current developments.[3]

Figure 4.1 Patrick Geddes, 1886

What Geddes would have thought of post-war planning, and specifically its approach to the office economy, can only be guessed. But it is quite possible that he would have found the *ad hoc*, reactive and ill-informed measures described here as quite absurd in the context of his ecological approach. Somehow, profoundly, British town planning lost sight of the notion of bottom-up change. Instead, by the 1950s, it had adopted much of the scientism of the age, in which a top-down approach, or social engineering, became prevalent. Like engineering, planning adopted the perspective that it could 'fix' cities: more planning would make cities more efficient, more ordered, more just; and as such, it became an extension of the state's social planning, with very little reference to underlying economic drivers.

Just as the supply of office space in London has lurched from phases of over-supply to under-supply during the post-war years (as described in Chapter 5), so urban planning has moved in and out of phases of direct control and *laissez faire* regimes. What is clear is that from the Second World War through to the late 1980s, there was a failure within the machinery of planning policy to fully appreciate the emerging role of the office economy. In short, it failed to learn from Geddes about the interconnectedness of activities: the role of the office economy within the national economy, its internal dynamics, its locational needs and its requirements for built space. The central problem was that there was no feedback mechanism, nothing to explain how changes within the office economy were driving changing demand for built stock within city centres.

Policies generally, at central, metropolitan and local levels were reactive, restrictive, inflexible and unresponsive to the diverse characteristics and changing requirements of office occupiers. Manners and Morris concluded that the achievements of policy "were on balance quite modest", and that:

> Undoubtedly, the central lesson of intervention in the 1960s and 1970s must be that old and unadapted policy vehicles should not be employed to secure new policy ends.[4]

They continued:

> The original rationale for intervention was rooted in certain perceptions of the Central London office 'problem'. Subsequent policy changes were merely tacked on to that original initiative, and on to a set of attitudes and initiatives towards manufacturing industry.[5]

Indeed, the post-war growth of the office economy came to be seen, in spatial planning terms, as a 'problem'. More significantly, it took on a political dimension as policy sought to protect manufacturing and productive activity from the encroachment of service activity in offices. Only in more recent times, with the setting up of the Greater London Authority (GLA), has a more strategic and supportive public policy context emerged for offices.

4.2 The first steps to constraining growth

Concerns over the proliferation of commercial and administrative buildings at the expense of industrial activity in the central area of London were being raised as early as 1940, in the Royal Commission on the Distribution of Industrial Employment (*Barlow Commission*).[6] While the planning policies of individual boroughs paid relatively little attention to office employment in central London,[7] the *County of London Plan* of 1943 set the tone for much of what was to follow. In his foreword to the plan, the Right Honourable Lord Latham, Leader of the London County Council (LCC), stated that

> Our London has much that is lovely and gracious. I do not know that any city can rival its parks and gardens, its squares and terraces. But year by year as the nineteenth and twentieth centuries grew more and more absorbed in first gaining and then holding material prosperity, these graces were over-laid, and a tide of mean, ugly, unplanned building rose in every London borough and flooded outward over the fields of Middlesex, Surrey, Essex, Kent.[8]

The plan itself, which was to form the basis of post-war planning in London, barely referred to office activities, and asserted that central London "must continue to be of many and varied uses and, in the main, redevelopment can take place without serious disturbance to the present natural zoning".[9] Figure 4.2 shows one of the key maps, with very little emphasis on the office economy.

An identifiably anti-office stance first appeared in the LCC's *Draft Development Plan* of 1951 (approved in 1955). The main strategic consideration of this plan was the decentralisation of industry and commerce and the expansion of residential development to address the city's shrinking population, through the use of zoning and plot ratio controls. While the retention of industry was later to fall back into favour as a desirable generator of jobs, herein lay the genesis of an obsessive concern within public policy with 'the office problem'. This obsession was to dominate the development and implementation of office policy, at local, metropolitan and central government levels, for the next 35 years and was to show a singular lack of understanding of the evolving role of London generally, and in particular the rise of the office economy.

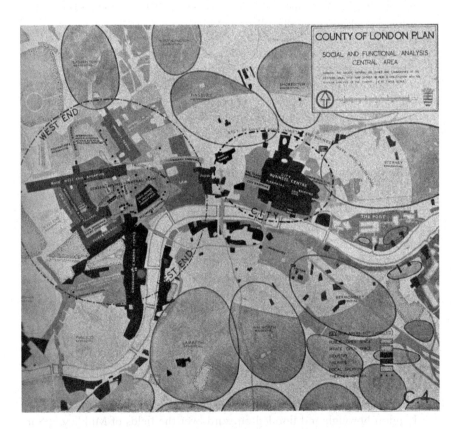

Figure 4.2 County of London Plan, 1943, social and functional analysis

Source: Reproduced from Forshaw & Abercrombie (1943).[10]

Donald Foley highlighted the long-term concern that "the London region was likely to grow too vigorously for the social and economic well-being of Britain". Since the recommendations of the Barlow Commission, it had been "central government policy that London's growth should be deliberately held back and that the growth of other relatively depressed regions should be stimulated", and it "became an underlying policy that central London in particular should be de-congested"[11] Post-war plans and policies assumed that there would be little new overall population growth in Greater London, and that "rigorous control of physical development could follow as desirable and feasible". So it was that population forecasts and policies to restrain growth

> merged to provide a planning doctrine blessed by central government and accepted by countless officials and planners for the next decade or two.[12]

Although damage to the fabric of London during the Second World War was extensive, post-war austerity measures constrained redevelopment activity. The operation of building licence controls, introduced by the post-war Labour Government through the 1947 Planning Act as part of its austerity programme, suppressed the ability of developers to build. Together with 100% development charges, they proved effective in discouraging development, in the early years.

Developers found loopholes in the legislation, most spectacularly, in the Third Schedule of the 1947 Act. This dictated that a building could be enlarged by up to 10% of its cubic content, a mechanism for allowing owners to modify buildings without the need to pay the prohibitive development charge. But, as Marriott observed, this did not mean that a

> new building was 10 per cent larger than the old, or that it housed 10 per cent more workers. Far from it. There was usually an increase much greater than 10 per cent. Old buildings have higher ceilings, thicker and more numerous walls, more space for passages, staircases, cupboards and directors' lavatories than new buildings. Thus the space actually used for work could be substantially increased by building a modern office box within the cube plus 10 per cent of the old.[13]

Marriott goes on to cite the case of the New Scotland Yard development in Victoria where the developer was able to provide a new office building with a net floor area of 400,000 sq ft (c37,000 sq m) "giving a ratio of site to floor space of 7:1. The maximum allowed by the LCC's zoning rules was 3.5:1". Developers maximised their opportunity, and abuse of the Third Schedule contributed to the 1950s construction boom.

But as business gradually returned to something resembling normality, there was inevitably rising pressure to redevelop bomb-damaged sites, and gradually the planning system yielded. Indeed, the seeds of the first post-war development boom were sown by Housing Minister Harold Macmillan in December 1952, when he introduced the bill that became the 1953 Town and Country

Planning Act, saying that "the people whom the Government must help are those who do things: the developers, the people who create wealth whether they are humble or exalted".[14]

As soon as licence controls were abolished in 1954, development applications increased markedly. Between 1948 and 1954, the LCC issued planning approvals for the construction of around 21 million sq ft (c2.0 million sq m) of new office space in central London. In addition, permissions for a further 6 million sq ft (c560,000 sq m) of conversions to office use were granted.[15] But many of the consents were sought to secure value rather than real development opportunities. The abolition of licence controls, however, was a prelude to a spectacular development boom. Between 1954 and 1958, the annual rate of construction completions increased nearly tenfold, from 700,000 sq ft (65,000 sq m) to 5.9 million sq ft (c550,000 sq m) in central London.[16]

Until this boom in development, office employment had not formed any part of central government's employment policies. But, as redevelopment led to a growing concentration of office employment, and as the *Development Plan's* strategic focus on the decentralisation of industry and commerce appeared not to be working, concern began to grow in central government over the impact of offices on London's employment structure and on congestion.

In 1954 (before the publication of the LCC's *Development Plan*) the Minister of Housing and Local Government, Duncan Sandys MP, asked the LCC to undertake a survey to determine land use allocations and to provide a factual basis for an attack on congestion. The ground was laid for a determined and sustained attack on the expansion of office activity, and Ministerial comments on post-war development plans hinted at the types of measures that were to follow. For example, in approving the LCC *Development Plan* in 1955, the Minister inserted a policy statement aimed at preserving residential uses in the central area. This stipulated that on the expiry of their permits for temporary use as offices, buildings would revert to their former, residential use. The attitude hardened, and the attack on offices gathered momentum.

In 1956, Sandys approved the *County of Middlesex Plan*, suggesting that "It would afford considerable relief to the congested central areas of London if more business concerns could be persuaded to establish their offices in the outer parts" of Middlesex.[17] Then, in 1957, the LCC published *A Plan to Combat Congestion in Central London*. While conceding that it was essential for London's position as the administrative and commercial capital of the country to be maintained (remarkably parochial ambitions given the City's already-established role in world commerce), the plan argued that policies needed to be amended. In particular, it sought restrictions on the areas in which large office buildings could be built.

As well as a concern over increasing congestion, 'the office problem' was also related to a growing concern over the increasing polarisation between 'productive activity' and the service sector. Manufacturing activity in London was declining severely during the 1950s, and policies aimed at decentralising office activity from central London were, in part, based on a fear that continued and largely uncontrolled growth of offices in the centre would encourage further loss of manufacturing jobs.

Manufacturing employment was seen to be more 'worthy' of protection than service jobs. In this sense, policy frameworks were addressing the historic role of the city centre rather than a more progressive stance that would have recognised the inevitable rise of services generally, and the office economy in particular. The plan to combat congestion did little in practice to halt the rise of offices. Indeed, London was proving

> far more attractive to various growth activities than had been envisaged – particularly headquarters offices, specialised services, and such manufacturing as electronic and electrical goods . . . it became increasingly unrealistic to assume that the plans for London could continue to be based on a policy of containment.[18]

In central London alone, the construction of new office space was particularly impressive, increasing from 153 million square feet in 1957 to 175 million square feet in 1962; that is, 22 million square feet of new stock in just five years.[19]

Little wonder that the perception was growing in central government circles that the metropolitan authorities were unable to prepare a coherent restrictive policy on the location and development of office activity. All the while, pressure was building among those property developers and speculators who had amassed their property portfolios during and immediately after the war. As controls on construction effectively lapsed, the years from 1953 to 1963 were ones in which, with the support of clearing banks and insurance companies, developers "were unleashed on the London environment to enjoy a decade of uncriticised and virtually untaxed profiteering".[20]

Central government came to perceive that, despite its potentially strong grip on development through the local development control process, the LCC was failing to control the spread of development. Eventually, its frustration surfaced in the form of a full-scale review of London government:

> A multitude of motives went into Henry Brooke's decision in 1957 to set up the Herbert Commission to examine the root and branch reform of London government. But there is little doubt that the LCC's shortcomings in the planning field . . . played a major part in it.[21]

But if the Conservative Government thought that the Herbert Commission (properly known as the Royal Commission on Local Government in Greater London) would streamline strategic planning, then they were to be disappointed.

4.3 Controlling the office economy

The 1960 *Herbert Commission* report led directly to the Local Government Act of 1963, which replaced the LCC with the Greater London Council (GLC). The GLC remained the strategic administrative body for Greater London from 1965 until its abolition under the Thatcher Government in 1986. At the same time (and in direct parallel with a move ten years earlier) the Act recommended that the GLC should carry out a survey of Greater London, so that it could submit to the Minister a *Greater London Development Plan* (GLDP). The results, and a *Written Statement*, were not to be published until 1969.

Impatient, the government chose not to wait for the metropolitan authority to establish itself and its policy regime. Its long-held perception that London-wide planning lacked the teeth and co-ordination to contain the office problem led it to intervene directly, for the first time, in the office development system. This was a bold change in direction for public policy on land use control, especially given the fact that it was based on at least two questionable assumptions.

The first assumption concerned the actual scale of growth in office employment. It seems that there was a significant gap between perceptions and reality. For example, the employment estimates on which the introduction of policy measures were based were themselves incorrect. The LCC had fuelled the debate over numbers in the First Review of its Development Plan in 1960. It had argued that the number of people employed in central London had been increasing by 15,000 each year, or nearly twice the national rate of growth.[22] The Town and Country Planning Association added its weight to the argument, stating: "the increase was almost entirely in office employment".[23] However, these comments were misleading. The 200,000 jobs thought by the government to have been created in central London offices between 1951 and 1964 was in fact a "very inaccurate" estimate.[24] On this basis, government took critical decisions about employment policies in order to justify its attempts to retain blue-collar work in central London.[25]

The inaccuracy of these estimates emerged when the results of the 1961 Census were belatedly published in 1966. These showed that the actual rise in office employment was significantly less than had been estimated: in fact, between 1951 and 1961, the average annual growth in London's office employment had been fewer than 6,000 jobs, or just over one-third the rate on which the previous policies, thinking and planning had been based.[26] The figures showed that the bulk of employment growth in the South East had taken place not, as had been previously assumed, within Greater London, but in the rest of the South East region.[27]

A second questionable assumption that had an equally important impact was that office employment was itself responsible for worsening congestion problems in the central area. There is, however, little evidence to verify that this was in fact the case. The very specific focus on macro supply dynamics, such as employment trends, distracted attention from other related issues such as traffic management, which are more difficult, expensive and sensitive to tackle. Consequently, instead of being "seen as an integral part of an overall urban growth and adjustment strategy, relating closely to transport, housing, manpower questions and other policies",[28] office activity was beginning to resemble a scapegoat for London's perceived ills.

The assessment of aggregate trends in employment lacked any clear understanding of the causes and structure of the increased demand for office employment. Most of the growth was taking place in industries that were tied to central London through a complex network of economic linkages, which were shaping London's role as a global centre of finance and commerce. And the introduction of restrictive office policies was unnecessary not only because estimates of office jobs were exaggerated, but also because the growth was taking place in activities that were tied to the centre.[29]

The government's preoccupation with controlling office development was undaunted by questions over the validity of its policy assumptions. On the contrary, frustration at the continued, seemingly uncontrolled growth of office space, and the apparent ineffectiveness of the metropolitan authorities, led to direct intervention. In 1963, and without waiting for the outcome of the GLDP process, the government took direct action.

In March 1963, while introducing the Town and Country Planning Bill, the Minister of Housing and Local Government, Sir Keith Joseph, noted that

> whereas before the war there were about 87 million square feet of office space in Central London of which 9½ million square feet were lost by war damage, the war damage loss has been much more than replaced in the years since the war.

Sir Keith observed that the newly constructed office space had been "on a scale which far exceeded the war damage loss and by 1962 there was in Central London one-third more office space than in 1939". He further noted that "office employment in Central London has been steadily rising to the tune of about 150,000 new jobs during the last ten years".[30]

Arguing that if "nothing were done about it, this trend could be expected to continue for at least some years", Sir Keith went on to state

> I scarcely need to emphasise to the House the implications for housing and transport of this growth of employment in a small central area. Finding land for housing within London or even within reach of London without violating the green belt is difficult enough, even for the amount of employment already there.

The Minister then pointed out that "Office work takes up well over half the total of jobs in the central area. No other single employment produces this figure". Further, "It is increasing at a much faster rate than any other use".

Showing an astonishing lack of understanding of the dynamics of the office sec-tor, or the underlying economic drivers of its growth, the Minister concluded that

> much office work, unlike many other forms of central area employment, can be carried out elsewhere at lower cost, without real loss of efficiency, and in very much pleasanter conditions.

And the Minister jumped to the conclusion that there could not be "any doubt that the proper target for any attempt to reduce the rate of increase of jobs in Central London must be in offices rather than in other forms of employment".[31]

The subsequent Town and Country Planning Act 1963 sought to control offices. One way in which it did this was to close the 10% loophole within the Town and Country Planning Act 1947. Sir Keith argued that the

> intention of the original legislation, which was passed in the times of Socialist Government, was to provide a 10 per cent tolerance, but it was expressed in terms of cubic capacity, not floor space . . . which enables the developer to drive a coach and horses through the original intention. With modern building methods, a 10 per cent. increase in the cube can be turned into a 40 per cent. increase in the floor space, and it is floor space that determines employment potential. The present law makes it difficult to see what would happen if the local planning authority tried to prevent an increase of this order.[32]

The next measure saw the principle of office decentralisation established as part of central government policy for the first time, and the establishment of a pub-lic agency – the Location of Offices Bureau – to make it happen. The second prong of the attack was to ration the amount of office space that could be built through tight development controls – the Office Development Permit system.

Decentralising jobs

The Location of Offices Bureau (LOB) was established in 1964 with the declared aim of encouraging the decentralisation of office employment from central London. Its remit also included the promotion of publicity and research, and the provision of information. During its tenure, there is little doubt that the scale of decentralisation from London increased markedly. Figure 4.3 shows that the number of jobs that were relocated between its inception in 1964 and closure in 1979 amounted to some 105,000, involving 374 businesses.

The extent to which these moves could be wholly attributed to the activi-ties of LOB has been hotly debated. Some argue that LOB's impact was only marginal, because "the horse had bolted and the government locked the stable

Year	Number of jobs moved from central London	Number of moves from central London	Mean size of move from central London
1964	4,500	16	281
1965	6,650	24	277
1966	7,800	26	300
1967	9,200	27	340
1968	7,500	21	357
1969	6,000	26	230
1970	6,100	23	265
1971	6,400	26	246
1972	4,800	27	178
1973	9,200	31	296
1974	9,200	33	279
1975	8,400	32	263
1976	7,350	26	283
1977	3,500	9	389
1978	5,700	11	518
1979	3,180	16	199
Total	**105,480**	**374**	**300**

Figure 4.3 Relocation from Central London, 1964–1979

Source: Location of Offices Bureau/Jones Lang Wootton.[33]

door".[34] For others, its success in decanting office jobs from central London could be attributed in large measure to the operation of market forces.[35] The LOB's annual reports provide some evidence to support these claims.

The results presented in the *1978/79 Annual Report*, for example, are typical. During the year in question, 10,428 jobs were moved out of the whole of London by 142 clients of LOB. Only half of the jobs (5,005), and one quarter of the clients (36) moved between regions, i.e. beyond the South East. Furthermore, "relatively few of these moves involved complete relocation of a company's offices".[36] Indeed, 3,000 jobs were relocated by five banks and insurance companies – mainly administrative staff. Of the remaining 5,346 jobs, only 54% originated in the central area.

While there was evidently some limited decentralisation of employment, there was negligible de-concentration of economic activity. Cost cutting is a primary reason for relocation, and this could be achieved relatively easily at the time by moving a large proportion of clerical and administrative staff, rather than the client-facing departments and decision-makers.

For 1978–1979 at least, LOB's success at redistributing employment was limited and, in addition, appeared to complement the established process of relocating administrative functions to cheaper, 'back office' locations. So what of the government's other anti-office policy – restricting the amount of space that could be built?

The government restated the problem by pointing out that outstanding planning permissions, together with existing use rights, could result in over a quarter of a million further jobs in London, half of them in central London.[37] Without clarifying the causes of this development pattern or the economic context that might have been leading to the growth in demand for space, the statement then boldly announced:

> call a halt to this rapid growth. First we must have a standstill on new offices in London. To this end, the Government will shortly be introducing a Bill under which, in stated areas, any new offices will require Office Development Permits.[38]

This reaction to the perceived problems associated with a further 250,000 office jobs in central London showed a startlingly simplistic view of the growth dynamic in the office economy and the city's evolution as a global city. In Marriott's words, it presented "the crowning gift to the developers" by restricting supply and thereby filling vacant stock.[39]

Restricting development

From 4 November 1964, the Government's Secretary of State for Economic Affairs George Brown introduced the White Paper, Offices, which led to the 'Brown Ban' coming into force. Henceforth, any proposals for new office developments amounting to more than 2,500 sq ft (232 sq m) would require an ODP. Furthermore, the ODP was subject to "named user" restrictions and would be granted only where the developer could establish a definite need for the office space in the proposed location. The Board of Trade (which was responsible for the control of office development before it was transferred to the Ministry of Housing and Local Government in 1969) required strict criteria to be met before granting an ODP. Where an occupier was specified, three further qualifying criteria were applied:

- that the activity for which the office accommodation was being sought could not be carried on outside the area of control;
- that no suitable alternative office accommodation could be found, and
- that the development was essential in the public interest.[40]

It is clear also, that the 'public interest' catch-all clause could, potentially, be used against most schemes that passed the first two criteria. The named user clause clearly favoured the further concentration of large occupiers, because it "operated with an in-built bias in favour of large national and international concerns by comparison with smaller firms, without any objective reference to real user needs".[41]

Figure 4.4 shows the number of ODPs granted during the full period of their implementation. The figures are not directly comparable for the whole

Year	Number of permits	000s sq ft	000s sq ft relinquished
1965	33	1,242	587
1966	69	1,229	676
1967	84	2,483	1,052
1968	127	6,601	4,051
1969	154	6,308	2,237
1970	266	10,921	4,827
1971	198	12,246	4,212
1972	139	10,302	3,788
1973	126	7,677	2,567
1974	–	–	–
1975	64	4,883	448
1976	75	6,545	1,119
1977	57	5,674	328
1978	52	7,753	333

Figure 4.4 Office Development Permits, Central London, 1965–1978

Source: HMSO.[42]

Year	Complete and vacant	Under construction	Outstanding planning permissions	Total
	Million square feet			
1964	5.1	14.1	17.9	37.1
1968	3.8	5.4	2.5	11.7
1969	3.0	4.4	2.9	10.3
1970	1.4	4.3	6.2	11.9
1971	1.9	5.2	9.5	16.6
1972	1.6	10.9	9.0	21.5
1973	1.0	12.7	10.5	24.2
1974	1.1	14.1	12.3	27.5
1975	1.2	4.5	6.3	12.0

Figure 4.5 Office space in the pipeline in Greater London, 1964 and 1968–1975

Source: Estates Times.[43]

period because the exemption limits for ODPs were raised on three occasions. In 1970, the 2,500 sq ft (232 sq m) limit was lifted to 10,000 sq ft (929 sq m); this was then raised to 15,000 sq ft (1,393 sq m) in 1976, before reaching its final limit of 30,000 sq ft (2,787 sq m) in 1977.

Figure 4.5 shows the amount of space in the Greater London development pipeline in 1964 and between 1968 and 1975. Complete and vacant floorspace stood at over 5 million sq ft (465,000 sq m) before the introduction of ODP controls, with a further 14 million sq ft (1.3 million sq m) under construction.

By 1968, the impact of the controls was beginning to show. Complete and vacant space had fallen by a quarter; space under construction had fallen by nearly two-thirds, and outstanding planning permissions had fallen by 86%.

The suppressing effect of ODPs on development allowed the GLDP *Written Statement* of 1969 to claim that the system had prevented developments that "could have caused a runaway demand for office workers".[44] This self-congratulatory remark is strange for its apparent suggestion that employment grows to fill the available space, rather than being a response to economic growth, but was symptomatic of the supply-side policy framework of the time.

In 1970 when the ODP exemption limit was lifted from 2,500 sq ft (232 sq m) to 10,000 sq ft (929 sq m), the number of outstanding planning permissions doubled, and a year later construction levels began to rise. The complete and vacant space continued to fall over the period, reaching less than 25% of its 1964 level by 1975.

Changes in government policy are also reflected in the scale of bank borrowing. Between February 1971 and February 1974, bank advances to the property sector rose from £362 million to £2,584 million, an increase of 614%. Similarly, following the Brown Ban, new construction orders fell to an average of £1,987 million between 1965 and 1970, before rising in the years following relaxation to an average of £3,283 million during 1971–1973.[45]

As with the LOB, the impact of ODPs has been hotly debated over the years. As long ago as 1971, the highly respected academic planner Peter Cowan went on record as stating that "there is no longer any doubt that the Office Development Permit system is weak and ill-considered".[46] Irrespective of its validity as a supply-side mechanism, it is clear that one of the most significant though less understood impacts of the ODP system was its effect on the businesses occupying office space in central London. Companies were being forced to occupy whatever space was available. As new space was in increasingly short supply, they were forced to occupy second-hand space that itself was of a poor design standard and increasingly obsolete for the changing operational needs of occupiers. As Gerald Manners and Diana Morris argued:

> employers failed to find accommodation in places which they preferred . . . working conditions for the office workforce were poorer than they might otherwise have been, with floorspace per worker remaining relatively small and the age of the occupied office stock somewhat older than might have been the case without government intervention.[47]

The approach to the LOB and ODPs demonstrated the failure of strategic planning and its obsession with supply-side economics. It demonstrated how far planning had moved away from the ecological approach propounded by Geddes, preferring instead a piecemeal, ill-informed approach to a perceived problem. Cowan *et al* confirmed as much:

> In the long series of bans . . . one particular feature has been common to each measure – none of them has had the effect initially intended. This is possibly because there is a fundamental resistance to the idea of a static or declining London. It is one thing to say that London's influence is growing too fast; it is a completely different thing to happily accept its decline.[48]

In the mid-1960s, the first glimpses of a less restraining regime began to emerge. In 1964, the Ministry of Housing and Local Government's *South East Study* concluded that "considerable growth in London was inevitable, as well as probably important for the overall economic health of the country". In 1967, the South East Economic Planning Council's *A Strategy for the South East* "proposed a rough, sketch-like scheme that could accommodate the growth".[49]

4.4 Rolling back state planning

In 1969, the GLC produced its GLDP *Written Statement*, claiming that ODPs had prevented developments that could have led to a runaway growth in office workers. Nevertheless, the *Statement* recognised that while total employment in London was falling overall, the number of office workers was rising and would continue to rise.[50] Its rather weak response to this growth in demand for space was that "it should be met to the extent that it can, for those activities which need to be in London, within the context of a declining labour supply overall".[51] There was also a rare recognition of the changing requirements of occupiers. The additional space to meet demand was required "to relieve the cramped conditions in out-of-date buildings and to provide the more spacious conditions which modern business methods require for achieving the maximum output".[52] Donald Foley wrote that the *Statement* represented an

> important movement away from containment as a principle for organising Metropolitan London and a shift toward an acceptance of growth, along with an increasing concern for maintaining a viable economy and ensuring workability and livability.[53]

Following the publication of the *Written Statement*, which had taken four years to prepare, it was subjected to a public inquiry – the Layfield Inquiry – which at the time was the longest such inquiry in history, clocking up three years of endeavour to its findings in 1973. Layfield was highly critical of the GLC's document, and there was a further three years' delay before the plan was finally approved by the Minister on 9 July 1976. A full 11 years after its establishment, twice as long as it had taken to prepare the *London Development Plan* of 1951, the GLC had a finished and approved strategy for governing London. Little wonder that Sheppard concluded that the "GLC's main failure was in the field of strategic planning: the principal purpose for which it had been established".[54]

The policy framework that emerged in the GLDP was focused on locational priorities. The need for more efficient land use in central London was emphasised in preference to extending central London activities; consequently, growth of those activities considered appropriate to central London would be achieved through further decentralisation. For those that remained, the plan stated: "some areas of central London are more suitable for the development of offices than others".[55] It sought to direct office growth to 'preferred office locations', or areas in which offices "can be located with benefit".[56] However, this was little more than lip service to the control of office development, because the authority lacked "both the necessary executive powers and the financial muscle" to implement its policy.[57] The Council could not directly intervene in the office market, and in this sense was less effective than central government.

The combined effect of central and metropolitan government lining up against the growth of office employment in London was to constrain occupiers geographically and in inappropriate space. The policy framework was myopic in its approach, and what was perhaps fundamental to the weakness of London-wide policies was that they operated in isolation: questions about office location policy

> were discussed and debated in the absence of viable solutions to metropolitan (economic and physical) planning problems as a whole. Yet . . . office policy must be seen as an integral part of an overall urban growth and adjustment strategy.[58]

Just as importantly, however, the policies appeared to be working in isolation of the needs and aspirations of the occupiers themselves. There was no real understanding of the nature of the office economy, how it was changing or in what direction future changes in office requirements might move. Certainly none of these were made explicit in policy statements. Indeed, the scale of policy makers' ignorance of the changing needs of occupiers and the manner in which the central London economy was evolving was breathtaking. There was evidence around had they gone looking for it, and there were a significant number of prescient forecasts in this area, such as the following from 1971:

> The central clustering of London's office activities will remain the most important single element in the country for the exchange of money, ideas, and information – the invisible transactional society, which will become even more dominant between now and the year 2000.[59]

However, overriding concerns with perceived congestion, with the concentration of office activity at the expense of manufacturing and other 'productive' economic activity, and with preserving the 'essential character' of central London, precluded any notion of an activity-based approach to policy-making. As we saw in Chapter 2, there was plenty of evidence which set out the dynamics of the expanding office economy, the emergent knowledge economy and particularly its locational characteristics. Yet UK policy makers, obsessed with supply-side economics, failed to grasp the significance of what was happening in city centres and sought instead to control 'the problem'.

It was inevitable that this constraining pressure would eventually be released in a new cycle of demand. The change of government in 1979 and the arrival of Mrs Thatcher at 10 Downing Street, signalled a hands-off approach to planning policy, as with many aspects of the wider economy. Just two months after the election, in July, the government announced that it had

> reviewed the control of office development. This was introduced as a temporary measure in 1965 against the background of an office building boom and was primarily conceived as a counter to congestion in London.[60]

Without appraising how successful the ODP system had been in relieving congestion, the government declared that it was "satisfied that office development within London . . . can be adequately controlled through the normal planning system".[61] Neither statement suggested a fresh understanding of the office economy, nor a more positive approach to its development; simply that supply-side policies would be repealed. Later in July, the government announced: "Special control of office development ends on Monday 6 August".[62] Fifteen years after directly intervening in the office market for the first time, central government was retreating. The relative indifference displayed towards office policy in the pre- and immediate post-war years by central government had come full circle by the return of the Conservative administration in 1979. And the anti-office stance of metropolitan government was somewhat stifled by its own battles with central government that eventually led to its abolition in 1986.

As central government rolled back controls over offices and encouraged a generally more *laissez faire* approach to economic management, it fell squarely on local authorities to devise their own policy approaches to the built environment. It was not long before the dreadful flaws in such an approach to the management of a global city began to become patently clear, and nowhere was the folly and potentially damaging effects of this approach more exposed than in the experience of the Corporation of London.

Planning in the City of London

As the local authorities took centre stage in the office policy process, there was a softening of approach, although many years of anti-office policies paid dearly in terms of a very poor understanding of the contemporary role of offices in the economy and the changing character of occupiers and their demands on built space. By 1982 the Corporation of London was beginning to understand the link between stock and demand:

> It is becoming increasingly apparent that prospective tenants are giving greater attention to the quality of the construction and design of the building. Poor standard of development is much harder to let than good quality of work. Funding institutions are now readily investing in refurbished schemes of good quality.[63]
>
> One section of property that is causing concern is the late 50s/early 60s speculative blocks. These units need drastic refurbishment if they are to be fully effective in their central locations. However . . . to bring them up to scratch means a very costly operation and no short term solution readily comes to mind.[64]

But such sentiment failed to transpose into a realistic policy framework, partly because the City continued to operate within the shadow of the strategic planning authority. The position of the latter was made abundantly clear in 1984 by George Nicholson, then Chair of the GLC's Planning Committee. Nicholson reinforced the GLC's position on offices with the following statement:

> There are no forecasts indicating that there is a real demand for office space much in excess of past take-up rates. Indeed, many firms are cutting back

their corporate staffing and are occupying smaller suites and not the quarter or half million square foot monsters so beloved by developers.[65]

Whilst it is not difficult to forgive the fortune-telling weaknesses of the GLC (after all, plenty of others have got the future wrong), it is difficult to conclude that this position was not driven by political considerations rather than a naivety about the operational needs of the evolving office economy. The fact is that the strategic planning authority for Greater London failed completely to anticipate any quantitative increase or qualitative shift in requirements for office space. The sharp increases in demand and shortages of space that were recorded in 1986 and 1987 were the result of long-term change. Yet take-up rates and aggregate employment trends were thought sufficient to guide Nicholson in his decisions on the level and type of office development to be permitted in the capital.

This was a classic example of how top-down planning, more concerned with social justice than with economic development, had catastrophically failed. It served to compound the failure of government planning with its LOB and ODPs. Patrick Geddes would have recoiled at the lack of bottom-up intelligence, or feedback, and the arrogance of planning policy that reflected political concerns rather than the underlying economic reality.

The fact is that, less than two years before Big Bang, there was absolutely no appreciation, in spatial planning terms, of the recent changes in the office economy, let alone any insight into the radical shake-up that was about to happen. At the citywide level, and as late as the mid-1980s, there remained a fundamental misunderstanding of what was around the corner, and this reflected itself in the local planning process of the City of London.

The disconnect between the evolving nature of the office economy and the planning process took a dramatic turn when the *City of London Draft Local Plan* was published in November 1984.[66] The plan fell woefully short of any coherent understanding of the relationship between City business and City property, and attracted a level of interest and criticism rarely raised over a local plan. The central weakness of the plan was that despite its claims to represent a world financial centre – with 55 million sq ft (c5.1 million sq m) of office space, 220,000 office workers and a residential population of just 4,000, all concentrated into one square mile – it continued to reflect the essentially anti-office stance of the metropolitan planning authority.

In the *Draft Plan*, 70% of the City's core area was designated a Conservation Area (Figure 4.6). Demolition and replacement there were to be strictly controlled in order to preserve the special character and architectural heritage of the City. The *Draft Plan* sought to resist a reduction in the diversity of uses affecting employment opportunities, and to restrict large-scale office development to the fringe areas. This it sought to achieve by exercising discrimination relative to location and existing uses, restricting new buildings to existing sites, preventing land assemblage by developers and designating Special Policy Areas. One of these, the eastern City area, sought to restrict the loss of wholesale and

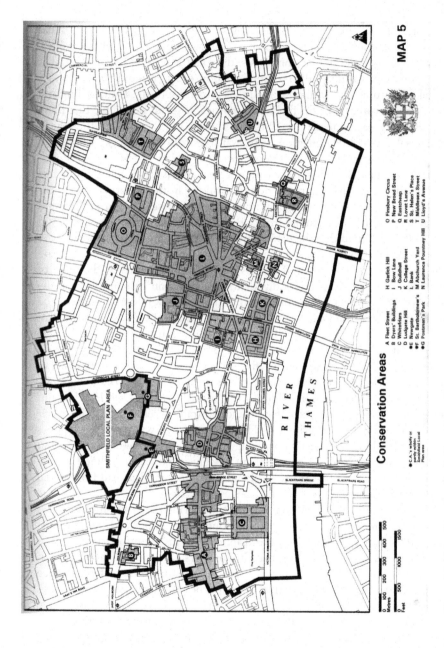

Conservation Areas

* C.A.'s wholly or
partly within
Smithfield Local
Plan area

A	Fleet Street
B	Dyers' Buildings
C	Whitefriars
D	Ludgate Hill
*E	Newgate
*F	St. Bartholomew's
*G	Postman's Park

H	Garlick Hill
I	Bow Lane
J	Guildhall
K	College Street
L	Bank
M	Abchurch Yard
N	Laurance Pountney Hill

O	Finsbury Circus
P	New Broad Street
Q	Eastcheap
R	Lovat Lane
S	St. Helen's Place
T	Middlesex Street
U	Lloyd's Avenue

MAP 5

Figure 4.6 City of London Draft Plan 1986, Conservation Areas

Source: Reproduced from Corporation of London (1984).[67]

industrial premises; to restrict the loss of "small" and "very small" units of business accommodation, and to discourage developments entailing the unification of several sites.

High buildings were seen in the *Draft Plan* as incompatible with the scale and character of the City; and demolition of certain non-listed buildings that contributed to the special character of a conservation area would be resisted. Echoing George Nicholson's comments, the *Draft Plan* went on to argue that new technology would lead to a rationalisation of space requirements, and a general fall in unit size. It also claimed that proximity to the Bank of England would remain paramount for financial activities. It seems incredible that these statements were being made just 23 months before Big Bang, when the lessons of financial services deregulation were available from New York, which had gone through a similar process of deregulation in 1975.

The Corporation invited comments on its plan. Many detailed comments were submitted, with a significant number coming from concerned City businesses. One of the most detailed (and published) of these was that put forward by the Centre for Policy Studies that outlined five major criticisms of the *Draft Plan*.[68] First, it argued that the plan demonstrated little apparent understanding of the City's economy or the economic and technological changes affecting the type of buildings required. This was clearly true and struck at the heart of the credibility of the plan. Secondly, that it was based on an uncritical acceptance of outdated policies contained within the GLDP. While true, it should be remembered the Local Plan worked within the framework of the strategic plan and was therefore constrained by its policies. Thirdly, the Centre argued that it did not indicate the Corporation's desire to promote the financial economy of the City. Again, this is true, and it reflects a very naïve understanding that was apparent at all levels of government in terms of understanding the contemporary role of the office economy. Fourthly, it argued that the *Draft Plan* had not justified the restrictive policies that would frustrate development. This point was related to the second criticism. The final criticism was that the proposals illustrated the shortcomings of a planning system that was cumbersome "when faced with a fast-changing environment".[69] This point is underlined by the long gestation periods that many local and metropolitan plans had from initial draft to final statement.

Whilst the Corporation of London was considering the criticisms of its plan during the summer of 1985, American property developer and head of First Boston Real Estate, G Ware Travelstead announced his intention to build up to 8 million sq ft (743,000 sq m) of space in London's Docklands. Travelstead was quoted as saying that the scheme represented "a massive boost of confidence in the future of London", and that "London needs offices that meet today's and tomorrow's needs if it is to continue to thrive".[70] Travelstead's move, while opportunistic, was a direct challenge to the City: if it did not want to accommodate the new financial players, then other locations could. It should also be recalled that at the time Paris and Frankfurt were making overtures to challenge London's dominant position in financial services in Europe. Suddenly, the City was under threat, and it was no surprise that the Corporation radically revised many of the central tenets of its earlier statement.

The *Draft Plan* was substantially re-written before a final version was deposited in May 1986. Michael Cassidy, then Chairman of the City Planning and Communications Committee, wrote in the foreword that, in formulating the *Local Plan*, the Corporation had

> sought to strike a balance between providing adequate office accommodation for the financial services industry, which predominates the City, and at the same time maintaining the unique character and quality of the townscape of the City.[71]

Indeed, the *Local Plan* had absorbed some of the key criticisms of the *Draft Plan*. For example, in an important development, the Plan recognised the significance of the space requirements of the City companies: 'Policy Econ 2' aimed to "encourage office development which meets the requirements of new office technology and which is flexible to cater for variations in office layout and use".[72]

The new attitude of the Corporation towards development was welcome. Nevertheless, two critical conclusions are inescapable. First, the Corporation had left itself open to the widespread opposition to the initial *Draft Plan* by failing to consult its key constituency. The occupiers knew what was happening to their businesses (it didn't take the Centre for Policy Studies long to ask them). Discussions would have avoided many of the fundamental flaws of the *Draft Plan*. Secondly, if consultation with occupiers during preparation of the *Draft Plan* had occurred, then the Corporation could have anticipated the major effects of Big Bang on demand for office space. As it was, they were caught off guard and faced with a hostile development industry eager to satisfy the pent-up demand. The Corporation's only option, made all the more necessary by the revitalised Canary Wharf scheme (under its new owners Olympia & York), was to take the pressure off pent-up demand by rushing through a swathe of consents that were to trigger one of the biggest development booms in the City's history. But, yet again, it had reacted to, rather than anticipated, the changing dynamics of demand.

At the same time as the Corporation was re-writing its local plan, central government was once again gearing up for a major intervention in the planning process – this time to the benefit of the office economy. Government had signalled its intent with respect to the planning system, and particularly the Town and Country Planning (Use Classes) Order 1972, in its 1985 White Paper *Lifting the Burden*. The Use Classes Order (UCO) contained 18 categories of use for buildings and land, which had remained unchanged since 1948.

The UCO controlled development by requiring that consent be granted for a building's use to change from one class to another. Applications for consent were typically time consuming for applicants and consumed large amounts of manpower within local planning authorities. In addition to these process issues, it was also becoming clear that the classes had not kept pace with the changing needs of occupiers, particularly in the emergent high-technology industries and the service economy.

Building for the tech sector

Following industry consultation, the government published *The Town and Country Planning (Use Classes) Order 1987*. The new UCO reduced the old 18 classes to 12 but, perhaps most significantly, introduced a new Business Use Class, or B1. This combined the old classes of offices and light industrial, together with laboratories, R&D and studios. From this apparent tinkering with a piece of planning legislation sprang one of the most profound changes in the office property market of the whole post-war period. The change in legislation had a major impact in city and town centres, where it allowed the conversion of light industrial buildings that had stubbornly clung on to their expensive, central locations simply because conversion was too difficult, and often times because the local authority sought to protect 'productive' jobs (and buildings) against the further spread of office jobs. But perhaps the even greater impact of the new legislation was that it resulted in the proliferation of an emergent building type and the transformation of large tracts of land, often on the edges of and beyond of built-up areas. The new building was the 'high-tech building' and its setting the 'business park'.

During the early 1980s, the technology industries were growing rapidly. Such companies were looking for a new type of building that allowed them to accommodate office and clean production activities into a single, flexible and simplified building type. The traditional office was, physically, too restrictive. They wanted to put office space, laboratory space and elements of light production, assembly and storage all in the same building. On top of this, they wanted to be able to change the configuration of the space as the needs of their businesses dictated. Many such companies were experienced in the North American market, where business parks had become commonplace. These kinds of buildings were not available in traditional office centres.

The development industry responded by providing high-tech buildings, typically simple, two-storey buildings that allowed for a mix of office and production/warehouse. The early examples were characterised by their brightly coloured cladding. However, these early buildings were little more than smartly designed light industrial units. The earliest examples, starting in around 1980, included Abingdon Business Park, Kembrey Park in Swindon and Aztec West, near Bristol). To cope with the growing demand and to overcome the restrictive nature of the UCO, certain planning authorities began to grant planning consents with special conditions, and others began to reach agreements outside the UCO.

As the market matured, a more sophisticated building type emerged that looked more corporate but still provided genuinely mixed space. The early phases of Stockley Park typified this kind of building, as did Frimley Business Park, Winnersh Triangle in Reading and Capability Green at Luton. These parks were typified by higher-quality master planning, and the buildings included atria and generally higher-quality materials. It is significant that Stockley Park, one of the earliest and perhaps *the* most successful park of its genre, was strongly influenced by research into occupier needs by design firm DEGW, particularly of founding partner John Worthington, undertaken for Stuart Lipton of Stanhope Properties. The research undertaken for Stockley was highly influential in convincing government of the need to revise the UCO as well as setting the standard for the new business park market. The success of this partnership between developer and advisor is all the more remarkable given that the same team was, simultaneously, having a similar impact on the City of London.

Pressure continued to grow on the increasingly outdated UCO, and in June 1987 the new Order was introduced. This formalisation of what had by now been happening by stealth for several years swept in a third phase of evolution in which the high-tech, mixed-use schemes yielded to straightforward office campuses for corporate head offices from a host of different sectors. The impact of the legislation was dramatic. Of course, change of use also meant change of value: the new legislation meant that suddenly developers could convert land and buildings that had previously been consented to accommodate light industrial uses into the new business use class. The temptation to pile in and benefit from the uplift in value led to a flood of new developments.

The new legislation underpinned one of the largest-ever speculative development booms. By 1990, nearly 800 developments were calling themselves business parks, around 48 million sq ft (c4.5 million sq m) had been built, a further 16.5 million sq ft (1.5 million sq m) was under construction and a massive 148 million sq ft (13.7 million sq m) had been granted consent.[73] Within the short span of under ten years, a new form of demand emerged; enabling planning legislation was enacted, and the market moved from starvation to glut. Combined with the development boom associated with Big Bang, the overhang of surplus office space in the early 1990s was, perhaps, the largest ever, before or since.

While the causes and features of the development boom were complex, there is little doubt that a planning process lacking bottom-up intelligence was a contributory factor. Batty and Marshall's observation is relevant here.

> For much of the twentieth century, city planners and the plans they produced assumed that cities were in equilibrium and the focus was almost entirely on implementing some form of blueprint depicting a desired end state.[74]

During the coming decades, market intelligence would improve immeasurably, and city planning would adopt a far more systemic approach to policy formulation, not least in its understanding of the office economy.

4.5 A new approach to strategic planning

Mrs Thatcher's Government abolished the GLC in 1986 and, along with it, strategic planning in London, which was absorbed into the government's own responsibilities. Perhaps learning from the experience of the Corporation of London's Draft Local Plan in 1984, and recognising that a complex global city such as London needed at least some form of strategic land use planning framework, while overcoming ideological concerns over interventionism, the government set up the London Planning Advisory Committee (LPAC) in 1986. This new body was composed of representatives from each of London's 33 boroughs, and its remit was to provide the Secretary of State with planning advice during the preparation of strategic planning guidance for London.

The Committee published a report on the principal *Policy Issues and Choices for the 1990s* in January 1988,[75] which was then subject to consultation before publication of the *Draft Strategic Planning Advice* in June.[76] Finally, in October, the *Strategic Planning Advice for London* was published.[77] This last document summarised the advice that LPAC wished to give on the scope and content of the Secretary's Strategic Planning Guidance; and it dealt with planning problems that LPAC believed would probably not be included in the Guidance.

Policy Issues and Choices noted that employment in London had fallen from 4.5 million in 1966 to 3.5 million in 1984, arguing that there should be no return to the decentralisation policies of the 1960s and 1970s. Rather, in order to stem the loss of jobs, the report argued that the broad goals of the 1990s should be to foster economic growth and regeneration. LPAC argued that strategic guidance should promote and support London's international and national roles through accommodating those sectors that export their services and products outside London.

This could be achieved in part by encouraging the expansion of large-scale business uses in a "core zone in central London, in West Docklands, and at a limited number of key public transport interchanges".[78] This, they argued, could involve a growth of around 300,000 jobs by 2001 (broadly equivalent to the size of the City of London workforce). Finally, after nearly four decades of denial over the role of offices in modern city economies, London's strategic planning body had begun to promote office economy jobs.

A major failing of the report, however, was that it offered no strategic advice on the potential demand for office growth in central London: "The Committee has attempted to estimate industrial and commercial requirements in London. It has concluded that the essential information required is not available".[79] It is curious, to say the least, that after so long spent by central and metropolitan government seeking to restrain office growth, such an admission could be possible. However, despite the absence of relevant information, the committee did argue that central London's international and national role as a centre for business, financial services, transport and communication networks "should be supported", and that "London must be able to offer high quality office sites in central London and a number of Strategic Centre locations".[80] The message was clear: the broad thrust of LPAC's approach was one of encouraging economic development and providing the office economy with the space it needed.

The Secretary of State's response in the form of the *Draft Strategic Planning Guidance*, was a much shorter document (17 pages, plus map) that stressed the need for planning to focus on land use planning "to facilitate development while protecting the local environment".[81] The Guidance emphasised the need for planning to provide "a framework for development control. This guidance therefore deals with matters that are directly related to the use and development of land".[82] The clear message in this statement was that borough councils were expected to avoid many of the broader social and economic issues that had been brought into borough plans within the framework of the GLDP and were, instead, to concentrate on the core subject of land use planning. Significantly, this meant a sharper focus on the office economy.

Planning for offices

The Guidance pointed to the role of the private sector in the regeneration of London and argued that "Boroughs should adopt a positive, flexible and realistic approach to business development throughout London".[83] Such a philosophy was in sharp contrast to the anti-development stance adopted by the GLC in the 1970s and early 1980s. The draft Guidance, in effect, was directed at shifting the balance in land use planning away from one in which an applicant needed to justify a scheme (within the normal framework of development control), to one in which the local authority would have to justify a refusal. Five key statements in the Guidance illustrated this shift in balance:

* development in London should not be restricted to a tightly drawn core zone;
* UDPs should reflect the changing needs of industry and current or likely future demand for such development;
* boroughs should not seek to distinguish between activities within the B1 Use Class;
* boroughs should not reserve land for general industrial use when there is no realistic prospect of such use materialising, and
* a local planning authority is not entitled to treat an applicant's need for permission as an opportunity to obtain some extraneous benefit or advantage.

Clearly the government was taking a more positive stance towards the growth of the office economy in London than had been the case for decades. The Guidance set the tone for the 1990s and provided a lead for strategic policy makers to follow. By the mid-1990s a far more accommodating policy framework for the development of the office economy was emerging than at any time since the war. For example, a study commissioned by LPAC urged a new strategic urban framework for London and presented a more sophisticated understanding of the role of the office economy in the late twentieth century than anything that had preceded it in planning policy terms: "Wealth creation in world cities is to be found in the agglomeration and network of interlocking and mutually supportive businesses serving world markets with products and services based on knowledge, information, technology and creative skills".[84]

Furthermore, in LPAC's first five-yearly review of its advice to government (Regional Planning Guidance [RPG] Note 3), a new strategic context specifically for offices was included, recognising that London's future as a global city was dependent upon the financial and business services sectors. It was not the prescriptive kind of guidance of bygone years in which strategic policy had sought to directly steer office development; rather, it was aimed at reducing uncertainty "by improving the availability of strategic information to planning authorities, and to help the market operate more efficiently by lessening the risks of developers and investors misreading it".[85] This was a major step forward in terms of providing the kind of information-rich environment that was a pre-requisite for addressing the volatility of the development industry, caused at least in part by the paucity of market-wide information.

The early to mid-1990s were, of course, dominated by the recession, and consequently, there was little pressure to deal with increasing demand for office space in public policy. Indeed, RPG3 specifically sought a shift of emphasis from providing further office supply to managing the surplus. However, the next major step in London's governance was taken in 1998, with a referendum asking whether there was support for the creation of a Greater London Authority, a London Assembly and a directly elected Mayor. The ensuing Greater London Authority Act of 1999 paved the way for Ken Livingstone, a stalwart of the old GLC, to assume the first of his two terms in office in 2000, to be followed by future Prime Minister Boris Johnson.

Following a brief squat in a non-descript office in Victoria, the newly elected Greater London Authority (GLA) moved into a smart new headquarters on the South Bank of the Thames adjacent to Tower Bridge. The GLA's planning remit included the preparation of a Spatial Development Strategy for London, into which borough-level Unitary Development Plans were expected to dovetail. The SDS replaced the guidance previously prepared by LPAC, and its aim was to provide an overall land-use strategy to support the Mayor's social, economic and environmental policies.

The *Draft London Plan* of 2002 committed to a review of the long-held opposition to high rise buildings, and also looked to earmark a series of sites around mainline rail termini for high density development.[86] The draft plan was adopted in February 2004.[87] The London Plan represented a sea change in attitude towards the office economy compared to its predecessors – some of which were signed off by the same Mr Livingstone now installed as Mayor. The plan began by recognising that "London has changed dramatically" over the past 20 years, particularly due to "the globalisation of many economic sectors, and the dominance of the finance and business sectors, frequently interlinked with dramatic changes in technology".[88] Twenty years after George Nicholson denied that London needed any further office space beyond that which existed, finally there was an explicit statement in citywide planning that London was a global city that needed modern office space.

The GLC has maintained its pro-office growth stance since the 2004 Plan. It monitors the market through the occasional *London Office Policy Review*. The 2007 edition referred to the GLA's "Goldilocks" planning policy

> a 'not too hot, not too cold' approach in which the emphasis has been on providing for growth with a London-wide strategy that seeks to balance geographical and temporal fluctuations in supply and demand.[89]

The next edition was published in 2009, following the Global Financial Crisis. The report referred to London navigating "turbulent waters", and then asked: "how can the London Plan encourage the office market to help protect and enhance London's world city?" The authors argued that, other things being equal, "London is positioned for growth", which means the need for a "planning framework that provides vision and certainty, within which the business community – and specifically the property industry – can operate with confidence and certainty". It went on to suggest

> In short, the London Plan requires a bold and ambitious statement of intent with respect to the city's most important economic driver – the office economy. Such a statement should encourage investment in London's future office market with an enabling framework for both CAZ and selected OL [Outer London] centres.[90]

The 2017 LOPR was published in the wake of the UK's decision to exit the European Union speculated on London losing ground internationally. The authors suggested that for London

> to fall from its current top spot to, say, 10th, within a decade or so would imply a seismic economic adjustment for a major city on a scale that has no historical precedence. Even a fall to, say, somewhere between 5th and 10th would imply an enormous shift in activities.

The report was more concerned about underlying trends within the office economy, most notably changing working styles and the prospect of automation and artificial intelligence, arguing that these could "dwarf any Brexit impact and could, ultimately, mean that spatial policy might be dealing with new locational preferences and building typologies".[91]

4.6 Planning for uncertainty

We saw in Chapter 3 how explanations of the office economy really got going only in the 1980s and 1990s. There was the odd exceptional piece of writing beforehand, but no collective body of knowledge and writing that could be said to relate to the office economy as a component part of geography or economics. It was little surprise that spatial planning failed to come to grips with the main driver of economic growth in cities over the second half of the twentieth century.

From the early days of the post-war period, offices and the office economy were seen as 'a problem', driving jobs growth that created congestion and additional demands on infrastructure and other services. Successive metropolitan and central governments pursued restraining policies. It seems quite strange, from the perspective of the twenty-first century, that past policies actually sought to restrain growth in office jobs. In the early years, there was certainly a political dimension, which regarded office work somehow as less worthy than manufacturing. Service activity was long regarded as a 'nice to have' rather than an integral part of the economy, and this fed through to spatial approaches to office activities.

By the mid-1980s, the economy was on the threshold of the 'technological revolution' which was to bring about immense change to the economy, to the businesses and to the buildings they occupied. White-collar work suddenly grew very rapidly, matched only by the speed of manufacturing's decline. We then experienced the deregulation of financial services which transformed financial and professional services. The internet, laptops, social media and agile working all followed. And at long last, in the late 1980s, planning responded with policies that were more in tune with the direction of travel in the office economy, and facilitating in the rebirth of London as a Global City.

Despite this more positive picture, the office economy faces tremendous uncertainty as it grapples with the consequences of technological innovation and disruption. Planning in such a context becomes extremely difficult. It is likely that future iterations of the London Plan will have to be sensitive to an extraordinary array of influences as the office economy and its internal dynamics evolve rapidly.

Notes

1 M. Batty & S. Marshall (2009) The Evolution of Cities: Geddes, Abercrombie and the New Physicalism, *Town Planning Review,* Vol 80, No 6, pp551–574
2 *Ibid*
3 *Ibid*
4 G. Manners & O. Morris (1986) *Office Policy in Britain: A Review,* Geo Books, Norwich, p145
5 *Ibid*
6 Town and Country Planning Association (1962) *The Paper Metropolis,* TCPA, London, p25
7 *Ibid*
8 J.H. Forshaw & P. Abercrombie (1943) *County of London Plan,* Macmillan, London, pIII
9 *Ibid*, p22
10 *Ibid*, Plate VI, facing, p24

11 Donald L. Foley (1972) *Governing the London Region: Reorganisation and Planning in the 1960s,* University of California Press, Berkeley, p94
12 *Ibid,* p96
13 O. Marriott (1967) *The Property Boom,* Pan Books Ltd, London, p199
14 Cowan (1969) *Op cit,* p165
15 TCPA (1962) *Op cit,* p22
16 R. Barras (1979) *The Development Cycle in the City of London,* Centre for Environmental Studies, Research Series 36, London, p41
17 TCPA (1962) *Op cit,* p27
18 Foley (1972) *Op cit,* p96
19 *Ibid,* p15
20 Jenkins (1975) *Op cit,* p215
21 *Ibid,* pp226–227
22 P. Cowan (1971) Employment and Offices. In: J. Hillman (Editor), *Planning for London,* Penguin Books, London, p65
23 TCPA (1962) *Op cit,* p20
24 I. Alexander (1979) *Office Location and Public Policy,* Longman, Harlow, p64
25 P. Hall (1969) *London 2000* (2nd edition), Faber & Faber, London, p77
26 Cowan (1971) *Op cit,* p66
27 Manners & Morris (1986) *Op cit*
28 G. Manners & O. Morris (1981) Does London Need an Office Policy? In: R. Barras (Editor), *The Office Boom in London,* Proceedings of the First CES London Conference, p6
29 A. Evans (1967) Myths About Employment in Central London, *Journal of Transportation and Policy,* Vol 1, pp214–225
30 K. Joseph (1963) *Town and Country Planning Bill,* Vol 673, 12th March 1963, https://hansard.parliament.uk/commons/1963-03-12/debates/a068f80e-8547-4842-b3da-2b760f6e7bc4/TownAndCountryPlanningBill
31 *Ibid*
32 *Ibid*
33 Location of Offices Bureau, reprinted in Jones Lang Wootton (1987) *The Decentralisation of Offices from Central London,* JLW, London, pp3–4
34 Marriott (1967) *Op cit,* p23
35 Manners & Morris (1986) *Op cit,* p116
36 Location of Offices Bureau (1979) *Annual Report 1978/79,* LOB, London, pp11–13
37 HMSO (1964) *Offices: A Statement by Her Majesty's Government,* SO Code No. 63–9999, 4 November, HMSO, London
38 *Ibid*
39 Marriott (1967) *Op cit,* p22
40 Board of Trade (1967) *Annual Report by the Board of Trade for the Year Ended 31st March* HMSO, London
41 Manners & Morris (1986) *Op cit,* pp127–128
42 Office Development Statistics in the Fourth Quarter and the Year 1972 *Trade and Industry,* 15 March 1973, pp578–579 HMSO, London; Office Development Permits in the First Quarter *Trade and Industry,* 8 June 1979, pp449–450, HMSO, London
43 Estates Times, 1st November 1973, p9 citing Annual Report by the Secretary of State for the Environment and the Secretary of State for Wales, 1974 and 1975
44 Greater London Council (1969) *Greater London Development Plan Statement,* Planning Department, London, p20
45 P. Scott (1996) *The Property Masters: A History of the British Commercial Property Sector* (2nd edition), Routledge, London, p184
46 Cowan (1971) *Op cit,* p68
47 Manners & Morris (1986) *Op cit,* p128

48 Cowan (1971) *Op cit*, p154
49 Foley (1972) *Op cit*, p96
50 GLC (1969) *Op cit*, p20
51 *Ibid*
52 *Ibid*
53 Foley (1972) *Op cit*, p99
54 Sheppard (1999) *Op cit*, p349
55 GLC (1976) *Greater London Development Plan Written Statement*, Planning Department, London, Paragraph 7.6
56 *Ibid*, Paragraph 4.13
57 Manners & Morris (1986) *Op cit*, pp108–109
58 Manners & Morris (1981) *Op cit*, p61
59 Cowan (1971) *Op cit*, p72
60 Department of the Environment Press Notice 270, 5th July 1979
61 *Ibid*
62 Department of the Environment Press Notice 320, 30th July 1979
63 Corporation of London (1982) *City of London Information Report: Offices*, Department of Architecture and Planning, London, p16
64 *Ibid*, p17
65 G. Nicholson (1984) The GLC Answers on Offices, *Building Design*, 11th May, pp14–15
66 Corporation of London (1984) *Draft Local Plan*, Department of Architecture and Planning, London
67 *Ibid*, p135
68 Centre for Policy Studies (1985) *Comments on the City of London Draft Local Plan of November 1984*, CPS, London
69 *Ibid*, p36
70 L. Mallett & J. Gardner (1985) US 'Star Wars' Scheme Set for Docklands, *Chartered Surveyor Weekly*, 13 June, p776
71 Corporation of London (1986) *Local Plan Written Statement and Proposals Map*, Department of Architecture and Planning, London
72 *Ibid*, p24
73 A. King & S. Bryant (1990) *Living with B1: UK Business Parks 1979–1999*, Applied Property Research, London
74 Batty & Marshall (2009) *Op cit*, p563
75 London Planning Advisory Committee (1988a) *Policy Issues and Choices: The Future of London in the 1990s*, LPAC, Romford
76 London Planning Advisory Committee (1988b) *Draft Strategic Planning Advice*, LPAC, Romford
77 London Planning Advisory Committee (1988c) *Strategic Planning Advice for London: Policies for the 1990s*, LPAC, Romford
78 *Ibid*, p19
79 *Ibid*
80 *Ibid*, p21
81 Department of the Environment (1989) *Draft Strategic Planning Guidance for London*, DoE, London, p2
82 *Ibid*
83 *Ibid*, p4
84 Coopers & Lybrand Deloitte (1992) *London: World City Moving into the 21st Century*, HMSO, London, p3
85 M. Simmons (1994) Planning for Office Development: A Strategic Approach, *Property Review*, December/January, pp213–216
86 Greater London Authority (2002) *Draft London Plan*, GLA, London

87 Greater London Authority (2004) *The London Plan: Spatial Development Strategy for Greater London,* GLA, London

88 *Ibid*, p1

89 Ramidus Consulting Limited (2007) *London Office Policy Review,* GLA, London, pXIV

90 Ramidus Consulting Limited (2009) *London Office Policy Review,* GLA, London, pXI

91 Ramidus Consulting Limited (2017) *London Office Policy Review,* GLA, London, ppVII–VIII

5 Building

A triumph of hope over experience

We saw in Chapter 2 how, during the nineteenth century, the office economy became a more distinctive part of the milieu of activities in the city. Alongside expanding industry, specialist office buildings emerged, somewhat grandiose in the early days, before evolving into the commercial buildings more familiar today. And London grew, denser in the centre while spreading outwards to house its rapidly growing and increasingly wealthy residents. As Marion Bowley observed, "Industrialisation, urbanisation and the growth of population all went together", but at the same time they "all required new investment in buildings".[1] She elaborated, buildings for "industry and trade are as much part of the basic capital investment required for the effective functioning of the economic system as machinery, railways, etc."[2]

Pioneering urban planner Patrick Geddes (see Chapter 4) conveyed a picture of London's growth as "Perhaps likest to the spreading of a great coral reef".[3] In one sense he was correct: over a long period, London grew larger and larger. But in another, he conveys the wrong image: one of slow, imperceptible growth.

Instead, London has passed through numerous phases of intense, rapid change as it responded to changes in society and the economy, interspersed with periods of slower growth. And these phases are not orderly, predictable, successive waves, but complex, interlocking and overlapping. Some phases of growth were driven directly by economic expansion and internationalisation (for example, the interdependent expansion of the docks and the City of London). Some were driven by the need to house, feed and entertain the growing numbers of people required to support the expanding economy (for example, the development of the residential and retail areas in the West End). Others were spurred by new technologies (for example, suburbanisation in the early part of the twentieth century following the proliferation of motor cars).

Many such phases of development left a trail of ruined developers and builders and half-completed projects throughout the eighteenth and nineteenth centuries and the first half of the twentieth century. But the post-Second World War period was different. While the same underlying causes of economic expansion and restructuring remained, the structure of the property supply industry changed radically; the scale of enterprise made a step change, and development became concentrated on the purpose-built, speculative office block.

Up to this point, especially in the nineteenth century, many development booms had had a broader mix of uses, mostly housing, retail and leisure activities. But this changed in the twentieth century as the office economy grew, and as "new office blocks became a familiar presence; new banks, company headquarters, insurance offices were built upon a massive scale, with intense and dramatic architectural effects".[4]

This chapter examines the physical development of the modern office economy, largely during the nineteenth and twentieth centuries, with a focus on the period of development following the Second World War and the rise of the institutional investment market.

5.1 The early developers

Nicholas Barbon (c1640–c1698) was a physician, financier, free market economist and one of the earliest speculative property developers. Born in London, Barbon studied medicine at the universities of Leiden and Utrecht in the Netherlands, receiving his Doctor of Medicine in 1661. Three years later, he was nominated an Honorary Fellow of the Royal College of Physicians. In the aftermath of London's 1666 conflagration he pioneered home fire insurance.

Figure 5.1 Nicholas Barbon

During his later years, Barbon wrote extensively on economic theory, and his *A Discourse of Trade* (written in 1690), met with critical praise from modern economists including John Maynard Keynes and Joseph Schumpeter. Barbon was also elected a Member of Parliament in 1690 and 1695. Quite a list of achievements; but Barbon the builder concerns us most here.

When the Great Fire engulfed the City of London, Barbon seized an enormous opportunity to become one of, if not *the* most influential builder of his time. His determination to succeed was ruthless, and his methods for procuring and developing land have been more politely described as unscrupulous, bending people to his will, and often riding roughshod over royal declarations and Acts of Parliament, to achieve his ambitions. This was most notably the case in what today is called Midtown, especially around what is now Bloomsbury and Strand, where he, in effect, connected the City of London with the City of Westminster. However, it was in nearby Holborn where Barbon's most infamous development took place.

Barbon started his largest project to date, Red Lion Square, in 1684. At this time, the land in the area was used as fields and paddocks, but Barbon had visions for a 17-acre (6.9 ha) redevelopment to provide new housing. Red Lion Fields lay to the north of High Holborn and to the west of the Inns of Court on the west side of Gray's Inn Road.

However, as Barbon started to divide up the fields to create a new London square, he was fiercely challenged by a group of lawyers from the nearby Inns of Court, many of whom objected to losing their 'rural' outlook. Naturally, they took him to court but, perhaps unexpectedly given their professional skills, lost the case. Relations between the two parties continued to deteriorate, so much so, that on 11 June 1684, a pitched battle broke out between roughly 100 lawyers and site workmen, with the main offensive weapons being bricks and other building materials. Naturally, Barbon led his workmen from the front. Several injuries on both sides were sustained but, eventually, Barbon and his workmen prevailed. Work resumed and Barbon's project proceeded, becoming a very fashionable district, popular with lawyers, doctors and other professionals.

Barbon had established a template, or modus operandi, for large-scale, speculative developers. He had also established a reputation for developers as a rapacious and single-minded class, sometimes with questionable ethics. To some degree, this reputational profile survives in the popular psyche to the present day.

Bowley writes that "until at least about the middle of eighteenth century the building owner himself or his architect, or agent, employed each craft directly".[5] Before Barbon – and indeed, for a long period after – most buildings were custom built for individuals on their own land, using their own labour and materials, whether residential or commercial. The buildings were relatively simple in form and straightforward to construct. Commercial buildings were typically general purpose in nature, occupied by small enterprises who did not make particular demands on building form. The notion of speculative development was a novel one, and independent developers who acted between landowners and occupiers of buildings were relatively rare. Barbon changed all of this.

Barbon the developer was a response to the increasing scale and complexity of development. He was paving the way for a new role: the speculative developer. The developer emerged to assemble land and capital, to organise the labour, to oversee the delivery of the physical building and to earn a financial gain upon completion. The developer became an intermediary between the landowner and the occupier of the building: he assumed a level of risk and responsibility and was rewarded accordingly. As the template became established, developers began to purchase land and develop in their own right. But, whether as an intermediary or as a direct player, as projects became larger and more complex, so the need for working capital and access to credit became paramount. In the early days, wealthy individuals and building materials firms were active in this sense, but increasingly there emerged a critical role for banks and insurance companies.

Developing the West End

Barbon's activities in and around Gray's Inn Road were part of the westward drift of London during the seventeenth century. As noted in Chapter 3, the West End was developed as the better off sought calm and dignity away from the intensity of the City, where much of the newfound wealth of the time, emanating from England's international trading, was being amassed. The money paid for large houses and luxuries, expensive clothes and exotic foods. Leisure time became accepted, and leisure time needed filling with leisure activities.

Nicholas Barbon was one of the more colourful seventeenth century developers, but there were many others, often wealthy individuals. One such developer was Francis Russell (1587–1641), more widely known as the 4th Earl of Bedford. In the 1630s, he built a house for himself north of the Strand and then, in an attempt to earn a return on the land, commissioned architect Inigo Jones to build a square. The result was London's first piazza, from which emanate both Russell Street and Bedford Street. Similarly, in the 1660s, Henry Jermyn (1605–1684), the 1st Earl of St Albans, laid out the area now known as St James's, including St James's Square and, of course, Jermyn Street.

Some developers were wealthy but untitled. Tailor Robert Baker (1578–1623) made his fortune selling, among other things, 'piccadill' collars (stiff, detachable collars popular at the time). He acquired land, which he then enclosed, and built himself a new home, Piccadilly Hall, on what is now Great Windmill Street. Following his death, his widow Mary (perhaps one of London's first female developers) began to build on the surrounding fields. The lane that stretched westwards from Piccadilly Hall towards the settlements of Knightsbridge and Kensington began to be developed in the mid-seventeenth century. It was then named Portugal Street, but would be renamed Piccadilly in the eighteenth century.

Richard Boyle (1694–1753), 3rd Earl of Burlington, was born into a very wealthy aristocratic family, and from a young age he was a patron of the arts who developed a passion for architecture. He is often credited with being one of the originators of the neo-Palladian style in eighteenth century London. He laid out the Burlington Estate, resulting in an area encompassing Saville Row, New Burlington Street, Old Burlington Street, New Burlington Place, Burlington Gardens and Sackville Street. Nearby, the developer Richard Lumley (1650–1721), the 1st Earl of Scarborough, developed Hanover Square from 1714 under a lease from the landowner Sir Benjamin Maddox (remembered in a neighbouring street), naming the square after the first Hanover sovereign's royal house.

At the same time, more grandiose buildings began to appear, reflecting London's global status in trade and commerce. For example, as paper became the dominant means of conveying security, permanent exchanges were established. The Bank of England moved into its George Sampson-designed building on Threadneedle Street in 1734, and the Corn Exchange (Figure 5.2) was established in 1751.

Running north of Piccadilly is Old Bond Street. The street is sited on land bought in 1683 by a consortium of Georgian investors led by Sir Thomas Bond (1620–1685). They bought the land, including the prestigious Clarendon

Figure 5.2 The Corn Exchange, Mark Lane (1809)

House, from the 2nd Duke of Albemarle. The house was demolished to make way for Albemarle Street, Dover Street and Bond Street. New Bond Street, to the north and linking through to Oxford Street, was built 40 years later in 1720. The combined street quickly became very popular with London's wealthy and socially elevated residents.

Of course, Bond Street became synonymous with wealth and luxury, a reputation that it has maintained to the current day. And Bond Street epitomised the newfound wealth of the eighteenth century. Shops expanded rapidly, with a burgeoning range of products and specialisms, and shopping became a leisure activity (for those who could afford it). Helen Berry provided some fascinating insights into eighteenth century retail. She observed that "As by far the largest and most diverse commercial centre in England, London was a magnet for polite society from all corners of Britain and abroad to spend their money". And while its streets "were still filled with mud, wide and handsome pavements now made leisurely browsing a more civilised and leisurely pastime". The leisure aspect is reinforced by the important role of 'browsing', in that the "cleanliness and convenience of the environment, and civil sociability of shopkeepers, helped to make browsing a polite activity".[6]

Shopkeepers became expert in finding ways to attract wealthy customers, not only in the quality and range of their products but also in their presentation. The shopping experience was added through the "extra pleasure of sensory stimulation" in so far as "artificial lighting was used to allow shoppers (who could pass by six-deep upon the broad pavements) to gaze at the brightly lit silver, china or glass within, long into the night". Berry continues: "The booksellers placed the most expensive books in their windows, the print sellers their most eye catching artists".

The eighteenth century saw a transformation in shopping and shopping habits. The shops themselves evolved in format as shopping increasingly became a leisure activity rather than a necessity, and as it became a sociable activity, and one in which one could convey one's wealth, status, taste and cosmopolitan outlook. Retailing became a strong force in re-shaping the urban environment. Shops became highly concentrated; they grew in size, and they became destinations in their own right. This was all achieved before the office had any measurable impact on the built environment, a situation that was to change in the following millennium.

5.2 Victorian grandeur

The emergence of the modern office economy in London during the Victorian period was described in Chapter 2. Even from this brief synopsis, it was clear that the scale of growth and change in the latter half of the nineteenth century was vast: a city being remade. By the end of the century London was a global city at the heart of a vast and productive Empire: "the public spaces, the railway termini, the hotels, the great docks, the new thoroughfares, the rebuilt markets, all were the visible expression of a city of unrivalled strength and immensity".[7] Nearly half of the world's shipping was controlled from London; it had become "the centre of international finance and the engine of Imperial power; it teemed with life and expectancy".[8] The City had become "the progenitor of commerce, and the vehicle of credit, throughout the world; the City maintained England, just as the riches of the Empire rejuvenated the City".[9] Amidst all of this growth and expansion, it is easy to forget that, for many, London remained a city of poverty and degradation.

> Old women squatted in the streets selling herbs, apples, matches and sandwiches. There was a floating population of ragged barefoot children who slept in alleys or beneath bridges. There were costermongers with their carts selling anything from coals to flowers, fish to muffins, tea to crockery. There were also epidemics of surprising speed and savagery.[10]

This was, after all, Dickensian London. While residing in Tavistock House, between 1851 and 1860, the great writer penned *A Tale of Two Cities*, *Bleak House*, *Hard Times* and *Little Dorrit*. The cheek-by-jowl existence of extreme wealth and poverty has remained a powerful feature of London: today the inflated salaries of bankers and lawyers in Canary Wharf contrast sharply with the relative poverty in the neighbouring Poplar.

James Burton (1761–1837) was perhaps the most prolific developer of Georgian London. Born in Strand, the young James was schooled in Covent Garden, before a period of private education, during which he studied architecture. In 1776 he became an articled surveyor, and in 1782 he began his first speculative project. He became prolific, with a small sample of his work including: large parts of Bloomsbury, including Bedford Square, Bloomsbury Square, Brunswick Square, Guilford Street, Tavistock Square and Tavistock House (for his own occupation); large parts of Regent Street, including St James's and Waterloo Place; and Regent's Park, including Clarence Terrace, Cornwall Terrace and York Terrace.

Burton was a pioneer in many ways, but even he was outshone by a near contemporary, Thomas Cubitt (1788–1855). While Burton was born in London, the son of a well-to-do property developer father, Cubitt was born in Buxton, Norfolk, the second of six children to Jonathan and Agnes Cubitt. His father was a carpenter, and Thomas learnt his trade as a ship's carpenter before setting up business on Gray's Inn Road in 1810.

What distinguished Cubitt professionally was that he was the first to establish the practice that we now call general contracting. In 1815 he made what for the time was the highly innovative and risky move from a business model involving engaging with multiple craftsmen to carry out particular jobs, to one where he applied his considerable organisational skills to building a business that employed the full range of skilled building craftsmen directly on a full-time basis.

Figure 5.3 Thomas Cubitt

Within four years of setting up in business, Cubitt was laying out the Calt-horpe Estate, near Gray's Inn Road. He was a skilled developer and urban designer, and he became one of the earliest and most successful and influential developer-builders.

Thomas was joined by his brothers: civil engineer Lewis, who built parts of Eaton Place and Lowndes Square and designed King's Cross Station; and contractor and politician William, who developed Cubitt Town on the Isle of Dogs in the 1840s and 1850s to house the rapidly growing numbers of workers in the nearby docks and shipyards. Cubitt's large, permanent staff required a constant flow of work, and this necessity led to a number of large-scale agree-ments with the Grosvenor Estate.

In 1826 Lord Grosvenor obtained an Act of Parliament enabling him to build on the land that is now Belgravia (named after a village on the northern outskirts of Leicester, where the Grosvenor family once owned an estate). Up to this point the land was used extensively, with market gardens and frequent duck shooting and cock fighting. Cubitt successfully applied his business and financial acumen to securing an agreement with Lord Grosvenor in which he gained virtually exclusive rights to build on the land, and he found financial backing from the City to fund his work.

Between 1826 and the mid-1830s, Cubitt bought building leases on adjacent plots, building as market conditions allowed. Cubitt based his plan on four

Figure 5.4 Belgrave Square, north-east side, today

Source: Oosoom at en.wikipedia (https://en.wikipedia.org).

squares – Belgrave (Figure 5.4), Eaton, Chester and Lowndes – that were linked by wide roads and crescents.

> The expert can detect subtle differences in each block as fashions in architecture changed from classical to Italianate. But the remarkable thing is that the extraordinary unity of the whole scheme was maintained over a period of some ten years of development.[11]

In an example of literal interdependence between the growth of the docks and the expansion of homes in the west, Cubitt used earth excavated from the new St Katherine's Dock (also a Cubitt family scheme) to raise the land level above its marshy surroundings.

Cubitt's skills as a builder, designer and businessman are shown by the fact that Seth Smith, who worked on the north side of Belgrave Square, sank rapidly into debt, and Joseph Cundy, who took on Chester Square, went bankrupt. Neither was able to replicate the grandiose vision and bold spirit of the master builder Cubitt. As well as Belgravia, Cubitt also worked on Gordon Square and Tavistock Square between 1820 and 1826, large parts of Pimlico during the 1830s and Victoria in the 1840s and 1850s. This huge output made Cubitt fabulously wealthy, but he never lost his zeal and vision for London. As well as a builder, Cubitt considered himself what we would call today an urban planner, and he made several contributions to reforms aimed at improving living conditions. But it was with his craft that he established his strongest reputation:

> He made his name, and kept it, by the excellence of his workmanship and the quality of his finishes. It was a reputation achieved only by the most painstaking attention to detail and by his innovation of building up a strong workforce – numbering 1,000 men at its height – to whom he assured continuous employment and from whom he secured a high degree of loyalty.[12]

At the time of his death in 1855, Cubitt had done more than any other individual to change the London skyline – and the development process. Reflecting on his life, Queen Victoria (for whom Cubitt had built Osborne House on the Isle of Wight) wrote sombrely in her diary: "a better, kinder-hearted or more simple, unassuming man never breathed".

Cubitt was a nineteenth-century phenomenon. His vision was awesome, his business skills extraordinary and his technical understanding of development probably second to none. He had 'commercialised' development and defined a new approach that was to be the forerunner of today's development industry. Cubitt's work, and that of others, was providing homes for the growing population of increasingly affluent people engaged in the city's commerce. During his working life, London had become "the workplace of the new 'professions', as engineers and accountants and architects and lawyers moved ineluctably towards the city of empire",[13] and apart from sowing the seeds of the office economy, these "affluent 'consumers' created a market for new 'department stores' and new restaurants" and "there arose a revived and more salubrious 'West End' of theatres under such actor-managers as Irving and Beerbohm Tree".[14] But in the time after his death, property development and construction changed fundamentally, partly in response to the increasingly complex nature and scale of buildings, but also in response to changing social structures and professional specialisation.

In some ways, James Burton and Thomas Cubitt embodied a frontier approach to development that has survived to this day: vestiges of both remain in today's property industry. However, as the nineteenth century progressed, so construction of buildings and other infrastructure grew in complexity. At the same time, specialisation flourished because neither general builders/developers nor architects could master the growing range of specialist skills alone.

Consequently, the supply process evolved into two distinct professions: design and development, with the latter fragmenting into a minestrone of sub-professions (see Chapter 7). The problem was that segregation resulted in more than division of labour to cope with specialisation; it also resulted in the establishment of 'competing' professions and the loss of the benefits of combining resources and working as one entity.[15] The split also, inevitably, reduced communication between design and delivery, and ultimately led to a separation of interests and a deep-rooted suspicion of motives among 'factions'.

As the property profession evolved in the later years of the nineteenth century, so it began to deliver new forms of buildings. Most businesses in the early years of the office economy occupied buildings that had been built for functions other than offices: residential buildings, industrial buildings, warehouses and so on. And others were buildings with ground floor offices, or counting houses, and upper floor accommodation. We saw how, in the 1830s, the first speculative office building appeared in the City of London. As the nineteenth century matured, so specialisation in development and building types evolved quickly with the growth in commercial property companies. For example, according to Scott, "1864 saw the establishment of two public companies whose sole aim was investment in, and development of, City office property – City Offices Co. Ltd and City of London Real Property Co. Ltd".[16]

Similarly, clients of the development process were becoming far more adventurous in their requirements for purpose-built office accommodation. Kynaston provides a few examples. First, Royal Insurance in Lombard Street, which commissioned "successive large headquarters from John Belcher in 1857 and 1863, the latter (demolished in 1910) full of panelling, plastered stories and allegorical sculpture". Then there were the joint-stock banks, "almost all of whom between the mid-1850s and mid-1860s built or rebuilt on a lavish, even monumental scale". Secondly, Kynaston cites City Bank in 1856, on a corner of Threadneedle Street, with its "imposing Italianate palazzo style" and thirdly, John Gibson's commission of National Provincial's 1866 "magnificent classical banking hall in Bishopsgate . . . resplendent with Corinthian columns and marble pillars" which could "have eased the doubts of even the most neurotic depositor".[17] It is worth dwelling briefly on the last of these for its reflection of the grandeur so typical of the mid-Victorian period.

The new building, to replace that shown in Figure 2.5, is shown in Figure 5.5. The picture emphasises height and grandeur, solidity and trust, and it was a successful attempt to establish what we might call today the bank's 'competitive position': it was at least as good as its rivals. Victoria Barnes and Lucy Newton provide a wonderful insight into the genesis of the design of the building, exploring the symbols and messages that the leaders of the bank wished to convey to the market. They note a *Bankers' Magazine* article, reporting that the "magnificent building" was "an important addition and ornamentation to the architecture of the city", highlighting the impressive statues and panels of the exterior, and the "banking room, commonly called 'the shop' [which] is 118 feet in length".[18] Ward-Jackson summarised the building thus: "the most extravagant of the City's Victorian joint stock banks, makes it mark principally through the profusion of its sculptural adornments".[19]

According to Black, the National Provincial Bank's new headquarters was a magnificent architectural statement.

> Gibson's design was for a single-storey building, bolder and grander than the Bank of England itself . . . with Corinthian columns running the full height to the cornice. Gibson broke with the established pattern of multi-storey building, signalling by the single-storey structure that this important and wealthy institution, new among the established metropolitan banks, had no need of upper stories for speculative letting space. That the facade was given over entirely to the expression of the banking hall further underlined the innovative nature of the building.[20]

Figure 5.5 National Provincial, Bishopsgate

Source: Stephen Richards (https://creativecommons.org/licences/by-sa/2.0/deed.en).

Kynaston also refers to the new limited liability discount houses which, "like-wise put a premium on appearance". He cites the National Discount Company's "palazzo-style building in Cornhill, completed in 1858 with distinctive cinque-cento motifs and detailing"; and the General Credit and Finance Company's "stunning effort" at 7 Lothbury in 1868 (Figure 5.6). Not everyone favoured the grandeur now common in financial houses, and Kynaston cites the *Bankers' Magazine* which, in 1863, suggested that they were "succumbing to a love of show, leading eventually to excessive expenditure and sometimes to embarrassment".[21]

So, by the 1860s the office economy was maturing rapidly. As we saw in Chapter 3, the firms were growing in size and complexity. And their buildings consequently became larger and more complex (and more ornate!). The combination of these factors encouraged the emergence of a mature property development sector, producing a range of purpose-built and speculative buildings.

But the increasingly complex nature of the emerging organisations demanded a more sophisticated and fitting kind of physical environment. Again, Marian Bowley captured the essence of the situation well.

> The significance of such investment is not always appreciated. Buildings required for industry and trade are as much part of the basic capital investment required for the effective functioning of the economic system as

Figure 5.6 General Credit & Discount, 7 Lothbury

Source: © Erik Brown, 2020

machinery, railways, etc. In the nineteenth century it was necessary to provide the buildings for an expanding economy changing from an agricultural to an industrial basis.[22]

In 1889, London gained its first directly elected government, with authority over the whole city (except the City of London) for the first time. It might have been hoped that a centralised, democratic, citywide body could bring some order to the chaotic nature of London's expansion – a more strategic view, a longer-term perspective, more resources for comprehensive planning, investment in infrastructure. Not a bit of it was delivered.

In the face of a Parliamentary Bill introduced by the Liberals, effectively to revive a seventeenth century plan to centralise the administration of London under the control of an expanded City Corporation, the Conservatives "seized their opportunity to create a Council that was initially hardly more formidable than the Metropolitan Board of Works",[23] which it replaced:

> It possessed no planning or welfare responsibilities whatsoever, while its housing powers, like those of the Board, were confined to slum clearance.[24]

Consequently the physical and economic growth of London continued in the fashion to which it was accustomed, that is, chaotically. Nevertheless, as the new century dawned,

> London's position at the hub of the world's largest empire gave work to half a million in the docks and in import-related occupations. It also guaranteed middle class employment in Whitehall but above all in the City, where tens of thousands of Pooterish clerks meticulously documented the transactions of the thousands of companies, great and small, involved in worldwide mercantile and shipping enterprises.[25]

5.3 Towards the machine age

London's population rose by one million in the nineteenth century to reach 6.5 million by 1901, accounting for one-fifth of the national population. London's population soared to 8.7 million by the outbreak of the Second World War, a level that, astonishingly, it was not to reach again for another 80 years! In 1901, it was the largest city in the world, and it would remain so until it was overtaken by New York in 1925.

The last years of the nineteenth century and the beginning of the twentieth were, in many ways, extraordinary years, and the impact of the technological revolution of this time was profound. Take, for example, the impact of the internal combustion engine. As the new century dawned, public transport on London's roads was still largely horse-dependent, although electricity was having a growing impact on the Underground and trams. Following unsuccessful trials with electric batteries, the petrol-driven, internal combustion engine became the focus of development. Then in 1896, the repeal of the Locomotive Act ended the 2 mph speed limit on horseless carriages, thereby spurring the development of the motor bus, the first of which ran in 1898. But there were major reliability issues and the development of the petrol engine continued apace.

The largest bus operator in London at the time, the London General Omnibus Company (LGOC), bought out its main rivals, and competition was effectively ended.

> the LGOC had the resources to design and build its own vehicles rather than buying buses from commercial manufacturers. . . . The first purpose-built bus was the X-type, a 34-seat double decker, introduced in 1909. Further developments resulted in the B-type in 1910, built at the LGOC works in Walthamstow using mass-production techniques. It was relatively quiet, trouble-free and easily maintained and by 1913, 2500 had entered service. On 4th August 1914, one week after the outbreak of World War One, London's last horse bus, operated by the Thomas Tilling bus company, was withdrawn from service.

Of course, alongside the development of buses, London saw its first motorised cabs. In 1897, which were powered by electricity. Designed by engineer Walter Bersey, who was General Manager of the Electrical Cab Company, the cabs became known as 'Berseys', and had a top speed of 9 mph. However, cost and reliability issues meant that the Bersey was withdrawn from service during 1900. Just three years later, in 1903, after costs scuppered a plan to import the Ford Model B from the USA, London's first petrol cabs were introduced. Instead of American cabs, London welcomed French taxis. The term 'taxi' was adopted around 1906 when the fitting of taximeters to display fares was made compulsory. The taximeter was invented by German Friedrich Wilhelm Gustav Bruhn in 1891, and was the device that accurately recorded the time and distance travelled, thereby generating an accurate fare for the journey taken.

Also around this time, one of the most iconic symbols of urban living got going: the humble traffic light. Its genesis was in fact much earlier: the first traffic signal was invented by John Peake Knight, a railway signalling engineer. It was installed outside the Houses of Parliament in 1868 and, unsurprisingly, looked like any railway signal, with semaphore arms and red-green lamps, operated by gas, under the manual control of a police officer.

The development of the signal was stalled due to a gas leak and explosion a month after installation, not reappearing until the internal combustion engine was common. Modern traffic lights were in fact invented in the USA, with red-green systems being installed in Cleveland in 1914. Three-colour signals, operated manually from a tower in the middle of the street, were installed in New York in 1918. The first lights of this type to appear in Britain were in London, on the junction between St James's Street and Piccadilly, in 1925. They were operated manually by policemen using switches. Automatic signals, working on a time interval, were installed in Wolverhampton, in 1926. The first vehicle-actuated signals in Britain controlled the junction between Gracechurch Street and Cornhill on the City, in 1932.

Major public buildings erected during this time included: County Hall on the South Bank (1911–1922), the seat of the London County Council; the

Methodist Central Hall in Victoria (1905–1911); the Old Bailey (1903); the Port of London Authority building in Trinity Square (1912–1922) and the War Office in Whitehall (1906).

The docks were still growing: the King George V opened in 1921. In 1919 the country's first civil airport opened on Hounslow Heath, offering daily trips between London and Paris. In 1925, the King opened the Great West Road, symbol of suburbanisation and the growing impact of the internal combustion engine. Soon dubbed 'the Golden Mile', the road attracted "scores of new factories . . . Smith's Crisps, Curry's cycles and radios, Maclean's toothpaste, and various American companies" including Firestone, Gillette and Hoover. The area also attracted Aladdin, Glaxo, Heinz, Horlicks, J Lyons, Marconi and Rockware Glass.[26] On the other side of London, Ford Motor Company began its long association with Dagenham.

The growth in population and commerce was accompanied by a growth in demand for leisure, shops, restaurants and hotels. Large swathes of the West End were redeveloped during this time, including Regent Street. The original John Nash buildings on Regent Street, from the 1820s, as far south as Waterloo Place near St James's Park, were entirely rebuilt between 1905 and 1927. Shops arriving included: Barkers in Kensington High Street (1937), Heals in Tottenham Court

Figure 5.7 Port of London Authority, Trinity Square

Source: David Williams (https://creativecommons.org/licences/by-sa/2.0/deed.en)

Road (1916), HMV in Oxford Street and Peter Jones in Sloane Square (1935), Selfridges in Oxford Street (1909) and Whiteley's in Queensway (1911).

The Dorchester Hotel and the Grosvenor House Hotel were opened in 1931 and 1929, respectively. The national sense of pomp and circumstance was sated with The Mall being designed as a parade route for state events; a re-facing of Buckingham Place and the construction of Admiralty Arch (1912), the Victoria & Albert Museum (1899–1909) and the Victoria Memorial (designed in 1901; unveiled in 1911 and completed in 1924).

Housing was also a key political agenda item between the wars. To meet population growth, as well as to address the famous 'homes fit for heroes' agenda, Parliament pushed through its Housing and Town Planning Act of 1919, under the stewardship of the Minister of Health, Christopher Addison. The results were spectacular.

> The act encouraged local authorities to build regardless of cost; any spend-ing above the product of a penny rate would be met from the Treasury. As a result, in the space of two years the LCC succeeded in building as many houses as in the whole of its previous history.[27]

Jenkins captured the essence of the newfound wealth and comfort of the city at large

> Twentieth-century Londoners need no longer walk to work, but could travel by tube or petrol-driven bus, cab or car. Unprecedented numbers had their own front door and a garden, and were served by sewers, mains water and central heating. They had electric, gas cooking and hot water. Their clothes were machine-stitched, their larders filled with imported food. Offices were served by typewriters and electronic cables. London had more telephone subscribers than the whole of France.
>
> Londoners could listed to recorded music on a gramophone and visit the local public library, swimming bath, technical college and cinema. Newspapers brought world news daily to front door.

Purpose-built office buildings also spread from the City to the West End fol-lowing the turn of the century. Among the earliest office buildings there were those along Kingsway which, with respect to its scale and ambition, was the Broadgate of its day.

An Edwardian Broadgate

In 1898 the London County Council agreed a scheme for the development of a road linking Vernon Place (north of High Holborn) to the Aldwych in the south. The scheme involved the complete reconfiguration of roads and the demolition of a densely populated area; 3,700 residents lost their homes (Fig-ure 5.8). It was one of the earliest schemes to manage traffic in a co-ordinated

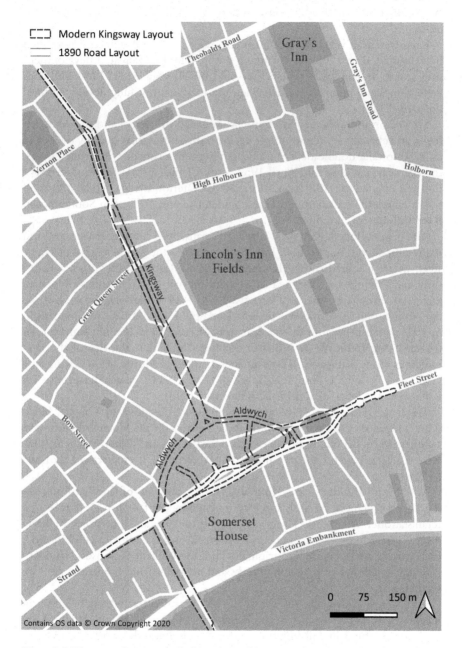

Figure 5.8 Kingsway shown over the historic road layout

manner, and incorporated an underground tramway line to link those of north and south London. It is the only underpass in London built specifically for trams.

Kingsway itself was built as a 30-metre-wide boulevard, which was opened by Edward VII in 1905, amidst much pomp. The LCC created development sites alongside Kingsway which they sold on 90-year leases. Most buildings were completed before 1914, although the road was not fully developed until the early 1930s.

Service on the underground tramway between the Angel and the Aldwych began in February 1906 with single decker trams; it was subsequently deepened in 1930 to take double decker trams. After an experiment with trolley buses, the tramway was closed in April 1952, and in January 1964 the southern section was opened to traffic as an underpass to Waterloo Bridge. The cutting and tracks where the trams emerged (at a gradient of 1 in 10 feet) in Southampton Row still survive (listed in 1998).

The re-developed Kingsway attracted many well-known businesses, and many offices continue to hold historic interest. At the southern end of Kingsway, where it joined Aldwych, were two grand, 'gateway' buildings, 1 Kingsway and 2 Kingsway, designed as an identical pair by Trehearne & Norman architects. Empire House, at 2 Kingsway (1913–1914) became the HQ of the Air Ministry in 1918, and was renamed Adastral House. It was renamed again in 1955 when it became the headquarters of Associated-Rediffusion London (Figure 5.9). Trehearne & Norman architects built the majority of the new buildings along Kingsway. These included Imperial House, 15–19 Kingsway (1912–1913); York House, 23 Kingsway (1913–1914); Alexandra House, 33 Kingsway (1915–1916) and 77–97 Kingsway (1913).

Waterman House, at 41 Kingsway (1909), was designed by Gibson, Skipwith & Gordon. Originally it was the headquarters for LG Sloan Ltd who were suppliers of rubber bands, playing cards, board games and Waterman pens (they were European agents for pen maker LE Waterman). It converted to a hotel in 2019.

WH Smith Building, at 7 Kingsway (1907), was one of the first buildings developed by Cubitt Nichols and was designed by Mr Langton Dennis for the retail stationer. At 40 & 42 Kingsway (1908–1909) was an eight-storey building on the east side by Edwin Lutyens in Portland stone with rusticated ground, first and second floors. It was originally built for William Robinson, visionary gardener, publisher of gardening books and magazines and proprietor of 'The Garden'.

Rather like Broadgate, the brand new Kingsway attracted large American businesses. Kodak House, at 61–65 Kingsway (1911), was designed by Sir John Burnet & Partners in collaboration with Thomas Tait (who worked on Unilever House). It was designed for the American Kodak Company and was later known as the Gallaher Building.

Victory House, 30–34 Kingsway (1919–1920) was occupied by another American business, FW Woolworth & Co Ltd. The firm's in-house architect

Figure 5.9 Empire house, Adastral House, Television House, Aldwych

Source: Stephen Richards (https://creativecommons.org/licences/by-sa/2.0/deed.en).

and construction department were in occupation between 1921 and 1929. Magnet House, 45 Kingsway (1920), was designed by R. Frank Atkinson and constructed for the General Electric Co (Figure 5.10). Extended over the years, the building was sold to Harry Hyams, who demolished the building and replaced it with a Richard Seifert building, also Magnet House.

One of the final additions to the original Kingsway scheme was Africa House, at 64–78 Kingsway. Also designed by Trehearne & Norman, Africa House was built in 1921–1922 and was occupied by International Combustion Ltd, makers of equipment for electricity generation who had recently moved their production plant to Derby.

The re-development of Kingsway was on a scale similar to that which started at Broadgate eight decades later. It involved the re-modelling of a whole area, with a range of buildings and public realm. It also took the development focus away from the City of London and its financial activities towards the West End, and towards a much wider array of industrial and service-based companies looking for the new office accommodation now being regularly delivered.

Figure 5.10 GEC's Magnet House, Kingsway (demolished)
Source: ShareAlike 2.0 Generic (CC BY-SA 2.0).

As we saw earlier, over the latter part of the nineteenth century and the early years of the twentieth, the number of jobs in the City of London grew rapidly. But within this overall picture, there were the beginnings of quite fundamental change, and this was the decline in physical trading and the growth of the City of London as Europe's pre-eminent services centre. The number of people employed in manufacturing and ports activities in the City dropped rapidly. As Ranald Michie observed, during this period, the City shrank as a port function, but, very significantly, became a centre for *organising world trade*. The outcome of this was that office activities grew relative to port activities.[28]

The inter-war years

By the turn of the century, the City of London had become a specialist supplier of two fairly distinct groups of services. One, centred on the Port of London and arising out of Britain's premier position in international trade, was provided by the banking, finance and insurance world; the other was met by a group of industries and services supplying mainly domestic markets, which had a particular and special need to be located in the centre of London, e.g. the newspaper industry and certain sections of clothing and precision instrument trades.

As the world's trading routes were increasingly managed using telegraph and telephone, not to mention railways, so London's role as a break-of-bulk and trans-shipment centre declined. By 1913, London was handling just 19% of British exports and 33% of imports, measured in value; while other UK ports such as Liverpool and Bristol grew, as did European ports including Antwerp, Amsterdam and Hamburg.[29] And as global trading became more integrated, there emerged the need for a dominant city to operate as a "fulcrum and mediator of the whole process".[30] Key to London's emerging role was technology, because, as Michie notes,

> The result of this communications revolution was that it became possible to conduct a global trading business from an office in the City, maintaining constant contact supplemented by rapid visits and the receipt and despatch of samples and catalogues.[31]

Increasingly, merchants, brokers and agents from both Britain and abroad gravitated to the City, organising and controlling world trade, most of which never touched Britain's shores. Not only trade itself, but shipping, finance and a whole host of ancillary services fundamental to world trade such as insurance, law and accountancy located themselves in the Square Mile. But, in some respects, this period also proved to be a high water mark, particularly for the City of London.

> In 1908 it was estimated that at least a million people a day flowed into and out of the City at a time when the telephone was only slowly becoming an accepted means of conducting business.[32]

Michie notes that 200,000 fewer people were employed in the City in 1981 than the 1930s peak of over 500,000. As telecommunications became more established and more sophisticated, the tight cluster that was both the *physical* City and the *functional* City began to loosen. Indeed, "improvements in communications was a two-edged weapon":

> Not only did it give the City increasingly instantaneous access to information previously restricted to local markets and so allow it to direct operations at great distance, but it also reduced the necessity of a physical presence in the City itself.[33]

In addition to fundamental changes in telecommunications, and setting aside the war years themselves, the inter-war period was dominated by economic slump. Barely had the century got underway before it was plunged into the

First World War, which had a dramatic impact on the economy. The 1920s were beset by stagnation, with low growth, high unemployment and deflation, which increased the burden of debt and reduced spending, causing a vicious cycle of decline. UK debt reached close to 200% of GDP.

Having suspended the gold standard in 1914 owing to the pressures of war, Winston Churchill returned sterling to the gold standard in 1925 at the pre-war rate of $4.86 to the pound. This proved to over-value sterling, reducing demand for exports and placing even further pressure on growth and jobs. The US stock market crash of 1929 precipitated a global recession and further reduced demand for Britain's exports. The UK economy was further hit by the sharp global economic downturn in 1930–1931: real GDP fell by 5% in 1931. In September 1931, Britain once again left the gold standard, allowing the government to cut interest rates, target higher inflation and devalue the pound against the dollar. The result was a boost to exports and domestic demand. A significant increase in money supply between 1932 and 1936 nurtured private sector investment and consumption. Housing construction increased, leading to the growth of London's suburbs and a rise in living standards.

While the 1930s economy was marked by the effects of the Great Depression, leading to poverty and hardship, the second half of the decade witnessed a period of economic recovery. In parts of the UK (especially London and the South East), there was a mini economic boom with rising living standards and prosperity.

Indeed, building activity in the office economy picked up significantly during this period. As Scott observes, the 1930s property development boom "was largely funded by the financial institutions. Insurance companies and banks had a strong bias towards lending backed by collateral security at this time".[34] The scale of increased commitment was remarkable.

> The most important group of institutional investors in urban property during the inter-war years were the insurance companies. From 1922 to 1937 the funds of UK-based life assurance companies more than doubled from £804 million to £1,655 million.[35]

Similarly, property companies expanded rapidly "when cheap money became available in the latter half of 1932". During 1933, 14 new public property companies were launched, and in 1934 there were a further 11.[36]

As a result of the growing activity of institutions and property companies, several signature office buildings were created during the inter-war years. For example, at the Aldwych, Australia House for the Australia High Commission (designed by A Marshall Mackenzie and A G R Mackenzie, and opened in 1918) and India House for the High Commissioner (Sir Herbert Baker and A T Scott, 1930) are both superlative examples.

Many corporate office buildings were completed during this period (more of which in Chapter 4), and there were several public sector buildings that came to be synonymous with their famous and long-term occupants: the London County Council's (later GLC) County Hall on the South Bank (1922); the

Figure 5.11 Australia House, Aldwych

London Passenger Transport Building (1929) occupied by the forerunner of London Underground in Victoria; the BBC's Langham Place (1931); the University of London's Senate House (1932); Shell-Mex House on Strand (1931) and the Wellcome Building on Euston Road (1931) are a few notable examples

Following the Second World War, the office economy matured in the context of a series of commercial property construction booms. The precise timings of the booms are open to question, and there have been large and small booms. The following sections examine the largest post-war booms, those of the later 1950s and the mid-1970s. Chapter 6 picks up the story of development booms with an examination of that of the late-1980s.

5.4 Post-war rebuilding: the 1950s boom

Developers and investors lie at the very heart of the supply process: without their orchestration of the process and ability to fund speculative development, there

would be no commercial market as we know it today. The two roles are undertaken both separately and in combination. For example, there are developer-
traders who borrow to fund development and then sell on; there are pure
investors who have little at all to do with the delivery process but instead purchase standing investments, and there are investor-developers who both fund
schemes and then hold the completed asset.

We saw in Chapter 3 that, during the mid-nineteenth century, a significant
shift took place in terms of how office buildings were delivered. Before this
time, they were almost exclusively built-for-purpose, or for an owner; and
typically they entailed ground floor offices with living quarters above. But then
this approach was gradually replaced by a more 'speculative' process, in which
buildings were built exclusively for office activity and for the open market. The
shift in delivery process also entailed a shift in roles, from one in which building owners or their architects employed crafts directly, with master craftsmen
engaging individual workers; to a model in which the developer orchestrated
the whole development process, including the architect.

We also met earlier in this chapter some of the first speculative developers
such as Barbon, Burton and Cubitt, each at their most active in the seventeenth, eighteenth and nineteenth centuries, respectively. But for these early
developers, office construction was a 'side-line'; their main focus was on delivering neighbourhoods. They and their patrons (often the landed estates), were
responsible for delivering huge swathes of modern London. It was not until
the latter part of the nineteenth century that the scale of the office economy
demanded a specialisation in the delivery of its accommodation. Not only was
there a rapid rise in demand for office buildings, but they were becoming larger
and more complex in an early example of a technological arms race.

By the outbreak of the First World War, the supply process had taken on its
recognisably modern form, with clients and patrons (owners and investors) at
the pinnacle, followed by architects, then developers, then the professions, then
general contractors, all supported by the humble sub-contractor. The social
stratification implied by this structure was real, and it largely survives to this day
(and is discussed further in Chapter 7). This nascent structure was added to by
the rapid involvement of speculative property investors, literally in the rubble
of the Second World War.

After the Blitz

Jenkins wrote that people walking London's streets in the mornings following
overnight bombing would notice suited men with notebooks walking around
the ruins.

> Their interest was not in Corbusian utopias, it was in profit. Many had
> graduated from the pre-war London College of Estate Management, and
> won their spurs demolishing and redeveloping old West End mansions.

They were outriders of a new boom. Their method was to ring around estate agents the morning after an air raid, seeking dazed owners desperate to sell.[37]

Indeed, the image of this time has something of a frontier feel about it, and the mood was encapsulated well by Jack Rose, citing personal experiences.

An estate agent instructed to sell a site would telephone a prospective purchaser who, with no further details than the address, price required and the approximate size, would take a taxi to verify the location. Satisfied on this he would telephone his architect.[38]

The architect would then seek clarification from the local authority on plot ratio and permitted uses, and the developer would calculate construction costs, fees and interest charges. A deduction of 20% from the gross building area would provide the net area available for calculating rental income.

A profitable deal would emerge if this income, capitalised at 10%, equalled or exceeded the addition of the development costs and site cost. Within hours of the first telephone call a deal could be struck.[39]

It really was as crude as this. Speculators rushing around to grab pieces of land that they could then 'turn', and create, sometimes, personal fortunes. Social historian Anthony Sampson described the period in vivid terms.

In five years several of Britain's biggest cities have been face-lifted: they have changed from horizontal to vertical skylines. Steel and glass cliffs have pushed up between old churches; rows of Victorian shops have given way to big white office buildings; skyscrapers have grown up around St Paul's.[40]

In the aftermath of the Second World War, a huge rebuilding programme was required to help repair the ravages of the Blitz. Central London lost around 9.5 million sq ft (880,000 sq m) of offices and in the City, around one-third of the stock "was completely flattened".[41] However, as Hedley Smyth argued, an overriding brake on development at this time was "the insufficient accumulation of capital to justify expenditure on new offices and retail outlets".[42] In other words, there were many greater priorities than commercial property; for example, economic and social reconstruction. Consequently, the development market remained relatively subdued for around eight years after the war. Many of the workers displaced by the blitz stayed away, only returning once the economy of the city began its slow recovery.

Two further constraints suppressed development during this period of austerity and nationalisation. First, building resources were rationed through building licences, introduced during the war. Secondly, there was the 1947 Town and

Country Planning Act. This was introduced by the Labour Government to constrain developers by introducing a 100% development tax and a new system of development planning.

Some developers found ways around the system and got new projects under-way, but overall new development was sparse. Perhaps the most significant additions to office stock during this time were those built specifically for government departments. These were developed under 'lessor schemes' in which the government granted a licence to a developer with a suitable site. The other way for developers to avoid licence control was by pre-letting new developments to companies involved in export industries.

There are two generic actors who dominate the property supply process: the institutions and the property companies.

The developer entrepreneur

Property companies, or developers, come in a wide variety of guises. They range from the small to medium-sized company led by an entrepreneurial and often flamboyant individual, to larger companies with a more sober, or corporate outlook. They also split between trader-developers who buy and develop sites in order to sell on the developed asset (often to institutions), and investor-developers, who develop sites with a view to holding them for their asset value. While their history pre-dates the Second World War, they first came to the general public's attention in the post-war building boom. The names of Harry Hyams, Charles Clore, Jack Cotton and Joe Levy came to symbolise development at this time. Smaller, more entrepreneurial companies, almost by definition, come and go, but more enduring examples include Bride Hall, Chelsfield, Greycoat, Helical Bar and MEPC. The larger companies are typified by the likes of British Land, Grosvenor, Hammerson, Land Securities and Slough Estates.

It is the changing roles and interactions between the institutions and property developers that so fundamentally shaped the dynamics of the post-war commercial property market, not least through its enduring impact on the office economy. How did this iron grip of the supply industry come about? Why did it last for so long? And what is its future? For the answers to these questions, we have to go back to the immediate post-war period.

The immediate post-war period was the age of the developer entrepreneur, the focus of Oliver Marriott's *The Property Boom*. Marriott makes the point that while property development required vast sums of capital, the industry was unlike others such as "the motor-car industry, which had to be highly concentrated to achieve economies of scale and thus funnelled great fortunes into the hands of a few men such as Henry Ford or William Morris". By contrast, in the property sector, "a relatively large number of individuals became extremely rich via property development between 1945 and 1965". Marriott estimated that there were 108 men and 2 women "each of whom must have made on my calculation at least £1 million in this golden period".[43]

One of the more powerful figures from this time, but perhaps typical of the genre, was Jack Cotton, an "exuberant character"[44] who worked out of a permanent suite at the Dorchester Hotel, where he was "surrounded by Renoirs, vivacious secretaries, brisk surveyors, maps of London, press cuttings, and a stream of visitors. He sits talking, drinking and laughing at the long table in his drawing room, sometimes till two in the morning".[45] By contrast his weekends were spent at his house by the Thames at Marlow listening to the Crazy Gang and collecting miniature bottles.[46] Cotton's single-mindedness led to a string of extraordinary deals. The son of a Birmingham Jewish export agent, Cotton set up an estate agency business at the age of 21 and was soon developing sites.

> He acquired a public company in 1946, to develop London and Birmingham properties with insurance companies – which grew enormously in the following ten years. He formed a host of subsidiary companies with insurance businesses (notably Pearl Insurance and Legal and General) and various corporate businesses (including ICI, Shell-Mex, BP and Barclays Bank) to undertake development projects.

And in 1960, Cotton merged his company with that of his friend Charles Clore to create the biggest property development group in the world, with assets of £67 million, and with Cotton at the helm.[47]

While the Klondike nature of the mid-1950s disappeared, and the process became more sophisticated, the basic role of the property developer has not changed much since then. He (for it was such at that time) spots a site with potential for development and will risk capital to buy the land, or at least an interest in the land; obtain planning consent and design and manage construction in the hope that he will reap the rewards of a greatly enhanced value at the end. Actually there are a few points along the way when the developer may decide to take his profit and move on. For example, he might choose to sell the site as soon as he gains a planning consent and make a capital gain without having to manage the risk of actual development.

As the 1950s gave way to the 1960s, the fruits of the first post-war development boom began to emerge in the form of the weaknesses of the speculative process. Much of the criticism for the 1960s architecture was laid firmly at the doors of the men who orchestrated the process. Marriott highlighted the image of the developer as someone for whom the development process was no more than an opportunity to create profit:

> (the) developer is like an impresario. He is the catalyst, the man in the middle who creates nothing himself, maybe has a vague vision, and causes others to create things.[48]

Certainly the prospect of building something of real aesthetic or architectural merit did not seem to be high on the agenda. This led to the widespread criticism that developers were responsible for creating rent slabs (in property market terminology) or urban wastelands (in populist terminology). Marriott was blunt in his assessment:

> Only a minority of developers were particularly concerned with the aesthetic design of their buildings. Since the developers were often men of little formal education, and since they were promoting their buildings for entirely commercial reasons, this is hardly surprising.[49]

Commercially, developers did not need to worry about quality. In the booming 1960s, there was so much demand for property that quality just did not matter: in short, it was a sellers' market. Rents were growing faster than development costs, and developers could make serious money from mediocre buildings. An architect would produce plans, a specification and an artists' impression; a quantity surveyor would calculate a construction cost and the developer would select the construction contractor that submitted the lowest bid. As Rose observed, the developers favoured those architects who maximised rentable area, compromising in the process specification and external aesthetics,

> Sacrificing elaborate facades in favour of curtain wall construction and limiting floor to ceiling heights to the minimum permitted by building byelaws. Hardly a single building incorporated air-conditioning. Central heating boilers were oil-fired rather than gas fuelled and there were generally insufficient electric power outlets.[50]

Nobody was concerned that cutting corners might come back to haunt the property market in future decades, but when the long leases granted at this time began to come to an end in the 1980s, the alternatives were demolition or refurbishments costing more than the capital value of the buildings.

It is little wonder that, in the public psyche, property development became a byword for greed and avarice, and that developers came to be widely held solely responsible for some of the most despised buildings in the country. As Scott observed, the office property boom "led to the property developer being cast in the popular mind as the arch villain of British capitalism", the reasons being "flaws and loopholes in planning controls, which the profit-maximising developer naturally exploited",[51] particularly plot ratios and the Third Schedule (see Chapter 4).

Perhaps nowhere was this approach to speculative development more intensively played out than along London Wall, creating a short-lived monument to the ideology of architectural modernism, à la Corbusier. A plan for the comprehensive re-development of London Wall was unveiled by the City Corporation and the LCC in 1955. The result included six, now infamous, office blocks: Moor House (1961), Royex House (1962), St Alphage House (1962), Lee House (1962), Britannic House (1964) and Bastion House (1976). Some have been replaced by newer buildings; one of those remaining is shown in Figure 5.12.

Figure 5.12 Bastion House, now 140 London Wall
Source: © Erik Brown, 2020.

No other single phase of development has done more to reinforce this unfa-
vourable image of the developer than that of the 1960s and 1970s. The period
left a legacy of despised town centres, in which the shopping centres and office
blocks owed more of their architectural heritage to Victorian prisons than
centres of commerce and community. Members of the general public would
probably not distinguish between institutions and developers, nor would they
typically be cognisant with funding requirements on property development or

the internecine aspects of planning law such as plot ratios. Instead, the general public would blame the 'property developer' for putting profit before community benefit.

Emerging institutions

The institutions include the pension funds and insurance companies (and to a lesser degree Property Unit Trusts) who invested vast sums of money in UK property during the second half of the twentieth century on the premise that it would generate secure returns. They are the large, monolithic organisations whose primary motive for investing in property is risk management. They seek to create portfolios of property spreading, typically offices, industrials and shops, with geographical diversity, and which themselves will be part of a broader portfolio including Gilts and equities. They are the grey suits in whom we entrust our savings to build a secure future. Names such as Legal & General, Prudential Corporation, Scottish Widows and Standard Life exemplify this group.

Institutional investors began to put their money (that is to say, our money) into the commercial property market during the late 1950s. Their priorities as owners of the buildings and landlords to most British business were to balance their portfolios, by off-setting the risks of investing in Gilts and equities, with security and long-term growth, achieved through the strictures of the standard commercial lease. Property became synonymous with the phrase 'hedge against inflation'. It was in part this hedge that led so many of us to believe that the value of our pension savings would continue to rise. The institutions gave the property market a kind of credibility that the property companies had failed to create, and it was the mutual interest of both that perpetuated a unique system of provider–occupier relationship for nearly 50 years.

In the early post-war years financial institutions played a passive role in the development process by providing fixed interest mortgages, often secured against leased assets. But as the 1950s progressed, the institutions increased their influence, firstly by developing their methods of financing property developers and latterly by transforming the very nature of commercial property as an investment medium. Insurance companies were first in, followed by pension funds. They became progressively more directly involved in the development process, until they arrived at the pivotal and powerful position that they would occupy for the rest of the century. One of their most influential roles was to define an 'institutional specification' for buildings that prescribed key aspects of design, paying little reference to the needs of the market and militating against innovation.

Prior to the mid-1950s, property investment and development were normally funded via a commercial mortgage or by offering a sale and leaseback. But as Scott observes, in the light of the rising cost of finance

> the number of joint development companies, share option agreements and other equity participation arrangements between the financial institutions and the property entrepreneurs began to multiply. These arrangements brough the two sectors into a close relationship.[52]

Scott also observed that, in addition to partnership deals with developers, a large number of developments were undertaken directly by companies including Legal & General, Norwich Union, Pearl Assurance and the Prudential. Moreover, by the late 1950s "the financial world was beginning to appreciate the value of a funding link with a financial institution to the growth prospects of the property company", but by the early 1960s, the "bulk of British insurance companies, and other financial institutions which participated in property development funding, were switching from conventional fixed-interest finance to equity participation agreements".[53]

The property market thus came to be dominated by a relatively small number of large, faceless financial institutions. There was no single architect or grand design to account for the way in which the post-war property market evolved in the UK, but rather a complex web of interconnecting and related economic, political and social factors. Angus McIntosh and Stephen Sykes (1985) named a few:

> Inflation, national economic policy, entrepreneurial skill, the poor performance of Gilts and equities, town planning and taxation have all contributed in different ways to the growth of the institutional investment market.[54]

The lifting of building licences in 1954 also played its part. It happened at the same time as the government decided to re-introduce lending restrictions in an attempt to counteract rising inflation and a deteriorating balance of payments, which had the effect of pushing up interest rates. Meanwhile, the Capital Issues Committee clamped down on company applications to raise finance through rights issues. All of which meant that the number of opportunities to develop multiplied just as the sources of finance diminished; it was inevitable that property companies would look elsewhere for money. So it was that they turned to the insurance companies for loans on low, fixed interest rates.

As their indirect involvement grew, however, the opportunities offered by direct involvement became irresistible to the institutions. They wanted a slice of the action, and they were in a hurry. By 1960, the *Investors Chronicle* was able to remark that:

> The granting of straightforward mortgages are a thing of the past. Equity shares and fixed returns have been demanded by insurance companies as they have wanted to benefit from the 'anti-inflationary principle' too.[55]

Insurance companies would commonly take an equity stake in a property company by obtaining shares or an option to convert loans into shareholdings. The maturing relationship between property companies and institutions was further encouraged as a side effect of government measures to curb speculative building activity. Chancellor Selwyn Lloyd's credit squeeze, combined with the rise in bank rates from 4% to 7% between 1959 and 1961, served only to reinforce the increasingly complex relationship.[56] The commitment of financial institutions to the property sector had matured from one of indirect investment during the 1950s to a very direct role by the 1960s.

With greater commitment from the institutions came power and influence. In fact, they fundamentally altered the workings of the property market. The institutional lease and institutional specification became dominant. And as their grip tightened, they determined where, how and what got built. In the process the direct link between the producer (developer) and consumer (tenant or occupier) was broken.

The growth of direct involvement by financial institutions in property and property companies was rapid. By May 1962, "the insurance companies held ordinary shares in nearly fifty property companies, and by 1965, the figure had risen to over a hundred".[57] Before long, the institutions found a way to share in the development profit by forming joint ventures with developers, or by providing fixed interest finance in return for subscription rights to some of the companies' share capital.

The 1965 Finance Act, however, introduced double taxation of company profits, making it highly unattractive for institutions to receive their share of rental income in the form of dividends from a property company. The response of the institutions was quite predictable. They began to seek greater involvement in the development process itself and

> to buy property directly, often by entering sale and leaseback agreements whereby the institution had an equity interest in the property rather than in the company.[58]

At first, institutional developers were willing to grant long occupational leases of 25 or 35 years without rent review. Until, that is, inflation caused building values to rise rapidly and landlords introduced rent reviews to keep pace with inflation during the course of the lease. Institutions also granted mortgages on variable rates of interest to developers, adjusting the interest payable periodically, becoming "related to the rental value of the property".[59] However, the institutions realised that if they actually purchased the investment themselves and then leased it back to the developer, they would have far more control over their investment. So it was that the sale-and-leaseback became common practice in the early 1960s. And as money poured into institutions from the investing public, their demand for property could not be satisfied by 'let investments', so they became involved in the pre-funding, construction finance and eventually direct development.

There is no question that institutional finance played a key role in the re-building of post-war Britain. Without their intervention, economic recovery would have been much slower and the scars of the war would have remained longer. There is, however, a question over their longer-term dominance of development finance, and certainly, the contractual structures they put in place for the occupation of commercial space. The issues were thrown into sharp relief when the economy began to change in the mid-1980s, and occupier's expectations of real estate began to change radically.

Boom and bust

It was in 1953 that commercial property development was to get its first big post-war break. The new Conservative Minister for Housing and Local Government, Harold Macmillan, introduced the Town and Country Planning Bill to repeal the financial provisions of the '47 Act and abandon 100% development tax. As Oliver Marriott relates the story, Macmillan believed "that laws which the public found hard to understand were bad laws".[60] Further, that while "the development charge is alright in theory . . . it will not work in practice", and that "if there is something wrong in practice it is just possible there is something wrong with the theory".[61]

In a remarkable comment Macmillan argued that "the people whom the Government must help are those who do things: the developers, the people who create wealth whether they are humble or exalted".[62] Of course, this comment must be taken in context: it demonstrated the government's desire to kick-start reconstruction without dipping into its own reserves. However, it sowed the seeds for a wave of activity that Macmillan probably never envisaged.

Together with the repeal of building licences in November 1954, these measures opened "the floodgates for commercial property development".[63] There is no doubt that the pressure to develop was already being compounded by an expanding economy and growing pent-up demand. Consequently, the conditions were evolving for a new era in property development: that of mass commercial property speculation. The growth in property companies and floorspace consents illustrate this point. The number of property companies grew rapidly in number from 25 quoted companies in 1939 to over 150 by 1965.[64]

It is notable that, just as the development boom was getting underway, pressure from business occupiers for greater protection led to the introduction of the 1954 Landlord and Tenant Act. Concern over the shortage of office space following the war led to calls for greater occupational protection, and the Act was passed to give tenants security of tenure, the right to apply for a new tenancy and a right to compensation should a tenancy be terminated. The result was enhanced value for the property industry generally and a boom time for property companies in particular: the aggregate value of quoted property companies: escalated from £103 million in 1958 to £800 million in 1962.[65]

The sharp upturn in development activity is shown in Figure 5.13. First, approvals picked up in 1954 while, secondly, the resulting completions started to come through in 1956. The scale of the boom in the latter half of the 1950s was unprecedented, as the institutions expanded their involvement both through providing development finance and through direct investment. According to Scott, direct insurance company investment in land, property and ground rents averaged £576 million in real (1990) prices during 1955–1964, an increase of 81.4% compared with the average level for 1946–1954. Pension fund direct property investment rose even more dramatically, with average net investment during this period amounting to £138 million, a 496% increase over the average for the previous nine years.[66]

Year	Million square feet	
	Approvals	*Completions*
1949	3.6	0.1
1950	3.3	0.6
1951	1.7	0.6
1952	3.0	1.1
1953	n/a	0.9
1954	5.7	0.7
1955	5.9	1.9
1956	3.7	4.1
1957	4.7	4.2
1958	4.2	5.9
1959	4.3	4.6
1960	4.3	4.2
1961	4.4	3.0
1962	2.9	4.7
1963	2.0	3.2

Figure 5.13 Office floorspace, approvals and completions, central London, 1949–1962

Source: Cowan, *et al* (1969).[67]

Scott expands with data on construction output.

> During 1955, the first year following the removal of building licence controls, the value of commercial property construction output amounted to £1,032 million in real (1980) prices, compared to £109 million in the previous year. For the 1955–1964 decade as a whole the value of commercial construction output averaged £1,847 million, compared to an average figure of only £90 million for the previous nine years.[68]

The long lease: shackle or security?

Prior to the mid-1950s, investment property was typically let, at fixed rents, on leases of 99 or 999 years. But as inflation became a more-or-less permanent feature of the economy in post-war years, new approaches were called for. According to Scott,

> a growing appreciation among institutional investors of the potential which property offered as an equity investment, led to the introduction of the rent review, an upward-only adjustment of rent paid for a property to market levels, at intervals stated within the lease.

Scott suggests that the first rent review clauses appeared around 1955, with the average rent review cycle falling from c70 years at this time, "until the present pattern of five yearly reviews became established in the early 1970s".[69]

The Landlord and Tenant Act of 1954 introduced the standard institutional lease, which remained virtually unchanged for over four decades. Throughout this time, it offered security of tenure to occupiers in return for long-term, onerous commitments, while appearing to create immense value for investors and owners. Above all this, the central riddle of the market is why its prime customer was, for so long, the investor and not the occupier.

This is how it was for businesses who signed a commercial lease from the 1950s until the late 1990s. It is of course, only possible where the landlord is in total control of the market. Such control was secured through the establishment of a central contract, namely the Landlord and Tenant Act 1954. This was an update of the Landlord and Tenant Act 1927, which was concerned mainly with shop premises. The legislation, complete with its feudal terminology, survived massive changes in the world around it, only to be challenged when it was clear that new ways of delivering property would be required if the current actors in the process were not to lose their relevance.

Businesses who typically planned no further than three, possibly five years ahead, were asked to commit their businesses to a property for 25 years. Not only that, but every five years, the rent would be adjusted and could *only* go up. To cap that, there was privity of contract. This particularly invidious clause described the relationship between the original parties to the lease. It meant that once a tenant signed a lease, they remained responsible for the rent throughout its full 25-year term, even if – and here is the twist – they assigned it to another business and moved to new premises. As if that were not enough, there was also the full repairing and insuring component. This onerous clause effectively divested the landlord of all management responsibility and placed it squarely on the tenant's shoulders.

It is little wonder that the UK institutional lease was so successful with the supply side, turning the UK property market into a global magnet. By locking occupiers into long leases with severe disincentives to moving, the institutional lease provided a secure income to owners. In short, it became a predictable cash flow, which was enormously attractive to long-term investors. In this sense it helped create the highly liquid market in commercial property that provided an alternative to Gilts and equities, and that proved a magnet to foreign capital.

The problem with predictability is that it can lead to complacency. In this case, apart from the problems that it created for occupiers, a prevailing supply-led culture evolved in which the development supply industry treated its most important customer as the owner, or investor, rather than the occupier. This was the 'let and forget' era. Buildings were developed with a view to selling to an institution, and we grew accustomed to 'institutional specification' and the 'institutional lease', neither of which recognised the needs of the occupier. What the institutions needed was product uniformity to underpin valuation and tradability: it was a short step to the 'rent slab'.

5.5 The rise of the institutions: the 1970s boom

Following the boom years of the 1960s, a period of general economic malaise settled over the UK. Full employment in 1967 gave way to a steady increase in the jobless; output began to slow, and imports of manufactured goods began their inexorable rise and eventual decimation of so much of the country's productive base. As the 1960s drew to a close, there were further signs of an impending economic downturn in falling investment and growing government debt. It was into these economic doldrums, in June 1970, that Edward Heath's Conservative Government was elected to office for the first time. The government's response to the economic climate would prove to be pivotal to the fortunes of the property market and for the office economy. Through a chain of interconnected events, the decisions of Heath's Government led to one of the most dramatic crashes of the post-war property market, and certainly the one with the greatest knock-on effect for the UK economy.

During 1971, unemployment, inflation and public spending were all on the rise. Historically, Conservative governments would have responded with a series of liberal economic policies (principally, non-interventionism). But this time it was to be different. Instead of sticking to the liberal economic tradition, Prime Minister Heath's Government sought to manage the economy in a more fundamental way perhaps than any Conservative government had tried in history. Along with his Chancellor, Anthony Barber, Heath believed that he could reverse the recession in trade and industry through increasing the money supply. And, in a further move to boost investment in the manufacturing sector, they reduced interest rates to around 5%. Perhaps most far reaching of the moves was the decision to remove all lending restrictions from banks that had previously worked within upper limits to their lending. These moves to invigorate the economy were a classic dash for growth: but they also created the ideal conditions out of which an unsustainable property boom would spiral.

By early 1973, the UK economy was showing signs of overheating. Inflation was rising, and in the mini-budget of December, the government reacted by sharply tightening credit controls. Unfortunately, the economic planning framework failed to have the desired effect. For example, the expected boost in investment by manufacturers failed to materialise because "it was clear to them . . . that world trade was going into recession".[70] But the seeds of trouble were not all domestic. In autumn 1973, the OPEC nations proclaimed an oil embargo, which led to the price of a barrel of oil quadrupling by spring 1974. The *Financial Times'* activities all share index showed a fall of 21% during the crisis of 1973 and 1974, finally falling by almost 60%; while by December 1974, the equity market was just 27% of its peak in the summer of 1972.[71]

Suddenly, the economy was heading towards recession: high wage demands were fuelling inflation, which spiked sharply, while demand throughout the economy, including demand for commercial buildings, collapsed. The world economy remained depressed throughout 1975 and 1976. Between 1972 and 1976, there was double-digit annual inflation and unemployment rose to nearly 7%.

Towards the boom

One of the less visible problems was that, at least one of the government's fiscal policies was working: borrowing for investment. The only problem was that the money was being borrowed in the wrong places: most notably in the commercial property sector. Thus, in the summer of 1972, "the Bank of England noted . . . that the expected rise in borrowing from industrialists was only 3% while lending to property interests rose by 50%".[72] This combination of a switch away from manufacturing investment, and the lifting of restrictions on lending, provided the necessary finance for a largely uncontrolled growth in property lending amongst secondary banks. Property companies were flocking to the banks for short-term borrowing, and between 1971 and 1974, bank lending to property companies increased six-fold.[73] This in turn led to a massive wave of speculative development.

In a move that simply reinforced the drift towards a property boom, the government also relaxed Office Development Permits (which had controlled speculative development) and repealed the Betterment Levy (a tax on development profit). The results of these attempts to manipulate the economy were disastrous for the property industry. The full horror of the economic dynamic behind the development boom can be revealed in a summary of relevant statistics:

> the retail price index . . . reveals a 7.9% increase in 1971, 9% in 1972, a further 7.7% in 1973 and by 1974 20%. Bank lending to industry went down by 28% in 1970 and a further 19% in 1973, whereas loans to property companies and financial institutions . . . doubled between 1970 and 1973 . . . The money supply went from 11% in 1971 to 27% in 1973. The growth in bank lending generally went up by 19% in 1970, 29% in 1971, 31% in 1972 and 41% in 1973.[74]

As the banks, and especially the secondary banks, piled into property, the government and the Bank of England were slow to respond, perhaps underestimating the scale of the escalating crisis.

During the early 1970s, bank advances to the property sector rose substantially, from £362 million in February 1971 to £2,584 million in February 1974, an increase of 614%. "The rapid inflow of funds to the property company sector fuelled a boom in property share prices; in the year to May 1972 alone property company share values virtually doubled".[75] The role of the secondary banks was pivotal and helped define what was about to happen as they provided the bulk of the lending for the enormous speculation, during a time when interest rates spiked from 5% to 13% between 1972 and 1973. As the UK economy tipped into recession in the face of such a hike, development projects that "had been viable with interest rates at 5% looked distinctly unhealthy with interest rates at 13%, and falling rents".[76] As banks began calling in property loans early, demand for floor space plummeted and property values fell by as much as 40%.[77]

Although aware of the growing problems, the Bank of England took no action until the late spring of 1972, when it issued a letter to all banks advising them to restrict further lending on property. In April 1973 the Counter-Inflation (Business Rents) Order 1973 was introduced which imposed a freeze on rents and which lasted until the Counter-Inflation (Business Rents) Decontrol Order

No. 21 which came into effect on 1 February 1975. Although introduced as part of a broader counter-inflationary strategy, such measures could do little to halt the development boom. During 1973, in the City of London alone, work had begun on around 18 million sq ft (1.67 million sq m) of offices.[78] The dynamics were not sustainable, and a slump was just around the corner.

The crash

The market crashed, in a most spectacular fashion, leaving a litter of ruined property developers and secondary banks. The property industry made one of its rare forays into the public arena when Prime Minister Edward Heath (in sharp contrast to Macmillan's post-war remarks) referred to property development as the unacceptable face of capitalism. The end was indeed sudden and dramatic:

> The end of the financial boom occurred within weeks of 9th November 1973 when the Financial Times' activities property share index reported an all-time high for property share prices. By 13th November the minimum lending rate went to 13% and on the last day of that month the Stock Exchange suspended the shares of the secondary bank, London and County Securities.[79]

In December, the mini-Budget introduced further measures to dampen the economy, including the tightening of credit controls. As part of this broader strategy, the property industry was hit. The budget contained proposals for a Development Gains Tax (DGT) and a First Letting Tax. Although Labour replaced the Tories in February 1974, the proposals were carried over and were contained within Part 3 of the Finance Act 1974.

The DGT provision was contained within Section 38 of the 1974 Act. It provided that a development gain should be regarded as income and as such constituted profits that were chargeable to tax under Case VI of Schedule D. The First Letting Tax proposed that the first letting of a property after development should be regarded as a disposal of the property at its then market value for the purposes both of CGT and DGT. In short, the measure entailed a tax on a profit that had yet to be made.

The institutions circled the market like a flock of vultures over a bloated corpse. And they spent heavily. Figure 5.14 shows how an annual average investment of around £350 million between 1968 and 1972 suddenly leapt to over £600 million in 1973, reaching £1 billion for the first time in 1976. The data show how institutional involvement grew in the period following the crash.

As they bought up land and buildings, the institutions tightened their grip on supply and increased their power in the industry. So much so, that one perception was that the role of the developer, whether property developer or individual, had been relegated to that of project manager or dealer.[80] Between

Year	Insurance companies	Pension funds	Property unit trusts	Total
1968	119.4	93.0	40.2	252.6
1969	185.9	112.0	43.3	341.2
1970	197.5	97.0	24.8	319.3
1971	198.1	91.0	22.7	311.8
1972	131.1	121.0	38.9	291.0
1973	306.8	248.0	56.9	611.7
1974	405.1	305.0	14.5	724.6
1975	406.3	339.0	33.8	779.1
1976	449.7	520.0	71.2	1,040.9
1977	410.1	535.0	66.2	1,011.3
1978	549.1	590.0	109.4	1,248.5
1979	631.0	510.0	91.0	1,232.0

Figure 5.14 Investments by financial institutions, 1968–1979, £ million

Source: Central Statistical Office.[81]

1975 and 1981, the total property assets of the institutions grew from £7.2 billion to £27 billion; or from 16.4% to nearly 20% of their total assets.

The wreckage of the property crash was everywhere, for all to see, with the familiar proliferation of 'To Let' and 'For Sale' signs, together with a crop of bankrupt property and development companies. But what were the lessons to be drawn from the boom and subsequent crash? What or who was to blame? Could it have been avoided? The headlines were clear enough. A phalanx of centralised planning, including increased money supply, lower interest rates and removal of lending restrictions, was clumsily introduced by a Conservative administration. This in turn led to boom conditions encouraged by plentiful, cheap money. The incentive to build was further encouraged by the removal of planning and fiscal constraints. Combined with a dose of inflation, the recipe was complete. But is this adequate as an explanation?

In 1978, The Royal Institution of Chartered Surveyors published its own post-mortem on the causes and effects of the property boom and the subsequent crash. The report began by identifying three factors responsible for the boom conditions in the late 1960s that preceded the crash, namely: user demand; investor requirements and the development and construction industry. The report offered no evidence on the increase in user demand, but noted anecdotally that a "high level of user demand had built up during the 1960s", caused by "an expansion in the number of office jobs" which, in turn, had been caused by "the growth of the public service and by the trend in the private sector from manual to white-collar employment".[82] However, figures show a small fall in the office workforce in London during the latter half of the 1960s.

The Institution's second presumed cause of the boom in the 1960s and early 1970s is also suspect, as shown by the data in their own report. The report showed that institutional interest in property between 1969 and 1972 actually *declined*. The great increase actually occurred only in 1973 when the boom was about to crash. As for the banks, the report again shows that by the time the money gushed into the system, the boom was almost over. The report cites Bank of England data to demonstrate that bank advances on property were quite stable between 1968 and 1971, doubling in 1972, and doubling again in 1973. Institutional and bank lending were not primary causes of the boom conditions although they were very much features of the subsequent crash.

The final particular cause of the boom cited by the RICS was the collective behaviour of the development and construction industry. The report argued that *"Because* of the high level of demand" [author's italics], developers were willing to enter into loan arrangements that "depended on expected future increases in rents to service their loans".[83] In other words, they backed a hunch. Following a few years of economic malaise, the property industry backed an upturn: nothing unusual there. What was remarkable was the conclusion that the RICS drew from this: "it is extremely risky to borrow to the limit of available collateral".[84] This seems to be a naïve and self-evident statement, and one that avoids perhaps examining the structural weaknesses of a supply process that were really at fault. As for construction, the report merely points to activity in housing and public sector construction as having provided an incentive for building firms to expand, and to borrow for the purpose.

The report then turned its attention to the causes of the crash. The report argued that "apart from the economic factors", falling demand, construction costs, the rent freeze, interest charges, DGT and the White Paper *Land* (which introduced the Development Land Tax), were the specific causes of the collapse. It must be recognised, however, that falling demand and construction costs were not primary causes of this, or any other, crash. The inability of the supply process to recognise and respond in a timely manner to falling demand is a cause; the inability to reflect rising construction costs in enhanced values within a deflationary context is a cause. But falling demand and rising costs are not of themselves *causes*. The fiscal measures referred to, such as rent freezes and taxation, are direct causes because they bring about swift and severe changes in sentiment: "The rent freeze shook the confidence of investors, who for many years had believed that commercial rents would go on rising indefinitely, thus providing an income broadly in line with inflation".[85]

In seeking to draw lessons from the fiasco of the collapse, the RICS argued that a highly unusual combination of international and domestic events (including inflation, a credit squeeze, economic recession, the oil crisis and a series of legislative measures) were all to blame. Of these latter, the RICS report was particularly critical: "it is highly dangerous for any government to attempt to regulate the property market by seeking to limit the supply of property . . . in response to demand".[86]

What is missing from the Institution's account of the boom and crash is any analysis of how the property supply industry itself could have behaved differently. In its rush to blame extraneous factors (viz: "lenders too often abandoned normal standards of commercial prudence"),[87] the Institution failed to look closely at the structure of the supply process. In fact, there was no criticism of the property industry itself, and its lemming-like desire to gorge on cheap money. The institutional market is barely mentioned, with only a reference to their ability to take the long view and "ride out short-term fluctuations in the value of their property assets".[88] There was no mention of the need for more, indeed any, market research and information. There was no attempt to understand the dynamics of demand. The two main lessons to be drawn were, first, that governments should avoid short-term intervention in the property market and, second, banks and other lending institutions should be more prudent.

There was just one behavioural explanation of the boom and crash that failed to find its way from the main body of text into the conclusions and lessons sections. In a nod to Heath's 'unacceptable face of capitalism' comment, the RICS did state that

> the cycle of boom and collapse was perhaps the inevitable consequence of lenders and borrowers treating as a short-term 'dealing' market what should always be regarded as a medium- to long-term investment market.[89]

Despite the findings of the RICS, it was almost inevitable that another boom-bust cycle should get underway. And the fact that it did, with an entirely different set of contextual circumstances, served only to underline the fragility of the official explanation for the 1970s crash. Within ten years of the RICS report, the new boom was getting into full swing.

5.6 A riddle wrapped in a mystery

The long historical sweep of this chapter has shown how London has grown: short periods of intensive development activity and expansion, interspersed with longer periods of relative stasis. In earlier times the development phases were mainly to house London's growing population, but from the later Victorian era this has included phases of intense building for the office economy. More often than not, the building phases have ended in a crash, with the debris of developers scattered all around.

We have also seen in this chapter how the developer entrepreneur was replaced by the financial institution as the costs and complexity of development finance multiplied. But throughout the post-war period, we have experienced boom-to-bust and back again. Two particularly severe episodes have been described, but there were other less damaging examples.

A riddle wrapped in a mystery inside an enigma. This was how Winston Churchill described Russian foreign policy at the outbreak of the Second World War. Famous for his rhetoric, Churchill found a way of both simplifying a very complex set of relationships while at the same time conveying his own sense of perplexity. There can be no better way to describe the supply side of the post-war commercial property market – at least from an occupier's perspective.

The structure of the market was entirely supply-driven, and it created buildings that, with notable exceptions, bore little relation to the changing needs of the businesses that rented them.

Amongst the mysteries was the commercial lease – the contractual agreement between the landlord and tenant, or owner and occupier, of a building. Whilst recognising the imperatives of the early post-war years, and the important role of institutional finance in 'getting things moving' once again, the problem seems to have been that the system ossified. It became an end in itself rather than a means to an end. This was particularly the case in the office economy: buildings were effectively 'under-written' by corporate occupiers, who signed long leases and committed to quinquennial rent increases on standardised space. The secure and predictable incomes that resulted proved an irresistible magnet for investors.

From the occupiers' perspective, the situation was bearable through the 1960s and 1970s, when they were typically slow-moving, unchanging, long-term organisations with few demands on building infrastructure. The problem came when this ossified contractual structure was exposed in the tornado of digital change in the 1980s and 1990s.

Notes

1 Bowley (1966) *Op cit*, p326
2 *Ibid*
3 Geddes P (1915) Cities in Evolution: An Introduction to the Town Planning Movement Harper Torchbacks, New York edition, 1971, p26
4 Ackroyd (2000) *Op cit*, p717
5 Bowley (1966) *Op cit*, p334
6 H. Berry (2002) Polite Consumption: Shopping in Eighteenth Century England, *Tans of the Royal Historical Society*, Vol 12, pp375–394
7 Ackroyd (2000) *Op cit*, p717
8 *Ibid*
9 *Ibid*
10 *Ibid*, p719
11 Jenkins (1975) *Op cit*, p81
12 *Ibid*, p84
13 Ackroyd (2000) *Op cit*, pp718–719
14 *Ibid*
15 Bowley (1966) *Op cit*
16 Scott (1996) *Op cit*, p22
17 Kynaston (1994) *Op cit*, p245
18 V. Barnes & L. Newton (2019) Symbolism in Bank Marketing and Architecture: The Headquarters of National Provincial Bank, *Management and Organizational History*, Vol 14, No 3, pp213–244
19 P. Ward-Jackson (2003) *Public Sculpture of the City of London, Public Sculpture of Britain*, Liverpool University Press, Liverpool, p35
20 Black (2000) *Op cit*, p368
21 Kynaston (1994) *Op cit*, p245
22 Bowley (1966) *Op cit*, p326
23 Gray (1978) *Op cit*, p293

24 *Ibid*
25 R. Porter (1994) *London: A Social History,* Penguin Books, London, p398
26 *Ibid*, p401
27 Jenkins (1975) *Op cit*, p198
28 Michie (1992) *Op cit*
29 *Ibid*, p34
30 D. Kynaston (1995) *The City of London Volume 2 Golden Years 1890–1914,* Pimlico, London, p17
31 Michie (1992) *Op cit*, p39
32 *Ibid*, p13
33 *Ibid*, p41
34 Scott (1996) *Op cit*, p83
35 *Ibid*, p68
36 *Ibid*, p73
37 Jenkins (1975) *Op cit*, p245
38 J. Rose (1985) *The Dynamics of Urban Property Development,* E&FN Spon, London, p151
39 *Ibid*
40 A. Sampson (1962) *Anatomy of Britain,* Hodder & Stoughton, London, p415
41 Jenkins (1975) *Op cit*, p209
42 H. Smyth (1985) *Property Companies and the Construction Industry in Britain,* Cambridge University Press, Cambridge, pp131–132
43 Marriott (1967) *Op cit*, pp11–12
44 E. Erdman (1982) *People and Property,* Batsford, London, p114
45 Sampson (1962) *Op cit*, pp416–417
46 *Ibid*, p417
47 *Ibid*, p418
48 Marriott (1967) *Op cit*, p36
49 *Ibid*, p40
50 Rose (1985) *Op cit*, p153
51 Scott (1996) *Op cit*, p143
52 *Ibid*, p148
53 *Ibid*, p153
54 A.P.J. McIntosh & S.G. Sykes (1985) *A Guide to Institutional Property Investment,* MacMillan, Basingstoke, p19
55 Associations with Insurance Companies *Investors Chronicle,* 29 April 1959. Cited in: Smyth (1985) *Op cit*, p147
56 Smyth (1985) *Op cit*, p155
57 W. Lean & B. Goodall (1966) *Aspects of Land Economics,* The Estates Gazette Ltd, London, p94
58 Herring Baker Harris (1992) *The Capital's Property,* HBH, London, p31
59 McIntosh & Sykes (1985) *Op cit*, p20
60 Marriott (1967) *Op cit*, p15
61 *Ibid*
62 *Ibid*
63 *Ibid*, p11
64 M. Bateman (1985) *Office Development: A Geographical Analysis,* Croom Helm, London, p16
65 Marriott (1967) *Op cit*, p311
66 Scott (1996) *Op cit*, p133
67 Cowan *et al* (1969) *Op cit*, adapted from Tables VII,1 (p162) and VII,3 (p183)
68 *Ibid*
69 Scott (1996) *Op cit*, p286
70 Rose (1985) *Op cit*, p164

71 *Ibid*, p166
72 *Ibid,* p23
73 R. Barras (1984) The Office Development Cycle in London, *Land Development Studies,* Vol 1, pp35–50
74 Rose (1985) *Op cit*, p165
75 Scott (1996) *Op cit*, p188
76 Herring Baker Harris (1992) *Op cit*, p32
77 *Ibid*
78 *Ibid*
79 Rose (1985) *Op cit*, pp165–166
80 *Ibid*, p167
81 Cited in: Smyth (1985) *Op cit*, p211
82 The Royal Institution of Chartered Surveyors (1978) *The Property Boom 1968–73 and Its Collapse,* A Supplementary Memorandum of Evidence to the Committee to Review the Functioning of Financial Institutions RICS, London, p5
83 *Ibid*, p8
84 *Ibid*, p14
85 *Ibid*, p9
86 *Ibid*, p13
87 *Ibid*, p14
88 *Ibid*, p15
89 *Ibid*, p23

6 Building

Re-shaping a global city

The previous chapter examined the physical development of the office economy from the eighteenth and nineteenth centuries, through the Victorian and Edwardian periods, to the post-Second World War period, up to the 1970s. The chapter highlighted the commercial property booms and crashes of the 1950s–1960s and of the 1970s. In this chapter we wind forward to examine the more recent physical development of the office economy. The chapter begins with a review of the 1980s boom – one of the biggest in UK history – and then examines this in the context of the earlier booms, in order to ask whether 'boom and bust' is an inevitable or avoidable model for the office economy.

The second part of the chapter looks at the changing physical office economy since the 1980s, through the millennium and up to the current time. The main means of conveying this is a description of the changing business geography of London, describing how, over three decades, London's office economy metamorphosed from a small, tightly defined market into a more complex mosaic of activities, to maintain its global city role.

6.1 The Big Bang boom

In 1985 the Apple 'Mac' introduced desktop icons, Microsoft Excel was launched, the first '.com' was registered and Cat 1 cabling (used to transfer voice data in telephones) was introduced, all presaging the digital era. Politically, the government was facing down the miners and the print unions. The economy was in the process of de-industrialising, and the share of the economy deriving from services generally, and financial services in particular, was rising sharply. As the 1980s progressed, the economy grew steadily, and the latter half of the decade was a positive time for the economy. The recession of the early 1980s was receding in the minds of lenders, and prospects were for a prolonged period of growth.

This positivity was underlined on 26 October 1986, when the City of London experienced one if its most radical upheavals in its history. This was 'Big Bang', the day financial services were deregulated; electronic trading became the norm and the floor of the Stock Exchange fell silent. It was the day that cemented London's role as a global city, competing with New York and Tokyo, eclipsing its European rivals in the process.

A re-organising City

Big Bang abolished minimum fixed commissions on trades, thereby increasing competition; it ended single capacity, the separation between those who traded stocks and shares (jobbers) and those who advised investors (brokers), thereby opening the floodgates of mergers and take-overs; and it permitted foreign firms to own UK brokers, putting up a 'welcome' sign to international banks. Trading switched from the face-to-face format of 'open outcry' on the Stock Exchange floor to desk-top screens on investment bank dealing floors.

The pressure to deregulate financial services had been building for some time. Internationalisation saw the number of foreign banks represented in the City rise nearly fourfold since 1970, to almost 500. These included large US and Japanese integrated merchant banks and securities companies (for example, Goldman Sachs, Merrill Lynch and Nomura Securities).

In the background, the USA had abolished fixed commissions in 1974, and in 1979 the new Conservative Government abolished exchange controls – triggering for many the UK's financial and economic rebirth. Then in the early 1980s, the competition authorities were threatening to take the Stock Exchange to the Restrictive Practices Court. The then chairman of the Stock Exchange, Nicholas Goodison, took the view in 1983 that rather than engage in a legal battle that would most likely be lost, he would strike a deal with Trade and Industry Secretary Cecil Parkinson, which gave the Stock Exchange three years before its rule book should change and commissions abolished.

One reason for the three-year lead-in was to allow time for the introduction of the technology that was an essential part of the ending of single capacity. The new dealing system would be quote driven – the Stock Exchange Automated Quotation System, or SEAQ, which itself was modelled on the established American NASDAQ system.

Following the Parkinson-Goodison Accord and in the lead up to deregulation, there was a feverish match-making activity as brokers and jobbers sought merchant bank partners who had the capital to navigate the times ahead. Figure 6.1 shows a selection of the alliances that were formed in the run-up to Big Bang.

In 1983 there were 214 broking and 17 jobbing firms in the City (following a long period of consolidation), and these were quickly snapped up by UK merchant banks and international investment houses. At this point, banks could take a maximum 29.9% in any target business. Some were bought by UK clearing banks, but many more were snapped up by much bigger US, European and Japanese banks. Clearly, new organisations were in the making: larger, more complex, more international and more technologically intensive.

Big Bang in financial services often overshadows a wider set of changes in the City of London. For example, also in 1986, Lloyd's of London moved into its new headquarters on Lime Street. Designed by Richard Rogers, the

Investor	Broker	Jobber
Barclays Bank	De Zoete & Bevan	Wedd Durlacher Mordaunt
Baring Brothers	Henderson Crosthwaite	Wilson & Watford
Charterhouse J Rothschild	Kitkat & Aitken	
Citicorp	Vickers da Costa	
Citicorp	Scrimgeour Kemp-Gee	
Credit Suisse	Buckmaster & Moore	
Hambros	Strauss Turnbull	
Hill Samuel	Wood Mackenzie	
Hong Kong and Shanghai	James Capel	
Kleinwort Benson	Grieveson Grant	Charlesworth & Co
Midland Bank (Samuel Montague)	W Greenwell & Co	
Morgan Grenfell	Pember & Boyle	Pinchin Denny
N M Rothschild	Scott Goff Layton	Smith Brothers
National Westminster (County)	Fielding Newson Smith	Bisgood Bishop
Prudential Bache	P B Securities	
Rowe & Pitman	Mullens	Akroyd & Smithers
Security Pacific	Hoare Govett	C T Pulley
Shearson Lehman	L Messel	
Swiss Bank Corporation	Hill Chaplin	
Union Bank of Switzerland	Phillips & Drew	

Figure 6.1 A selection of Big Bang mergers

building achieved instant recognition from the fact that its services (lifts, stairs, drainage pipes and so on) were 'attached' to the exterior of the building. The rationale was to create internal space that provided ultimate flexibility to facilitate changing needs.

Big Bang proved to be the biggest spur to property redevelopment since the war. Deregulation was fundamental, and suddenly it seemed that the whole financial services industry needed re-housing in purpose-built buildings, with dealing floors the size of football pitches. The property industry responded with an orgy of development. Not only were many old buildings demolished and replaced, but whole swathes of land were developed. Perhaps the best known of these was Broadgate at Liverpool Street Station (of which more later), where railway land, air rights and new construction technologies were all exploited to permit the creation of a whole new district in record time. Other rail stations followed suit, with major developments at Ludgate, Cannon Street and Charing Cross. And, of course, there was Canary Wharf, a largely disused dock in east London.

Boom time (again!)

With Big Bang seeming to spur the City into a new and expansive age, 1987 saw confidence within the property industry running high. The Conservative Government had just won a third consecutive general election, with a landslide result. Work had begun on the Channel Tunnel; Sir Clive Sinclair

Figure 6.2 Lloyd's of London, Lime Street
Source: © Erik Brown, 2020.

launched his Z88 Portable Computer; *The Simpsons* was broadcast for the first time and London Docklands Airport opened for business. Indeed, the commercial property market was enjoying very benign conditions. The economy was expanding at more than 3%, and demand for space was rising. In central London, the amount of empty space had shrunk from nearly nine million sq ft (c840,000 sq m) to just two million sq ft (186,000 sq m) over three years. As a result, rents were rising steeply and investment yields were falling sharply. Analysts were in general agreement about the short-term future: "the prospects for the City office market are good and we do not anticipate an over-supply of office accommodation . . . within the foreseeable future".[1] Similarly: "We believe that latent demand is still underestimated . . . and occupiers will continue to take up the rapidly expanding supply of new developments through to the early-1990s".[2]

Even the effects of the infamous stock market crash on 'Black Monday', 19 October, appeared to have been softened by the underlying strength of demand for space. This was surprising. On the fateful day, the FT-SE fell 250 points (10%); on the 22 October, a further 110 points were wiped off, taking the market down to 1,833; and prices fell again on Monday the 26 October by another 150 points. In just one week, the FT-SE had fallen 28% from its 1987 high.

Despite one particularly dire warning that "the events of the last two weeks will bring about a property crisis comparable with that of 1974",[3] the market as a whole brushed aside the crash. One year on, while recognising that the real estate market had not been immune from events in the financial markets, property advisor Jones Lang Wootton (now Jones Lang LaSalle) stuck to the business-as-usual line: "an imminent collapse in tenant demand and consequent fall in rents which was forecast in some quarters at the beginning of the year did not materialise".[4] While JLW was one of the most respected firms in the property market, with one of the most forward-thinking research teams, this view was backward looking and failed to address closely what was looming over the horizon. Some analysts, however, were prepared to cast doubt over the strength of the market.

Indeed, the first warnings of the impending disaster began to emerge as early as the spring of 1988. Geoff Marsh, then head of Applied Property Research, gave one of the earliest public warnings. His central London office market report identified 45 construction starts in 1988 – *simply on schemes of over 100,000 sq ft (9,300 sq m)* – totalling nearly 16 million sq ft (c1.5 million sq m). Marsh concluded:

> the central London office market is heading for oversupply . . . Construction activity has actually increased significantly to levels which imply that demand has to double again from the already very high levels of 1987 and 1988.[5]

The development tap was turned to maximum flow, but who was listening? Construction completions in the City had been steady at around 2.5 million sq ft (232,000 sq m) between 1986 and 1988. Then, as property companies continued to gorge themselves on a seemingly endless supply of cheap money, in 1989 and 1990, completions leapt to 4.36 million sq ft (405,000 sq m) and 4.34 million sq ft (403,000 sq m), respectively; and in 1991 jumped to a massive 6.92 million sq ft (643,000 sq m).[6] Figure 6.3 shows data comparing the extent of lending to property companies in the 1970s and 1980s. Property lending as a percentage of all loans doubled between 1984 and 1989, while bank lending to property companies increased almost sixfold.

Three months after Marsh's forecasts, the *Sunday Times* asked: "Is [the] City Lost In Space?"[7] The article went on: "Are the first cracks beginning to appear

Year end	Total bank lending (£ millions)	Bank lending to property companies (£ millions)	Property lending as % of total
1970s			
Nov 1968	9,267	336	3.63
Nov 1969	10,651	327	3.07
Nov 1970	13,570	341	2.51
Nov 1971	16,792	501	2.98
Nov 1972	23,872	1,155	4.84
Nov 1973	36,413	2,330	6.40
Nov 1974	45,877	2,802	6.11
1980s			
Nov 1984	150,793	5,420	3.59
Nov 1985	168,440	7,111	4.22
Nov 1986	202,893	9,335	4.60
Nov 1987	244,749	13,333	5.45
Nov 1988	304,634	21,287	6.99
Aug 1989	395,579	29,608	7.48

Figure 6.3 Bank lending to property companies, 1970s and 1980s compared

Source: Phillips & Drew (1989).[8]

in London's office construction boom?" Drawing attention to the volume of new space that was due to come onto the market in the next two years, the article noted that this was happening at a time when City firms were believed to have laid off more than 12,000 staff in the previous ten months, and when the volume of business had roughly halved over the same period.

In the same month, August, broker Morgan Grenfell Securities warned of a "risk of oversupply" as "a risk remains that developers will still proceed with their proposed speculative schemes".[9] To round out the year, property advisor Winter & Co gave a stark warning:

> We anticipate growing problems for the market in 1989. Occupiers will shelve expansion plans in the face of expensive borrowing, an unstable Stock Market and a decline in consumer spending. Demand will begin to fall dramatically and supply of both new and second hand space will continue to grow.[10]

So how big was "the risk" of oversupply? Asked another way, should it have been obvious that the market was *certainly* heading for oversupply? A simple calculation puts the quantum of supply in 1988 and 1989 into stark focus. This calculation involves a rule of thumb to calculate the number of office workers that would be required to fill all of the space being developed. Assuming a conservative occupation density of 250 sq ft (23 sq m) gross per person, the construction starts for 1988 alone represented 64,000 office workers. Even allowing for the crudeness of the calculation, against a total City working population of around 250,000 workers, it should have been clear that the scale of development was unsustainable. And what was happening in the City was being replicated elsewhere. One observer did put the development boom into a wider geographical context, although not until late in 1989. Using slightly different calculations to those used previously, but coming to the same over-all conclusion, journalist David Brierley was blunt: "In the south east, where 95 million sq ft will be built, about 800,000 new office workers will be needed to ensure that no buildings are vacant – an impossibility".[11]

In a prescient article in the *Financial Times* at around the same time, Michael Brett compared the raging boom with the 1970s. Citing recent warnings by the Governor of the Bank of England over the level of bank lending to property companies (around £30 billion in October 1989), Brett noted that one undeniable similarity between the early 1970s and late 1980s was that "on both occasions the Governor of the day sounded public warn-ings on property lending and the banking system roundly ignored them".[12] Brett went on to note:

> In the 1970s cycle, the warning was sounded in 1972. Bank lending to property companies grew by 131 per cent in 1972 and by 101 per cent in 1973 to £2.3bn. In 1974 the commercial property market collapsed.

Within six months, over half of the value of property companies had been wiped out. Brett concluded his article with a direct comparison to 1973:

> we have just experienced a period of rapid growth in property values, bank lending to property is at a peak, interest rates have risen sharply . . . and the authorities are attempting to damp down the economy. And now, as then, we have considerable instability in the securities markets.
>
> It is not a re-run of 1973. But neither is it a background that gives much comfort to the banking and property industries.[13]

The Governor's warning in 1989 followed a 43% rise in property lending in 1987; a 60% rise in 1988; and a 39% rise in the first half of 1989; interest rates were again at an all-time high, and securities were again in turmoil.

And crash (again!)

In 1990, the bloodletting began in earnest. By the end of April, shares in property companies Sheraton Securities, Rush & Tompkins and Wiggins were all suspended, and both Finlan and Priest Marians had revealed problems with borrowings. In May, the receivers were called in to South Quay Plaza, a £50 million development in Docklands. It took another two years for the full horror of the debt situation to become fully clear through collapses of well-known development companies. As Vanessa Houlder observed:

> It is more than three years since the UK commercial property sector went into decline, but 1992 was undoubtedly the year when the cracks became chasms.[14]

Noting that nearly 32 million sq ft (2.97 million sq m) of offices in central London lay vacant, and that bank debts to property companies stood at £35 billion, Houlder referred to research by accountants Touche Ross suggesting that since April 1990, property businesses had been going bankrupt at the rate of one every 36 hours. And it was not just small, anonymous companies that were disappearing. The roll call of well-known company collapses in 1992 was staggering. In March, Olympia and York, one of the largest property companies in the world, and Heron International proposed restructuring of their debts. In the same month Randsworth, a West End company bought by US investors at the top of the market in 1989, went into receivership with debts of more than £350 million.

In April Speyhawk revealed that it was technically bankrupt, disclosing that its liabilities exceeded its assets by £70 million. Heron International finally collapsed weeks later. In May, with debts of about £500 million, Mountleigh (a company that had been built out of an old textile mill by Tony Clegg) called in the receivers. Its shares were among the top ten performers of the 1980s, and Clegg had sold a 22.5% stake to Nelson Peltz and Peter May, two Americans with property interests in America, for £70.4 million in November 1989. They paid 200p per share, a 40% premium to the market. Mountleigh collapsed with debts of £400 million. Also in May, Rosehaugh collapsed with debts of £350 million; and Olympia & York recognised the scale of its problems and filed for insolvency protection in Canada and the USA. Within days, Canary Wharf was put into administration.

What went wrong? What led the market to stumble from apparently benign conditions at the start of 1988 to total chaos in 1990? The 1980s property boom was one of the greatest ever. Much of the activity was concentrated in the capital where more space was created in a shorter period than ever before. The boom even exceeded, proportionately, that of the rebuilding after the Great Fire, principally because a large proportion of the building was on land not previously used for commercial space, such as railway goods yards, disused docks and derelict land. The scale of the boom is illustrated by the fact that the City's stock of offices grew by a third between 1984 and 1991. The legacy of the 1980s development boom was breath-taking. When the crash came, 40 million sq ft (c3.7 million sq m) of space, equivalent to nearly 1,000 football pitches, was empty in central London alone, and there was a development pipeline of around 100 million sq ft (9.3 million sq m), representing a staggering 50% increase in total stock.[15] Vacancy rates topped 20% in the City. As well as the physical legacy, there was also the financial hangover. The boom was funded on a seemingly inexhaustible supply of money, and total lending to the property sector peaked at around £40bn. To put this figure in context, annual lending had historically run at between £1bn and £2bn.

The boom was, undoubtedly fed by a huge wall of money. And for the first time, much of the cash came from overseas. This was a globally-financed boom, not a UK boom. During the early stages of the Broadgate project, for example, the apparent risks of the venture (off-pitch location, scale, new building types, and so on) were too much for traditional UK investors, and the bulk of finance came from the Far East. Germany, Japan, Sweden, the Middle East and others piled in later. According to one estimate, overseas investment in UK property stayed below £0.5bn between 1980 and 1987. Then in 1988 it leapt to nearly £2.0bn; and in both 1989 and 1990 reached the dizzy heights of around £3.0bn.[16]

The forces that shaped the 1980s boom were very different from those that shaped the 1950s and the 1970s booms, described in the previous chapter. For example, in addition to the ready availability of cash, at least four other inter-related factors played crucially important roles, including:

- the expectation of exceptional levels of demand following deregulation of the financial services industry;
- changes to the Use Classes Order which led directly to new building types, especially the out of town business park and retail parks;
- rapid developments in office technology, contributing to building obsolescence, and
- the availability of air rights over railway lands (for example, Broadgate and Ludgate) and regeneration land (for example, Canary Wharf).

Indeed, it seems that all booms have a different 'profile' in terms of causes, duration, amplitude and impact. This time round, perhaps the greatest single cause was the actual and anticipated growth in office work as both the financial services industries and the technology industries expanded extremely rapidly. The planning and design communities responded with new building types, and the rush was on to supply large, highly specified buildings in city centres for financial and business services firms, and large, simpler but internally flexible buildings on business parks for technology, and a host of other office-based sectors. The overwhelming development hubris that followed was unsustainable and largely uncontrolled.

The post-war reconstruction boom coincided with a period of economic growth, but began with the removal of building licences in November 1954, and was led by speculative finance. This boom was then followed by a period of slow economic growth, which caused new development to decline sharply, before picking up again sharply to a development activity peak in 1973–74 as the national economy fell into decline and as the oil crisis took hold. Edward Heath's dash for growth to combat recessionary pressures in the early-1970s involved increased money supply, low interest rates and the abolition of lending restrictions. For the first time, the 1980s boom was underwritten with a surge of global finance. There were also other, less spectacular, booms in the post-war era that also had credit as a key enabler.

What is clear, is that while the profile of each boom was an outcome of the interaction of a unique and complex web of events and circumstances that coincided in time, before dissipating, the single factor that resonates across all of the post-war booms is the relative availability of finance to the property industry. And in the post-war era this has been particularly the case for commercial, speculative development involving offices, retail and light industrial property. Throughout the eighteenth and nineteenth centuries, through to the Edwardian period readily available finance was a key ingredient of boom periods. But during these times, the commercial element of the booms was relatively limited: most of the exposure was to residential markets. After the First World War and through to the 1970s, the public sector took a more significant role in the provision of social housing, while commercial property expanded rapidly. The post war booms took on a different character.

The frequency of development booms has led to a widespread belief that they are inevitable. Indeed, they are commonly referred to as property cycles, directly implying some spurious sense of periodicity and regularity. Without doubt, successive development cycles, or booms, have provided a wheel of fortune on which the property industry casino swung successively from boom to bust and back again. But it is important to be clear about one thing in particular: a boom and a crash are not necessarily part of a single process.

Developers and investors are always looking for signals that help them decide when, where and what to build. At the most basic level, development activity increases when there is a growing consensus that demand is growing sufficiently to justify new space: classic demand and supply. In this sense, it is not really the genesis of any particular boom that should be at issue. Far more important to unravel is the scale of *overbuilding*, and the direct cause of overbuilding really only has one culprit: easy credit (oh, and a dose of hubris).

So, how are we to regard development booms? Are they really cycles? Are they avoidable? Or are they inevitable? The second part of this chapter looks at our understanding of modern development booms and the efforts to explain them.

6.2 Boom-to-bust: inevitable or avoidable?

In Chapter 5 we saw how the building booms of the 1950/60s and 1970s/80s both got underway before unravelling in chaotic oversupply. In the previous section, we saw the same pattern occur once more as the boom of the 1980s, with the same chaotic outcome. Are these phases of boom-to-bust inevitable? Must we just endure them, due to some sort of inherent weakness in 'the system'? Or are they in fact avoidable? It is difficult to believe that a set of common unavoidable causal factors runs through the residential booms of the nineteenth century and the commercial booms of the post-war era.

So vexed has this issue been that a whole body of literature has grown up to help explain. A recent literature review of UK and US property cycle research over the past century cited 95 references.[17] With such a weight of accumulated knowledge, it would be fair to anticipate a number of keen insights. Indeed there are many explanations, often involving pages of impenetrable mathematics. However, one explanation is studiously avoided above nearly all the explanations, and that is insatiable avarice in the context of cheap and easy finance.

The cycle disciples

Weiss observed that while writers from the Duke de la Rochefoucauld had noted the boom-bust behaviour of the real estate market as early as the eighteenth century, it was only during the depression in the 1930s that "scholars began to systematically examine the historical pattern of real estate cycles".[18] It was then that the search began "for historical explanations and research methods that could help predict the timing and impact of future fluctuations".[19] It was this perceived link between analysis and prediction that was to so profoundly influence thinking about development booms for decades to come: *if only we can measure them, then surely we'll be able to predict them.*[20]

American economists such as Charles Wardwell, Simon Kuznets and Arthur Burns were among the first cycle disciples. Working in the 1920s and 1930s, they all claimed to find results that indicated cyclical characteristics in construction and business capital investment.[21] They shared a belief that the investment and building cycles were related to but distinct from shorter-term, more irregular business cycles and longer-term Kondratieff swings in prices.

There were, however, detractors from this generally held belief that cycles could be identified and predicted. Joseph Schumpeter, the Austrian-born economist who led the 'depression generation' of economists at Harvard during the 1930s and 1940s, provided a robust critique in the context of business cycles:

> the question of causation is the Fundamental Question. . . . there is no single cause or prime mover which accounts for them. Nor is there ever any set of causes which account for all of them equally well. For each one is a historic individual and never like any other, either in the way it comes about or the picture it presents. To get at the causation of each we must analyze the facts of each and its individual background. Any answer in terms of a single cause is sure to be wrong.[22]

Equally forcefully, Homer Hoyt, in his classic *One Hundred Years of Land Values in Chicago*, argued:

> Before we go any further, we should dispose of the notion that there is anything rhythmical about business cycles . . . Unfortunately for the forecaster, each business cycle seems to be unique, and its length and amplitude are determined by a combination of forces never duplicated before and probably never to be duplicated in the future.[23]

Despite these more cautionary voices, the cycle disciples gathered large amounts of data to 'prove' the cyclicality of real estate. One of the classic studies from this era was that by Clarence Dickinson Long (Jr). Long was born in South Bend, Indiana, in 1908, and took a PhD from Princeton University in 1938. He was a professor of economics at Johns Hopkins University (1946–1963);

served in the US Navy during World War II, and was a Democratic congress-man between 1963 and 1985. In an extensive study, Long assembled a monthly index of construction over the period 1868–1940, based on building permits.[24] Long's endeavours led him to identify short building cycles with an average four years' duration, and longer cycles of around 20 years. These findings were substantiated statistically and cross referenced with other investigations into the subject.

Such work by cycle disciples cast a very long shadow. During the post-war years, development booms generated enormous interest amongst academics and researchers, all seeking some explanatory or causal mechanism, from which a predictive model could be built. But how exactly do periodic phases of over-development with their unique causes and events, as described earlier, fit with the concept of cycles? The distinction between 'phases' and 'cycles' of over-development is more than a semantic one. The former describes an economic inefficiency that is periodically, and often traumatically, corrected, but which might be thought of as avoidable. The latter describes a form of regularity, with beginnings and ends, passing through an identifiable series of events during the process, and which might be thought of as inevitable.

Property cycles

Many of the early studies of property cycles were focused almost entirely on the residential market. There can be little surprise over this, given the late apprecia-tion of the role of offices in cities (see Chapter 3) and the bespoke nature of much industrial plant until the mid-twentieth century. However, the extensive literature on residential cycles set the framework for approaches to understand-ing phases, particularly of overbuilding, in the commercial sector. As early as 1933, Hoyt suggested that business conditions, commodity price levels, the cost of money, and population growth were the main causes of short-term cycles.[25]

Following Hoyt's work, Long identified the four-year short-term cycles and 20-year long-term cycles mentioned previously.[26] And it seems that the real estate industry has remained imbued with the notion that it is a prisoner within an inherently cyclical process ever since, cycles that lead the market to oscil-late between wild optimism and frenetic activity, and periods of falling prices, negligible activity and chronic oversupply.

But over time, the explanations have shifted focus. As already noted, the cycle pioneers emphasised prices (money and commodities) and population growth (this was at a time when commercial markets were not dominated with widespread speculative development). Transportation innovation has also been tagged. For example, Isard reported in 1942 regular transport building cycles and a causal relationship between real estate and transport building.[27] In the mid-1960s, Lewis also cited population, production, harvests, migration, rent levels and credit supply.[28] Later studies emphasised the role of institutional finance and credit generally[29] and technology.[30]

Lewis's focus on credit supply was picked up by Fleming, who observed "a building boom will only take place if abundant funds are available to finance it".[31] This was perhaps the first recognition that money supply lay at the heart of building booms. But, even in the mid-1970s, there was a lack of focus on development booms in the commercial sector. For example, Gottlieb undertook a major study across many countries, only to conclude that favourable economic conditions encourage the formation of new households, feeding demand and supply of new homes.[32]

Indeed, many observers continued to look for structural explanations rather than financial explanations. Many studies in the 1980s focused on 'endogenous forces'. This shifted the explanatory focus away from exogenous factors (population growth, cheap money, migration), to factors inherent to the industry itself. Chief among the villains identified here was the 'production lag'. This was most clearly set out by Barras, whose conceptual model of building cycle dynamics progresses as follows.[33]

1   Strong upturn in the business cycle coincides with a relative shortage in the available supply of property, following a period of low development activity during the previous business cycle.
2   Strengthening demand and restricted supply cause rents and values to rise, improving the profitability of development and triggering the first wave of building starts.
3   If credit expansion accompanies the business cycle upturn, then it can lead to a full-blown economic boom, while at the same time banks fund a second wave of more speculative activity.
4   A major building boom is underway, but lags inherent in the development process mean that little new space is completed, and rents and values continue to rise.
5   As the bulk of new buildings reach completion, the business cycle has moved into its downswing, money supply is tightening and interest rates begin to rise.
6   As the economic boom subsides, demand for property weakens, just as the new supply reaches its peak. Rents and values begin to fall and vacancy rises.
7   As recession looms, the fall in rents and values accelerates; the credit squeeze hits property companies with un-let buildings.
8   The outcome is a property slump with depressed values, high levels of vacancy and widespread bankruptcies in the property sector.

The reasoning of this explanation is clear, but it ignores the role of contingent factors that determine the timing and scale of change in economic activity that feed through to heightened confidence in the development sector. Let us now examine the supply-side weaknesses in more detail.

The first problem with the cyclical interpretation of the commercial property market is the belief that the market is *inherently* cyclical. Many commentators, notably Ian Alexander,[34] Richard Barras,[35] Barras and Ferguson,[36] Michael Bateman[37] and Tony Key *et al*[38] have discussed at length the inherent cyclicality of the property market. For example, in their conclusion to the *Understanding the Property Cycle*, Key *et al* rooted their analysis firmly in the relationship between broader economic trends and the property market:

> The property cycle is the compounded result of cyclical influences from the wider economy, which are coupled with cyclical tendencies that are inherent to property markets.[39]

In the foreword to this work, the then President of the RICS, Clive Lewis, made clear that the research team's brief was "to make clear the enduring forces, both external and internal to commercial property that have produced this sequence of cycles – to explain cycles as a recurring phenomenon, not as the product of chance conjunctions of circumstances". In this sense, the researchers fulfilled the brief.

At the macro level, periodic influences on the property market from the wider economy are not questioned here. As economic fortunes wax and wane in the short term, there are bound to be reciprocal variations in the level of demand for accommodation, resulting in growing and shrinking supply. But are cyclical tendencies *inherent to property markets*? There was nothing cyclical about the deregulation of financial services or the introduction of the new Use Classes Order during the 1980s boom. The fact that the development industry responded over-optimistically to these events does not reflect an inherent cyclical tendency, but rather a herd instinct in a free market context. More specifically, there was no reliable feedback mechanism to act as a safety valve on the growing development pipeline.

A second problem area is the dependence on the hypothesis that major property cycles are generated by endogenous forces in the supply industry, the key factor being an inherent production lag within the construction industry, i.e. the time that it takes for construction to catch up with expanding demand, only to find that the business cycle has taken a turn for the worse. A classic statement on this is provided by Barras, who found evidence for cyclical tendencies in the production process itself:

> The supply of new developments exhibits an inherent cyclical trend, created by the delay of three to four years between the start and completion of a development scheme.[40]

While there might be some truth in the essence of this proposition, it is also the case that building techniques and technologies have speeded up the development process and also made it easier to wind down. Fundamentally, as the 1980s boom showed, the industry simply builds too much space. Even if the economy had continued to grow at a very pleasant 3% per annum into the 1990s, there is no way that sufficient jobs would have been created to absorb the space being built.

The third problem with development cycles is that the evidence for commercial property booms, as distinct from wider real estate booms, is based on a relatively short period of time (i.e. the post-war period), and on the office market in central London in particular. Very rarely are data of a more general basis than this presented in the discussion of property cycles. Consequently, the property industry continues to be profoundly influenced by a circuitous argument that has as its foundation little more than a short-term historical experience.

This temporal restriction is troublesome to even some of the more ardent cycle disciples. For example, Key *et al* conceded in the mid-1990s that the "behaviour of the UK property market can be traced back over no more than thirty years"[41] and that the number of recorded cycles is "too few to construct a definitive classification of cyclical patterns".[42] Of the data on which the notion of development cycles is based, the authors concede that "Statistical analysis supports, but cannot confirm, the existence of property cycles".[43]

A fourth problem area for cycle theory is that it is based on evidence of symptoms, not causes: it is an abstraction. Many measures of the cycle are property indicators themselves. Key *et al* demonstrate this bias in analysis:

> Property cycles are recurrent but irregular fluctuations on the rate of all-property total return, which are also apparent in many other indicators of property activity, but with varying leads and lags against the all-property cycle.[44]

The all-property total return is an indicator of market sentiment not inherent to cyclicality: it is inadequate as a descriptor of cyclical behaviour. Yet the property industry continues its circuitous belief that by measuring cycles with its own self-defining measures, it will achieve the goal of predictability.

The final problem with cycle theory is the sheer improbability of it all. The property industry responds to signals from the broader economy and seeks to do what it is set up to do when conditions are perceived to be most favourable. And all this takes place within a 'business as usual' context. As Key *et al* accept, it is "plausible that there is a property demand cycle directly linked to the business cycle . . . longer cycles . . . linked to supply side production lags . . . and still longer 'swings' . . . generated by obsolescence and replacement cycles";[45] and that

> Theoretically, since property is subject to influences from all these factors, the final pattern in property performance would be the net result of several underlying cycles, sometimes overlapping and reinforcing each other, at other points cancelling out.[46]

The same authors also commit the 'related to but distinct from' sin. In their own words: "property cycles are driven not by economic cycles, but through mechanisms which give the property cycle its own dynamics and characteristics",[47] And, "It is [the] lagging effect which underlies property's tendency to occasional major cycles interspersed with minor ones, to some extent independent of the accompanying economic cycles".[48]

Besides all this, why is it that offices are more 'cyclical' than any other form of property development? Perhaps the risk-and-reward structure is such that the amplitude has little to do with inherent cyclicality and more to do with the structure of the supply industry. Geoff Keogh, Tony McGough and Sotiris Tsolacos[49] note the greater volatility of the office market compared to the industrial market. They attribute this to differences in the building, planning and construction periods in the two sectors that are longer in the office sector, thereby delaying the supply response. But again, this is a circuitous argument. The fundamental point is the sheer scale of development, not its timing. Furthermore, it ignores behavioural traits associated with the attractiveness of office development (compared to industrial) due to the much higher returns that are possible.

It does not necessarily follow that a market that is unstable and prone to periodic phases of boom and bust is cyclical. Certainly phases of overbuilding occur, but, as we have seen, the term *cycle* implies a regularity of periodicity and indeed common causation. Yet each of the post-war booms was varied in its length, amplitude and causes. Key and colleagues conceded that "it is essential to . . . accept that cycles in economic activity are not as regular or repetitive as the everyday use of the term might imply", and that the property market "consists of recurrent groupings of upswings and downswings which vary in length, scale and composition".[50] It is surely not mere semantics to suggest that such variation undermines the very notion of a regular and potentially predictable cycle of activity.

Recent analysis has pointed to the role of the global market in property finance. The 1990s cash glut was not a domestic phenomenon, but a global one. Money was pouring in from all corners. Davis and Zhu undertook a study of data across 17 countries to assess the relationship between finance and commercial property. The authors found that particularly strong links of credit to commercial property are found in the countries that experienced crises linked to property losses by banks in 1985–1995; there is evidence that credit expansion boosts the volatility of property prices, and that it is commercial property prices that "cause" credit expansion rather than autonomous credit expansion boosting property prices. They conclude that the results pose

> a challenge for bank managers and regulators to understand the dynamics of the commercial property cycle at a macro level and to detect deviations from fundamentals. Sound credit assessment of individual loans, diversification across projects and adequate capital are necessary but not sufficient to overcome the type of systematic shock that the downturn in the

commercial property price cycle can generate. Credit needs to be appropriately priced and rationed inter alia to allow for high cyclical volatility
which may be persistent over time and marked cross-country correlations
in property markets.[51]

The global integration of financial and real estate markets continues as an ever-
present threat. The Global Financial Crisis of 2008 was fundamentally caused
by irresponsible lending on residential property that borrowers could not afford.

Much of the explanatory research discussed here has been searching for its
Holy Grail in the so-called *inherent* instability of the market. It has done this by
linking longer-term development cycles to shorter-term business cycles, and as
we have seen, this has proved difficult. Furthermore, endogenous causes, such
as so-called production lags, do not account for why the volume of development gets a head of steam in the first place, only why it peaks when the business
cycle has moved on. Similarly, symptoms of over-development such as yields
and returns do little to explain causality, let alone provide evidence of cyclical
behaviour. Perhaps an explanation of development booms, or more specifically,
the spectacular ensuing crashes lies elsewhere.

According to analyst Ian Cundell, the notion of property cycles rests on two
separate, yet related, propositions. The first, more reasonable notion, is what
might be called the 'gravity proposition': what goes up must come down.
A boom will, with terrible inevitability, end in a bust. The bigger the boom,
the more spectacular the bust. The second is more contentious; what might
be called the 'anti-gravity proposition': what comes down will eventually go
back up. Cundell roots this proposition in Say's Law, which states that the price
mechanism ensures that demand and supply will try to even out, commonly
stated as 'supply creates its demand'.

> Claims that we are at the 'bottom of the cycle' are the vox-pop of this idea.
> We are at the bottom, ergo, the only way is up. The fundamental problem
> here is that if Say's law operated, then prices (rents) would adjust. But
> institutional inflation in the form of upward only rent reviews prevents this
> from happening in the office market.

But Say's Law has long since been dispensed with by economists, says Cundell.

> A ball will bounce in defiance of gravity – what stock market analysts call
> a 'dead cat bounce' – but the bounces get smaller and smaller. Eventually
> gravity asserts itself. Unless, that is, somebody or something gives the ball
> a kick.

This kick could be in the form of regulatory change, novel technology or a
novel virus. But they could also be gradual, like the loss of institutional memory causing past lessons to be forgotten. Notions of cycles which do not explic-
itly recognise this are as flawed by design as is Say's Law.[52]

The contention here is that the commercial property cycles of the post-war era were a factor of financing rather than building lags, output levels or technology. Whether this was ever the case is another matter. But, in terms of learning the lessons of the past, one of the greatest lessons is to control the zeitgeist.

The role of zeitgeist

In a commentary following the crash of 1990, Marsh felt comfortable in stating that many commentators on the London office market:

> still put their analyses firmly in the context of historic supply and demand cycles. Thus, a return to normality is simply a matter of time, before the merry-go-round of rental growth and a construction boom starts again.[53]

There was a general acceptance within the industry itself that 'the market' is cyclical. And, in as much as practitioners absorb the output of academics, their 'gut feeling' is supported by apparently credible research. The key issue here is that behaviour – development, funding, advice – are all profoundly influenced by an argument that is circular and that, in turn, is based on evidence that is not at all clear.

Following the growing institutional involvement in commercial property, a growing focus on the availability of finance as a principal cause of development booms began to emerge. For example, Barras recognised that favourable financial conditions fuelled two speculative property booms, one in the early 1970s and another in the late 1980s and early 1990s.[54] Baum was more specific by pinpointing the rapidly growing property portfolios under the control of financial institutions.[55]

Nevertheless, in the wreckage of the 1990s crash – generally considered worse than its 1970s cousin – the RICS could only re-tread old explanatory factors rather than move the debate on. Its analysis of the most recent boom-bust phase stated that "the cycle is a feature embedded in the operation of property markets, not the result of chance events or a single, simple link to one element of the economy", and it concluded that "the causes of cycles can be divided into those external to the property industry, and those that are a product of the way in which the property industry itself operates". In terms of the externalities, the report allocated 'blame' between occupiers (demand) and the investment markets. To be clear, the RICS was arguing that growth in the economy is a primary cause of development booms and that bond yields and inflation trigger activity in the investment markets. The RICS pinned its endogenous explanation firmly on "the development lag and inflexibility of the built stock".[56]

So, the 'official' post-mortem following the largest development boom ever experienced placed the blame on occupiers and construction (with a little assistance from investment markets)! It is difficult to find anywhere in the RICS report any reference to the glut of cheap cash that was pouring into the sector. The report fills many pages with analysis of rent and yield data, but these are mainly *outcomes*, not causes.

There is virtually no recognition in the RICS paper, or pretty much any other paper, of the role of zeitgeist. What is rarely discussed in property cycle theory is the raw emotion of property development: the excitement, the hubris, the avarice, the passion, the zeal for the deal. Whatever your chosen term, property development is without question a human endeavour that entails emotions that are rarely aroused, for example, in the production of widgets. When these emotions are fed with cheap money from one-step-removed lenders, the basic ingredients for a crash-and-burn scenario are firmly in place.

Perhaps the closest equivalent is the corporate deal. The plan, the hunt, the chase, the courtship and the consummation all drive towards closing the deal. Yet, many deals destroy shareholder value rather than enhance it because they are driven not by logic but by fear or vanity, overvaluations and mis-aligned motives, resulting in cultural conflicts, management distraction and loss of intellectual capital.

The ludic fallacy

The property industry emerged from the war very differently from its pre-war guise. First, aided by the need to redevelop large tracts of war-ravaged land and buildings, the era of the modern speculator had arrived (see Chapter 3). Developer entrepreneurs whose sole *raison d'être* was speculative offices proliferated. Secondly, they were shortly to be joined by pension funds and other institutions looking for secure, long-term investment opportunities. From the late 1980s onwards, these largely domestic funders were joined, indeed, swamped by global finance. Thirdly, the surveyor community began dabbling directly in development as well as conducting business-as-usual brokerage earning fees based on a percentage of transactional values from both developers and funders. Conflicts of interest abounded, but the motivation was clear. Given these three agents of change, who needs a cycle theory? Human frailty and the curious ability to make the same mistake time and again jump to the fore. The explanation, it is suggested, is not one to be found in complex equations of yields and returns, but rather in behaviour, or psychology.

Analysis of property cycles – particularly of commercial property – in the post-war period has suffered from the ludic fallacy. A term coined by Nicolas Taleb in *The Black Swan*,[57] a ludic fallacy is the assumption that statistical models apply to situations where they clearly do not; alternatively put, the misapplication of models in real-life situations. This can result in an overconfidence in modelling and a lack of awareness of the complex context in which it sits, with exogenous factors far too numerous and subtle to predict. Taleb distinguishes risks (which you can compute) and uncertainty (which you cannot). Cycle theory calculates risk but cannot account for uncertainty (for example, the removal of building controls in 1954 or Big Bang in 1986).

To illustrate this notion, we can reference rational expectations. This is a modern twist on classical economics. First discussed by John F Muth of Indiana University[58] in the early 1960s but popularised by economists such as Thomas Sargent and Robert Lucas in the classical revolution of the 1970s, the rational expectations hypothesis describes situations in which the outcome depends in part upon what is expected to happen. In other words, it is a utility optimisation model: agents behave in a manner which maximises their returns. In turn, the theory rests on the notion that perfect knowledge will lead to the avoidance of problems. In the context of property cycles, the theory leads to the construct that if agents (say, developers) have perfect foresight (i.e. knowledge of all potentially influencing factors), then they can avoid a precipitous market crash by bailing out in a timely fashion. But, of course, the classical notion of perfect foresight is an illogical construct.

The fundamental point here is that property has become a global investment medium. The influence of cash in the system is not a new one, as we have seen. Indeed, since the 1950s, there has not been a single commercial development boom that has not had, at its heart, easy credit. And the 1980s boom illustrated the critical role of global money. According to Chen and Mills, "global real estate investment has become an increasingly important component of efficient, global mixed-asset portfolios".[59] And researchers including Case et al,[60] Jackson et al[61] and Stevenson et al[62] have identified a high level of synchronicity in real estate cycles across international markets.

As we saw in Chapter 2, after a 'slow burn' growth in the hundred years up to the Second World War, the second half of the twentieth century witnessed a phenomenal growth in the office economy: an almost relentless growth of white-collar work reflecting fundamental shifts in the national (and international) economy, in which white-collar office work came to dominate. This growth in the office work (and the service sector generally) represented a fundamental shift in the national economy that required a novel solution to its accommodation needs: it needed the new accommodation quickly, efficiently and in significant bulk. The property industry responded with large-scale, speculative development.

Despite discontinuities during recessions, the overall pattern of growth of the post-war office economy was one of relentless expansion, and this economic restructuring actually needed a speculative development process to keep pace with the demand for purpose-built office space to replace old and obsolete stock.

Seen thus, the post-war commercial development booms were a part of a longer-term process of economic restructuring. As part of this process of change, the property industry expanded and provided the accommodation necessary to house the growing office economy. But the key question remains: do the post-war property booms reflect an inevitable, 'cyclical' process of supply, or have supply-side weaknesses played a central role in market instability? We have to look to the industry itself for at least a very significant part of the explanation.

It was the emergent *structure* of the commercial property industry, rather than its inherent or endogenous supply processes that made the pattern of post-war speculative development booms and busts not only possible, but almost inevitable. In particular, the periodic gorging on cheap cash made available by willing lenders, increasingly on a global stage, led developers and their advisors to do what they know best: create buildings! This claim clearly shifts the emphasis away from exogenous factors (the inevitable) towards factors based more squarely on the industry itself (avoidable). And the consistent problem here has been the pattern of funding for speculative development.

6.3 Building a new business geography

In Chapter 2 we saw how, during the late 1980s, London became a pre-eminent global city, alongside New York, Singapore and Tokyo. It was Europe's finance centre and global city. It boasted a prized time zone advantage, allowing trading with the Far East in the morning and the US East Coast in the afternoon. It had many pre-existing advantages that allowed it to gain early mover advantage over Paris, Frankfurt and Geneva. But to cement its global role, it also had to undergo a physical change to match the scale of change in its economy and institutions. In short, it needed to re-house its office economy. This meant more than technologically advanced buildings (more of which in Chapter 9); it meant a radical re-organisation of central London's established business geography.

What ensued was one of the periods of intense building activity referred to in the previous chapter. The problem, simply put, was one of capacity: the City of London was physically not big enough. Unlike New York, Singapore and Tokyo with their largely twentieth-century street patterns, the physical City remained strongly influenced by its medieval layout and would have struggled to accommodate the rapidly evolving office economy. The layout, combined with highly complex land ownerships, meant that assembling large parcels of land for comprehensive redevelopment could take decades. We saw in Chapter 5 how the Luftwaffe assisted in this respect, but by the 1980s, a new solution was required.

In the three decades following Big Bang, London's physical office economy grew rapidly. Figure 6.4 shows the 1980s development boom in the context of the following period. The graph shows that, in the five-year period 1985–1989, some 47.7 million sq ft (c4.4 million sq m) of new construction began, in one of the largest development booms ever. But in the period 1998–2002, another 45.9 million sq ft (c4.3 million sq m) was commenced. This was the 'dot.com' boom, when there was a frenzy of activity to deal with expectations over the internet economy. But there were two further 'mini-booms': the period 2005–2008 (in the run-up to the Global Financial Crisis) saw 24 million sq ft (c2.2 million sq m) of developments start; while the period 2015–2018 witnessed another 25.5 million sq ft (c2.4 million sq m) of offices started.

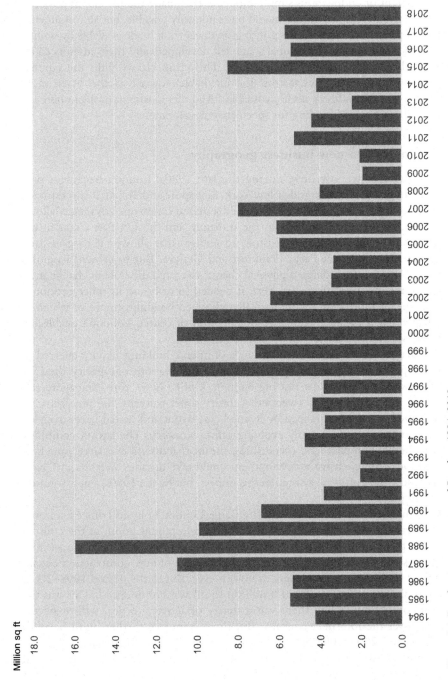

Figure 6.4 Development starts, central London, 1984–2018

Development starts do not provide a clear picture of growth, because they do not discount the space that is being replaced. To get a clearer picture of overall growth in stock, official figures from the Valuation Office can be used. These show that central London stock of office space grew from 145.4 million sq ft (c13.5 million sq m) in 1986, to 203.0 million sq ft (c18.9 million sq m) in the year 2000, and to 233.3 million sq ft (c21.7 million sq m) in 2019.[63] That represents a growth of around 60%, or almost 88 million sq ft (c8.2 million sq m). This breaks down to an annual average growth of about 3 million square feet (c280,000 sq m).

By London's historic standards, these are big numbers. They encapsulate the maturation of the office economy leading up to and beyond the turn of the twenty-first century. They reflect the enormous growth that was spurred by Big Bang in the 1980s and 1990s, and they reflect the rapid impact of the 'digital revolution' in the 2000s.

All of this growth could not possibly have taken place within the confines of London's traditional property markets. Something had to give. New capacity had to be found. And, as throughout its long history, the city responded.

A polycentric office economy

Up to the latter part of the 1980s, central London's business geography was really quite simple, with four tightly defined sub-markets. First, of course, was the City, the 'Square Mile', with its four 'EC' (East Central) postal districts, accommodating its financial services and associated businesses. Secondly, there was the West End, focused on Mayfair W1, where there were corporate head offices, advertising and media businesses, the property industry and professional services. Thirdly, there was WC1 and WC2, or Midtown as we know the area today. This was the beating heart of the legal profession, with the Inns of Court and the Royal Courts of Justice, and little else. Finally, there was Victoria, home to government and its army of civil servants, plus some remnants of the engineering industry and the petro-chem industry. That was it; all tightly defined. Outer London hosted a collection of back offices of mainly City firms but in practice was not considered a major office market, with the exception of the 'golden mile' along the A4 through Hammersmith and Hounslow that was home to the electronics industry and the first wave of computer industry. Most large property agents had both a West End and a City office; smaller firms tended to specialise in just one market.

Even within these functional clusters there were 'sub-markets'. For example, the City of London contained a number of concentrations of distinctive business activities, including banking, commodities, insurance, press, publishing, shipping and stock broking and jobbing. Such clusters were often associated with the historic centres of markets such as the Bank of England, the Stock Exchange, Lloyd's insurance market or the Commodity Exchange. To this day, the insurance sector remains tightly packed into the EC3 postal district in the south-east corner of the City.

In the early 1980s, there also remained vestiges of London's industrial past, including clusters of activity such as fashion (Bethnal Green and north of Oxford Street), furniture (Shoreditch), jewellery (Hatton Garden) and printing (Fleet Street). Of course, medicine was clustered around Harley Street and Wimpole Street. As is evident with the historic nature of some of these clusters, they are not immutable; they come and go with the economy and with fashion. One of the most dramatic phases of change in recent times was the mass exodus of the press industry from the street that was once synonymous with its name – Fleet Street. Another example was the furniture industry in Shoreditch that suffered the combined impact of a City creeping northwards and structural changes in its market.

All of this was fine, but it would not accommodate an emergent global city with the international investment houses being assembled in the run-up to Big Bang. Something more radical was required. What was needed was an entirely new approach to building design and construction, and a fresh vision of what, in spatial terms, constituted the central London office market.

Informed by the emerging requirements of global investment banks and professional services firms, developers in London began to think big. The logical solution to this problem was to look 'off pitch', and the first mover in this respect was Rosehaugh Stanhope, with its redevelopment of Liverpool Street station, later to become Broadgate. The key was to utilise redundant and operational railway lands: suddenly, albeit recognising major engineering complexities, unified parcels of land could be assembled to create the large footprint buildings required by the global office economy.

The result was a series of 'mega schemes' forming a necklace of integrated and masterplanned office campuses around the fringe of central London, focused primarily, but not exclusively, on major rail and underground hubs and including a large element of public realm, and a high retail and leisure content. The mega schemes were first named as such and examined for their combined impact on the London office market in the *London Office Policy Review* of 2009.[64] Figure 6.5 shows these 'first-generation' mega schemes.

The Square Mile spread north and east, exemplified by Broadgate around Liverpool Street Station. The South Bank began its remarkable revival with London Bridge City, which then led to More London in the 1990s (and, most recently and dramatically, to the Shard). Canary Wharf rose from the wreckage of London's Docklands, balanced in the west by Chiswick Park. The area around Paddington Station began to emerge as a distinctive office location, as did Regent's Place on Euston Road.

Between the late 1980s and 2008, the mega schemes delivered around 24 million sq ft (c2.2 million sq m) of new office space – enough to accommodate around 170,000 workers:

- London Bridge City: 800,000 sq ft (c74,000 sq m);
- Broadgate: 4,500,000 sq ft (c420,000 sq m);
- Canary Wharf: 14,000,000 sq ft (c1,300,000 sq m);
- Paddington: 1,600,000 sq ft (149,000 sq m);

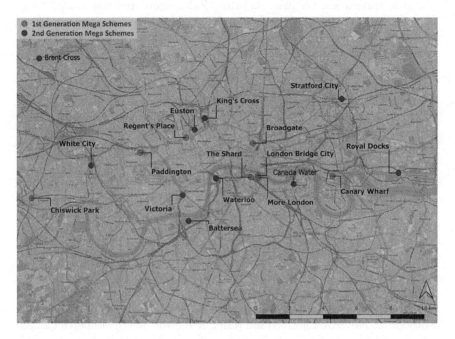

Figure 6.5 Mega schemes in the Central London office economy, 1988–2020

Source: © Ramidus Consulting Ltd.

- Chiswick Park: 1,100,000 sq ft (102,000 sq m), and
- More London: 1,900,000 sq ft (176,000 sq m).
- **Total: 23.9 million sq ft (c2,220,00 sqm)**

To put the nodes into a physical context, as we saw earlier, official data show that the stock of offices in Central London in 1986 amounted to around 140 million sq ft (c13 million sq m). The schemes provided, on average, 1.0 million sq ft (c93,000 sq m) per annum. The long-term contribution of the six schemes equated to around 18% of the total Central London annual development completions of around five million sq ft (c500,000 sq m) over this time.

But apart from adding greatly to the stock of London's office space, the mega schemes began to alter the distribution of business activity in London. Most notably, Canary Wharf attracted large swathes of the investment banking sector, including Bank of America, Barclays, Citibank, CSFB, HSBC, JP Morgan and Morgan Stanley. Indeed, it was reported in 2012 that Canary Wharf had overtaken the City in terms of the number of people employed in banking. The *Financial Times* revealed that, following the imminent arrival of JP Morgan, Canary Wharf would house 44,500 bankers compared to the City's 43,300.[65]

More London on the South Bank became a business services cluster, with large lettings to accountants Ernst & Young, Lawrence Graham, Norton Rose and PricewaterhouseCoopers. Similarly, Paddington attracted a large number of corporate occupiers, including Kingfisher, Marks & Spencer, Misys, Nokia, Orange, Prudential, Statoil, Visa and Vodafone; while Regent's Place attracted ATOS Origin, Gazprom, Lend Lease and Santander. Then came the Shard.

Since the Global Financial Crisis, 'second generation' mega schemes have been developed (see also Figure 6.5). Perhaps the largest and most complete of these is at King's Cross. With over 4 million sq ft (c372,000 sq m) of office space, King's Cross has attracted a range of international businesses including Google, Astra Zeneca, Havas PRS for Music, Universal Music and YouTube. The potential cluster impact of Google is significant, with its ability to attract other TMT companies. Battersea, a completely untested location in office market terms when it was first mooted, has attracted both the new US Embassy (from Grosvenor Square) and Apple Corporation for its new 500,000 sq ft (c46,000 sq m) European headquarters. Around 16,000 new homes are planned as well as an extension to the Northern Line.

White City is located off-centre, and the overall project is focused on the refurbished BBC Television Centre (leased to The White Company, Publicis and BBC Worldwide); White City Place (BBC, Cutting Edge, Gamma Delta, Huckletree, Novartis, OneWeb and Ralph & Russo), and the Imperial College campus. The last of these is an extensive scheme, the first phase of which was given planning permission in 2019, and will involve around two million square feet of scientific research facilities.

Victoria is not strictly a mega scheme, including as it does a number of discrete projects. However, it merits inclusion here in recognition of the scale of redevelopment undertaken by Landsec (formerly Land Securities) and the quality and scale of the buildings and public realm this has created. Cardinal Place (560,000 sq ft, c52,000 sq m) set a tone for the area's renaissance with innovative architecture, followed by 62 Buckingham Gate, the Zig Zag building (almost 200,000 sq ft, c18,600 sq m) and Nova (600,000 sq ft, c55,700 sq m).

Waterloo is also not a mega scheme, but the redevelopment of the Shell Centre (complete) and Elizabeth House above the station (under construction) will create an office location of nearly 3 million sq ft (c279,000 sq m). In east London, further large office centres are emerging at Stratford and the Royal Docks; while Canary Wharf is developing the neighbouring Wood Wharf. Waiting in the wings is British Land's 3 million sq ft (c279,000 sq m) mega scheme at Euston.

The mega schemes established the pattern of "off-centre" developments. When Broadgate was mooted in 1983, half a mile from the Bank of England and tipping into the neighbouring Islington, it was revolutionary. Today, building in Waterloo, King's Cross or Paddington is considered mainstream. The original seven mega schemes, all with their roots in the 1980s property boom, signalled that London's old locational ties had loosened dramatically. And while they continue to evolve (Canary Wharf, for example, has further development capacity), they have encouraged a string of other large-scale developments across central London.

The mega schemes thus diverted existing activity away from established market areas as well as attracted new growth that might otherwise have located elsewhere. More London is an exemplar: the large number of professional services firms now based there would traditionally have located more centrally. The developers have achieved this in large measure by providing suitable premises, designed and built to modern, exacting standards, in campus style, with public realm and traffic-free environments, with plentiful support activity in the form of retail and leisure activity.

The mega schemes also set new standards in building design. The buildings designed and built in the 1990s and 2000s were far superior to anything built in the 1960s and 1970s. Figure 6.6 shows two exceptional additions to the City of London: 122 Leadenhall Street ('Cheesegrater') and 30 St Mary Axe ('Gherkin'). The buildings exude quality, confidence and the boldness of a global city.

The experience of the mega schemes also suggests that infrastructure can be a crucial ingredient of "critical mass". The success of Canary Wharf can be traced to the construction of the Jubilee Line Extension, while Paddington received a huge boost when Heathrow Express opened in 1998. Similarly, while the idea of redeveloping the railway lands at King's Cross had been around for at least 20 years, it was the huge investment associated with Eurotunnel that gave the project credibility. However, such connectivity is not essential, as demonstrated by More London and Regent's Place.

As a result of mega schemes, and much more development besides, the business geography of central London has changed radically. The old model of City, Midtown, Victoria and West End has yielded to a far more complex tapestry of sub-markets (Figure 6.7). In the 'old model, the City was tightly and strictly defined by its EC postal districts, except the area to the west of Ludgate Circus, which was not considered to be 'real' City. The West End was further subdivided in sub-markets, including Victoria, Mayfair and St James's, and Bloomsbury. Midtown was not really a string corporate market but was home to the legal profession, which inhabited a wide range of old buildings.

In the polycentric model (Figure 6.8), the 'northern fringe' stretches from Regent's Place in the west, through Euston, to King's Cross (including London's nascent life sciences cluster). Victoria has been transformed from an enclave of government to a diverse and vibrant office market. The South Bank is unrecognisable from the pre-Big Bang era, with high-quality offices including campus environments from Battersea through Waterloo, to London Bridge

Figure 6.6 Digital Office buildings: the 'Cheesegrater' and the 'Gherkin'
Source: © Erik Brown, 2020.

with London Bridge City, the Shard and More London. Canary Wharf and its environs form a whole new London quarter.

The City has spread out and continues to push eastwards through the eastern fringe. The West End is the only market not to have been transformed. It has extended northwards to Paddington, and the area to the east of Oxford Street, centred on Mid City Place at the junction of Tottenham Court Road and Oxford Street, has become an identifiable office node. But the core area of St James's and Mayfair remains pretty much as it was.

Occupiers have become far more footloose: non-financial businesses have been moving into the City; hedge funds have clustered in the West End, and tech companies have located all over the sub-markets. There is also evidence to suggest that firms that previously might have located in Outer London or even further afield in the Thames Valley have chosen instead to concentrate in the centre, where quality buildings, good-quality public realm and highly skilled staff are most readily available.

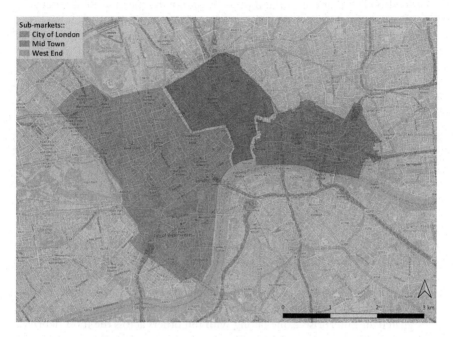

Figure 6.7 The 'old' central London office economy

Source: © Ramidus Consulting Ltd.

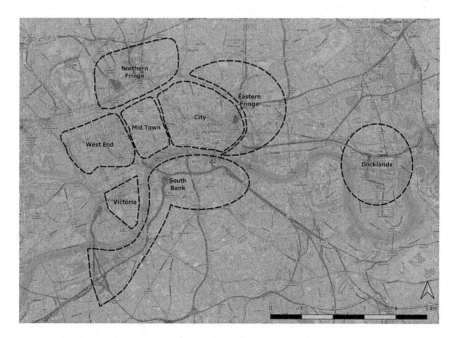

Figure 6.8 The 'new' polycentric central London office economy

Source: © Ramidus Consulting Ltd.

It is also clear that the London office economy has been satisfying demand from two very different demand profiles: large, corporate occupiers and smaller businesses looking for quality space and a service offering. Each of these is illustrated here, the first by reference to the evolution of Canary Wharf, and the second by reference to the growth of the tech sector around Shoreditch.

Manhattan-on-Thames

The fortunes of London's modern docks were relatively short-lived, from modest beginnings in the early nineteenth century to a peak 50–60 years later. By the 1980s, they had been in terminal decline for decades. East India dock closed in 1967, and the whole area fell into a vicious spiral of economic and social decay; the final docks to close, the Royal Docks, ceased operations in 1981.

To tackle the resulting unemployment and economic shrinkage, the Thatcher Government established the London Docklands Development Corporation (LDDC) in July 1981. The LDDC was charged with the regeneration of Docklands, an area of some eight square miles – the largest regeneration project in Europe. As part of a package of proposals, part of Docklands was designated an Enterprise Zone (along with ten others across the most deprived parts of the UK), opening in 1982. Enterprise Zones enjoyed a range of benefits, including: exemption from Development Land Tax; exemption from rates; corporation and income tax allowances for capital expenditure on buildings, and relaxation of planning restrictions.

The Isle of Dogs presented a huge challenge in urban regeneration terms. Three miles east of the City, in abandoned docks and with negligible public transport and hopelessly inadequate road links, the area was a wasteland. It was difficult to imagine at the time what could possibly be achieved, but slowly new buildings were created, and new businesses began to appear. But it was a fateful visit by an American banker, on a tour of the area in search of a site for a food processing plant in 1984, whose single-minded vision and boldness was set to change the fortunes of the area forever.

Michael von Clemm, then Chairman and Chief Executive of Credit Suisse First Boston, was making his visit at a time when he had also been struggling with City planners to resolve CSFB's urgent need for large, open trading floors to cope with the firms' activity in the Eurodollar market. In Docklands, he was shown an abandoned warehouse that had received bananas from the Canary Islands, and was known as Canary Wharf. The building he was shown included two 40,000 sq ft (3,700 sq m) open floors, and his idea was to fill them with financial traders. Immediately he contacted Morgan Stanley's chief executive Archie Cox to invite him to view the site and to join his plans to move there. It was at about this time that the then Governor of the Bank of England, Robin Leigh Pemberton, relaxed the controls over where banks could locate,

Figure 6.9 Canary Wharf
Source: © Erik Brown, 2020.

removing the requirement to be close to the Bank of England. Von Clemm then called developer G Ware Travelstead.

Despite reservations over poor transport and isolation, Travelstead accepted the challenge and joined the project. Under his guidance the project was transformed into a plan for a 10 million sq ft (c1.0 million sq m), 72-acre (29 ha) development including twin towers. He launched a major PR campaign, the like of which had rarely been seen in London, and the press reacted with predictable incaution, with many articles following the 'Manhattan-on-Thames' theme.

The one thing Travelstead did not have was committed financial backing from his anchor tenants – only intent. As the months dragged by, he faced increasing pressure to 'come up with the goods'. Negotiations dragged on and on, and discussions were held with many other financial institutions and developers, including Canadian developers Olympia & York (O&Y). Then, in mid-1987, CSFB and Morgan Stanley withdrew as financial backers to the project.

As the project floundered, O&Y swooped. Within a month of Travelstead's consortium falling apart, the project had new owners. Toronto-based O&Y was run by three brothers, Paul, Ralph and Albert Reichmann, and had immense financial stature. At the time of its takeover of Canary Wharf, it owned huge tracts of Manhattan and was said to have around $12 billion in assets. The deal gave O&Y virtual ownership: First Boston International retained a residual stake in the project, and CSFB and Morgan Stanley were committed as anchor tenants. Within three months O&Y unveiled their revised plans for the project, and the development programme was gearing up for major activity. The project was set fair, and construction began in earnest.

Figure 6.10 shows how the first crop of buildings came to the market in 1991 – the depths of the recession and post-boom crash.

But, of course, no matter how fast O&Y could build, and no matter how hard they sought to attract tenants, they were not immune from the travails of the wider London office market. As we saw earlier, tremors began to shake the London office market in 1989, and a year later it imploded. Just as new supply reached maximum flow, and tenants disappeared, Canary Wharf phase one reached completion.

The efforts to attract tenants were redoubled and rumours were abounding of huge incentives to attract occupiers to the project, including two-year rent-free periods, a long fixed rent, generous fit-out allowances and the assumption of responsibility by O&Y for premises left behind with obligations. Suddenly, the audacity of one the world's strongest developers began to many to look like outright foolishness. Inevitably, market talk was of how long Canary could last.

Then, like an earthquake following tremors, O&Y filed for insolvency protection in Canada and the USA in May 1992. Debts were estimated at $3 billion (£1.6 billion). The sense of shock within the property market was palpable. The developers fought a rear-guard action and were reported as saying that Canary Wharf would be unaffected by the move and would continue on schedule.[66] But the glue was undoing. Two weeks later the project itself was placed in administration, with debts of £576 million. Under the headline "Banks abandon Canary Wharf", the *Financial Times* reported "on Wednesday the 11 banks [the leading creditors] told O&Y that an insolvency filing was the only option".[67] The shock was palpable. Olympia & York had seemed invincible, and the stakes at Canary Wharf too great to fail. A torrid time followed with endless speculation over the future of the project and its chief sponsor. Various international personalities were linked to the project, including Sheik Maktoum al Maktoum of Dubai and Li Ka-shing of Hong Kong, reflecting the increasingly international nature of the UK office market.

Eventually, the 11 banks that had financed the O&Y project brought Canary Wharf out of administration the following year, 1993. After a slow start, the project saw visible signs of activity. Major lettings were achieved with the Personal Investment Authority and the European Medicines Agency. Morgan Stanley increased its commitment to the project with a further 500,000 sq ft (46,500 sq m) letting in 1994, and in spring 1995, BZW, the investment-banking subsidiary of Barclays Bank, committed to a similarly sized letting. In April, confidence in the project was sufficiently high to allow construction to restart.

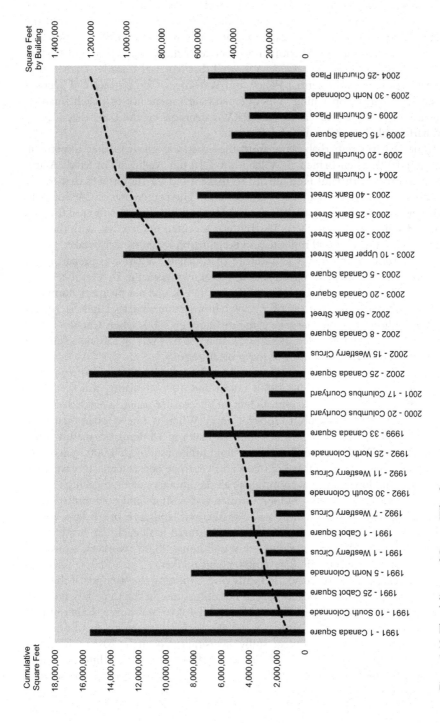

Figure 6.10 The delivery of Canary Wharf

In fact, such was the renewed success of the project that the Corporation of London once more began agitating about companies moving away from the City. BZW's decision to relocate was too much for some in the Corporation to bear. Referring to a "furious row" between the Corporation and Canary Wharf, the *Independent on Sunday* reported that Michael Cassidy (chairman of the Corporation's influential Policy and Resources Committee) had accused Canary Wharf of breaching "a gentlemen's agreement not to poach financial institutions".[68] He had "accepted" BZW's decision on the basis that Canary Wharf would not seek further defections.

The Wharf's remarkable rise from the ashes was given further impetus in August 1995 when Prince al-Waleed bin Talal bin Abdulaziz of Saudi Arabia joined forces with Paul Reichmann to make a bid for the project that only a year earlier had been valued at £500 million. The prince owned 24% of Euro Disney and was the largest single shareholder in Citicorp (a major creditor and, later, occupier of the project). The bid was accepted in October, with the bidders paying nearly £800 million, and Reichmann retaining a 5% stake.

In October 1998, it was announced that Canary Wharf was preparing for flotation, and these plans were realised the following spring. By October 2000, Canary Wharf overtook Land Securities as the UK's largest property company, with a market capitalisation of £3.94 billion, compared to Land Securities' £3.91 billion. The *coup de grace* came in October 2000, when Canary Wharf entered the FTSE100 index. By 2003, Canary Wharf was up for sale again, and a protracted battle for control of the project began.

Thus it was that Canary Wharf came full circle. Its 20-year history was a complete microcosm of the 1980s boom and 1990s collapse, followed by a measured recovery. It was reminiscent of the earlier *grands projects* that were perfected by Thomas Cubitt. Today, Canary Wharf is a new urban quarter in its own right. It is a large, 86-acre (35 ha) estate of 18 dramatic, mainly steel-and-glass, tall buildings, totalling some eight million sq ft (c750,000 sq m), with a further four million sq ft (c372,000 sq m) yet to be constructed. Its cavernous interiors and lower levels accommodate all the frontages of a typical high street: shops, travel agents, banks, numerous cafes, coffee shops and restaurants; fitness clubs. All of this is set within a size manicured landscape of gardens, sculptures and water features. And through and around it all each day bustle some 55,000 workers. When it is complete it will house 90,000 workers, equivalent to around one-third of the City employment market.

While the project was criticised early on for diluting the strength of the City of London, it has quite possibly saved London as one of the three key financial centres in the world, alongside New York and Tokyo. It is home today to many of the leading investment banks, several of which occupy in excess of one million square feet. And it has successfully diversified its occupier base with a range of other companies relocating there since 1991, including: Clifford Chance (law); BP and Texaco (oil); Ogilvy & Mather (advertising); Financial Services Authority and European Medicines Evaluation Agency (government); Reuters (media); and *Independent on Sunday*, McGraw Hill Group, *Reader's Digest* and the Telegraph Group (publishing). And there are numerous other smaller occupiers in the multi-let 1 Canada Square – the iconoclastic tower – and elsewhere on the estate.

Canary Wharf has become a spectacular success in terms of responding to the changing needs of the office economy with real estate that met a need in terms of building specification. It was also perhaps the zenith of speculative development in the UK. What it did not do was to address the underlying concerns of occupiers in terms of offering different terms. It replicated what was available throughout the market, and as the 1990s recession took hold, the nature of the standard lease once more came to the fore.

Accommodating the flat white economy

During the 1980s, when computer technology was evolving rapidly and the era of the personal computer arrived, the locus of the UK's technology sector was the Thames Valley, an area that stretches westwards from Hammersmith in London to Reading and beyond in Berkshire. The market (which has over the years been variously described as 'the M4 Corridor', 'the High-tech Corridor', the 'Heathrow Corridor' and the 'Golden Triangle') was first referred to as a discrete entity in the early 1980s, when its attraction to international technology-based companies became evident.

However, the genesis of the market stretches back much further, to the location of government defence industry research establishments in the area after the war. Atomic energy research at Aldermaston and Harwell, aircraft research at Farnborough and transport research at Bracknell are all examples. As they grew, they attracted small and specialised sub-contracting firms in, for example, engineering and electronics; which in turn drew more and more highly skilled workers to the region. The newly named London Airport was officially opened for commercial operations in May 1946, but for much of the post-war period, the high-tech nature of the local occupier base was largely incidental to the airport itself. It was not until much later that proximity of commerce and airport came to be crucial.

During the 1960s and 1970s, entrepreneurially minded staff within the large corporate structures began to realise the potential for the commercial application of their work, often setting up spin-off businesses to produce electronic hardware and computer software. Growing connectivity helped: the eastern sections of the M4 were opened between 1961 and 1971, most of the M3 in 1971, the M40 to Oxford in 1974 and the M25 in 1986. Indeed, it was during the 1970s that the Thames Valley property market began to take on its modern form, as a new causal mechanism of growth kicked in.

> The presence of specialized labour and skills in the research establishments and defence industries when blended with the skills and resources of the incoming multinationals produced a large enough pool of interacting skills and demands for a genuine fusion to take place. At some stage, therefore,

perhaps in the mid–1970s, sufficient agglomeration had occurred for internal and indigenous growth processes to take off.[69]

In other words, the Thames Valley had gained the critical mass to generate self-
sustaining demand. By the early 1980s, this indigenous growth was being sup-
plemented increasingly by foreign multinationals, many from the USA, some
from Japan, including well-known brands of the time such as 3M, Dell, Digital,
Fujitsu, Hewlett Packard, Honeywell, IBM, ICL, Racal, Rank Xerox, Thorn
EMI, Toshiba and Wang, who found the area around Heathrow a convenient
one from which to serve UK and European markets. By the mid-1980s, it had
become a 'high-tech corridor', to mimic Silicon Valley.

Such firms were at the centre of the technological advances of the time.
In 1981, IBM launched its PC having built its reputation supplying large
mainframe computers to corporations across the globe. This move acted as
the catalyst for a revolution in software, which spawned countless new firms
developing products for personal and business applications. IBM clones such
as Compaq together with Intel processors and hard disks ensured that prices
tumbled and PCs were ubiquitous by the end of the 1980s. In 1984 Apple
released its revolutionary Macintosh computer with its Graphical User Interface. The screen of the 'Mac' was clear, mimicking paper with its black characters on a white background, and came with the iconic 'mouse' for clicking
and dragging. The CD arrived in 1983, revolutionising how we listened to
music and stored information. The Sony Walkman, camcorders, satellite TV
and video game consoles matured and evolved, changing the ways in which
people engaged with entertainment. The fax machine arrived as did the humble telephone answering machine.

Rapidly evolving technologies, changing social habits, new retail and leisure
formats churn in the nature of work, new requirements on buildings. And it all
took place in parallel with the financial and professional services revolution in
the City following Big Bang. Indeed, many innovations overlapped the business and the personal environments. The PC is the classic example, but many
of the communications technologies co-evolved in both markets.

The development of the office economy has been dominated by the inexorable advances of digital technologies since the 1980s and 1990s, in both business and consumer markets. These advances reflected the underlying structural
change in the economy from a manufacturing to a services-based economy,
and particularly the consolidation of the office economy.

But as far as the business geography of London is concerned, something else
happened in the late 1990s that caused the established order to change: suddenly, 'tech businesses' were expanding not in the Thames Valley cluster, but
in central London. In the mid-1990s, the internet, mobile phones and laptops
emerged as ubiquitous technologies and, as the millennium approached, they
were joined by smartphones, personal devices and social media to bring about
truly revolutionary change in almost every aspect of life. At this point, internet-
based businesses started to cluster in central London.

In an age when seemingly almost everything is in some sense 'digitised', it can be difficult to comprehend a discrete digital industry. Broadly speaking, there are two generic components. First, there is the technology itself, with systems (like broadband networks), hardware (computers and servers), software and services (such as sales, installation and maintenance). The second component is 'digital content', including broadcast media together with advertising, design, music and publishing, and sometimes referred to as 'Telecoms, Media and Technology' or TMT.

The expansion of digital firms in Inner East London was rapid (Figure 6.11). Between 1997 and 2010, the number of digital firms in London grew from 271,062 to 392,334, up 45%. In the heart of the digital sector's preferred location, Inner East London, employment rose by 275%, from 12,931 to 48,586; while the firm count more than doubled.[70]

Investment in the digital sector has continued to pour into London. For example, total venture capital funding into technology companies across major European cities in 2018 saw £1.8 billion going to London, far ahead of Berlin (£936 million), Paris (£797 million), Stockholm (£224 million) and Barcelona (£183 million). At the national level, the UK's £2.49 billion eclipsed Germany and France (at £1.38 billion and £1.03 billion, respectively), and the distant Switzerland (£531 billion).[71]

By 2016, property advisory firm Colliers was able to report that tech and media occupation in central London had grown by 41%, to reach 5.2 million sq ft (c483,000 sq m), since 2009 (bearing in mind the impact of the Global Financial Crisis after 2008), and that its Tech City heartland had grown by 90%.[72]

| Year | Inner East London | | London |
	Firm count	Employment	Employment
1997	1,591	12,931	271,062
1998	1,802	23,488	286,027
1999	1,980	25,068	297,402
2000	2,096	20,728	265,751
2001	2,203	27,013	306,545
2002	2,207	27,183	322,108
2003	2,600	36,172	384,713
2004	2,539	43,867	406,271
2005	2,499	43,461	381,549
2006	2,680	44,110	381,662
2007	2,786	43,940	371,928
2008	3,246	47,583	385,554
2009	3,288	48,577	408,448
2010	3,289	48,586	392,334

Figure 6.11 Digital firms and jobs in Inner East London, 1997–2009

Source: Nathan *et al* (2012).

By 2020, Amazon, Facebook and Google each had London footprints in excess of one million square feet; Facebook occupied over 700,000 sq ft (c65,000 sq m) and Apple's new 500,000 sq ft (46,500 sq m) European headquarters was emerging in Battersea.

There emerged a patchwork of organically evolving, identifiable clusters of digital activity, including concentrations around Aldgate/Whitechapel, Clerkenwell/Farringdon, Fitzrovia, King's Cross, Covent Garden and South-bank. But the cluster which received by far the greatest attention was around Shoreditch, where Old Street and City Road met at Old Street roundabout – often referred to as Silicon Roundabout. From here, the cluster spreads north-wards to Haggerston and Hoxton; south to the City, Spitalfields and Aldgate; westwards towards Clerkenwell and Farringdon and eastwards into Bethnal Green.

Through to around 2005, there was very little that distinguished Shoreditch: it was a little-known back land of the City. Much of the area was character-ised by older buildings, often industrial and warehousing in character – classic zone in transition material on the edge of the CBD. Many of the buildings had housed furniture makers, jewellers, clothing designers and other manu-facturers in times past. Of course, rents were very cheap, but at the same time,

Figure 6.12 Old Street ('Silicon') Roundabout, Shoreditch

Source: Creative Commons (https://creativecommons.org/licences/by-sa/3.0/deed.en).

the location was fairly central, and the atmosphere was very non-corporate. The industrial buildings offered character to contrast with the corporate office buildings to the south; volume, subdivision into multiple small units, large windows and plenty of natural light, a slight untidiness, interiors stripped back to breeze blocks and bare ceilings with exposed services.

But then, for reasons that remain unclear, it began to attract digital start-ups. Add to the mix galleries, bars, upcycling shops, artisan bread and, *de rigueur*, coffee shops – and a sense of community, more importantly – a 'cool' community emerged for predominantly younger, tech-savvy, entrepreneurially minded workers. People could also afford to live close by and cycle to work. The number of companies grew as more arrived.

It is reported that in March 2008, Matt Biddulph of Dopplr created the moniker Silicon Roundabout in a tweet which ran "Silicon Roundabout: the ever-growing community of fun start-ups in London's Old Street area".[73]

The community was not designed by planners or developers; rather, it emerged organically from the like-minded businesses that clustered there. It is not clear how much collaboration and sharing took place, but feedback from businesses at the time highlighted a perceived inter-connectedness between the businesses. Soon Shoreditch was *the* tech location. It was the location of choice for start-ups, young entrepreneurs and aspiring Apples and Googles. The one thing that seemed to unify the companies flocking to the area was their impossible-to-spell names. Among the early arrivals were: AMEE, Dopplr, Kizoom, Last. fm, LShift, Moo, Poke, Redmonk, Skimbit, Songkick and Trampoline Systems. A mixture of tech start-ups and tech-enabled creative businesses.

In July 2010, TechHub was launched, an accelerator without the venture capital element, providing low-rent, flexible workspace for start-ups and entrepreneurs. Google was a founding partner. In 2013, TechHub entered a partnership with British Telecom and the launch of the BT Infinity Labs programme, and in 2014 it launched its first scale-up centre nearby. TechHub has supported numerous start-ups and scale-ups, with notable alumni including Monoidics (acquired by Facebook), Nexmo (acquired by US-based Vonage), SwiftKey (acquired by Microsoft) and Wercker (acquired by Oracle).

But Shoreditch became a prisoner of its own success. In the first instance, growing demand reduced available space, which in turn pushed up rents, forcing the cluster to begin to leapfrog into neighbouring areas. More mainstream, established businesses began to move in, able to afford higher rents. Cisco, Facebook, Google, McKinsey and Microsoft all invested in the area, as did various universities. In 2011 Google acquired a seven-storey building on Bonhill Street, with the aim of hosting events, activities and new companies. The building, known as Campus London, opened in March 2012. In 2017, internet giant Amazon moved into its new 625,000 sq ft (c58,000 sq m) headquarters in the nearby Principal Place. It was no great surprise that local rents surged to the same as those in the City.

Secondly, developers saw an opportunity. Suddenly, hoardings began to go up as sites were bought either for redevelopment or refurbishment. While some were sensitive to the ethos of the place and sought to replicate the area's intrinsic features, others saw an opportunity to provide 'Grade A'.

Thirdly, the cluster was noticed by government in 2010. Desperate for good news stories in the face of on-going Global Financial Crisis bad news, Prime Minister David Cameron spotted an opportunity in his 'rebalancing of the economy agenda' to shift the focus from 'greedy bankers' towards 'beardy nerds'. Cameron visited Shoreditch and spoke about the importance of the digital economy and his plans to support the cluster, including £15 million in government funding. Being an ex-marketing professional, Cameron immediately saw the potential for a re-brand, and so 'Tech City' was born, and behind it, an organisation to promote the area's strengths.

Tech City was an enormously valuable addition to London's office economy, boosting activity and global reputation. While the iconic Shoreditch continued, so digital activity spread across central London.

6.4 New generations of activities

The 1980s construction boom was the same as those in the 1950s–1960s and the 1970s, discussed in the previous chapter. But it was also different. And there is the rub. From a distance, successive booms and busts give the impression of inevitability – something that happens periodically, due to the economic cycle, catching investors unawares and leading to a necessary correction between supply and demand. But when each boom is examined closely, the causal patterns in each are different, or at least, largely different.

Government policy, the Oil Crisis, inflation, urban planning, construction lags and many other factors have all been put in the witness box. But of course, the one common theme running through all the booms is the availability of cheap cash from willing lenders, increasingly on a global stage, encouraging developers to indulge their desire to create buildings. Whether this makes booms more or less inevitable or avoidable is a moot point.

The cash available to developers during the 1980s boom was particularly cheap and particularly plentiful; and the crash of the early 1990s was one of the most severe on record. There has not been a crash on anything like that scale since. The severe market disruption of 2008–2010 was caused by a global economic problem, not a development boom. And the market disruption in 2020 was caused by a global health problem. Neither resulted in major oversupply. Whatever the causes of the 1980s boom, its effect was to equip London's office economy with a new generation of buildings capable of responding to the radical changes in the economy.

Financial deregulation in the City in 1987 was a catalyst to enormous change in the London office economy. While comparisons are difficult, the scale and fundamental nature of change between 1987 and 2007 was probably greater than that which happened a century previously. Between Big Bang and the Global Financial Crisis, a whole new building form emerged (large, deep plan, highly specified buildings), which could cope with the servicing needs of the technologically driven occupier requirements. Large, global businesses poured into London, both to work directly in the new financial markets and to support those firms with technology and other office needs. In the 1980s and 1990s, large investment houses led the way on letting new space, and Canary Wharf was their iconic home.

In the late 1990s and into the new century, the digital revolution added a second wave to the financial revolution. Suddenly, the workplace was radically altered under the influence of the internet, email, laptops, mobile telephony, social media and so on; the modest fax machine, which had appeared to great applause only in the early 1980s, was redundant. The result of the digital revolution for the office economy was a new raft of occupiers – the 'tech sector', to use one of its many labels. Once again we witnessed a re-ordering of business activities in London. The phenomenon of Shoreditch and Old Street was particularly publicised; but tech businesses appeared everywhere in central London.

The overall message here is about London's ability to re-invent itself as new generations of activities emerge. The changes that have taken place in the office economy over the past three decades have transformed the physical structure of London. The city has evolved through a periodic pulse of change and adapted to a new ea.

Notes

1 County NatWest & Baker Harris Saunders (1987) *The City Office Market: An Examination of Supply, Demand and Debt* CNW/BHS, London, p4
2 Savills (1987) *City of London Office Demand Survey,* Savills, London, p24
3 Winter & Company (1987) *The New Property Crisis: 1987/1988?* Winter & Co, London, p3
4 Jones Lang Wootton (1988) *The City Office Review 1984–1988,* JLW, London, Introduction
5 G. Marsh (1988) *Central London Office Market Forecasts,* Applied Property Research, London, p8
6 Herring Baker Harris (1992) *Behind the Façade,* HBH, London, p17
7 J. Bevan (1988) Is City Lost in Space? *Sunday Times,* 21 August, pD3
8 Phillips & Drew (1989) Cited in: *Estates Times,* 27 October, p56
9 Morgan Grenfell Securities (1988) *City Office Review,* MGS, London, p2
10 Winter & Company (1988) *The New Property Crisis: 1989,* Winter & Co, London, p5
11 D. Brierley (1989) Tremors Shake Property, *Sunday Times,* 22 October, pD3
12 M. Brett (1989) Learning the Lessons of the Past, *Financial Times,* 30 October, p38
13 *Ibid*
14 V. Houlder (1992) Cracks Became Chasms, *Financial Times,* 18 December, p9
15 G. Marsh (1991) *End of the Recession in the London Office Market?* Applied Property Research, London, December, p2
16 Debenham Tewson & Chinnocks (1992) *Money into Property,* DTC, London, p24
17 A. Jadevicius, B. Sloan, & A. Brown (2017) Century of Research on Property Cycles: A Literature Review, *International Journal of Strategic Property Management,* Vol 21, No 2, pp129–143
18 M.A. Weiss (1991) The Politics of Real Estate Cycles, *Business and Economic History,* Second Series, Vol 20, pp127–135

19 *Ibid*

20 R. Barras (2009) *Building Cycles: Growth and Instability,* Wiley-Blackwell, London

21 J.T. Black & W.E. Hauser (1993) Moderating Cyclical Overbuilding in Commercial Real Estate. In: *Land Use in Transition,* Urban Land Institute, Washington, DC, p44

22 J. Schumpeter (1939) *Business Cycles: A Theoretical, Historical and Statistical Analysis of the Capitalist Process,* McGraw-Hill, New York, pp11–12

23 H. Hoyt (1933) *One Hundred Years of Land Values in Chicago,* University of Chicago Press, Chicago

24 C.D. Long (Jr) (1940) *Building Cycles and the Theory of Investment,* Princeton University Press, Princeton, NJ

25 Hoyt (1933) *Op cit*

26 Long (1940) *Op cit*

27 W. Isard (1942) A Neglected Cycle: The Transport-building Cycle, *Review of Economics and Statistics,* Vol 24, No 4, pp149–158

28 P.J. Lewis (1964) Growth and Inverse Cycles: A Two Country Model, *Economic Journal,* Vol 74, No 293, pp109–118

29 M.C. Fleming (1966) *Review. Building Cycles and Britain's Growth,* by PJ Lewis, *Economic History Review,* Vol 19, No 2, pp435–436

30 R. Barras (1987) Technical Change and the Urban Development Cycle, *Urban Studies,* Vol 24, No 1, pp5–30

31 Fleming (1966) *Op cit*

32 M. Gottlieb (1976) *Long Swings in Urban Development,* National Bureau of Economic Research, New York

33 R. Barras (1994) Property and the Economic Cycle: Building Cycles Revisited, *Journal of Property Research,* Vol 11, pp183–197

34 Alexander (1979) *Op cit*

35 Barras (1984) *Op cit,* pp35–50; R. Barras (1981) *The Office Boom in London,* Proceedings of the First CES Conference, London; Barras (1979) *Op cit*

36 R. Barras & D. Ferguson (1985) A Spectral Analysis of Building Cycles in Britain, *Environment and Planning A,* Vol 17, No 10, pp1369–1391; R. Barras & D. Ferguson (1987a) Dynamic Modelling of the Building Cycle: 1. Theoretical Framework, *Environment and Planning A,* Vol 19, No 3, pp353–367; R. Barras & D. Ferguson (1987b) Dynamic Modelling of the Building Cycle: 2. Empirical Results, *Environment and Planning A,* Vol 19, No 4, pp493–520

37 Bateman (1985) *Op cit*

38 T. Key, F. Zarkesh, B. MacGregor, & N. Nanthakumaran (1994) *Understanding the Property Cycle,* RICS, London

39 *Ibid,* p81

40 Barras (1981) *Op cit,* p17

41 Key *et al* (1994) *Op cit,* pii

42 *Ibid,* p82

43 *Ibid*

44 *Ibid,* p9

45 T. Key, F. Zarkesh, & N. Haq (1999) *The UK Property Cycle – A History from 1921 to 1997* RICS, London, p26

46 *Ibid,* p26

47 *Ibid,* p29

48 *Ibid*

49 G. Keogh, T. McGough, & S. Tsolacoss (1995) *A Model of the Office Development Cycle in the UK,* Paper presented to the American Real Estate Annual Meeting Hilton Head Island, SC, March

50 Key *et al* (1994) *Op cit,* p81

51 E.P. Davis & H. Zhu (2004) *Bank Lending and Commercial Property Cycles: Some Cross-country Evidence,* Bank for International Settlements BIS Working Papers No 150

52 I. Cundell (2020) *Personal communication*

53 Marsh (1991) *Op cit*, p21

54 Barras (2009) *Op cit*

55 A. Baum (2001) Evidence of Cycles in European Commercial Real Estate Markets – Some Hypotheses. In: S.J. Brown & C.H. Liu (Editors), *A Global Perspective on Real Estate Cycles*, The New York University Salomon Center Series on Financial Markets and Institutions, No 6, pp103–115

56 RICS (1994) *Understanding the Property Cycle. Main Report: Economic Cycles and Property Cycles*, The Royal Institution of Chartered Surveyors, London

57 N.N. Taleb (2008) *The Black Swan: The Impact of the Highly Improbable*, Penguin Books, London, p127

58 J.F. Muth (1961) Rational Expectations and the Theory of Price Movements, *Econometrica*, Vol 29, No 3, July, pp315–335

59 L. Chen & T.I. Mills (2005) *Global Real Estate Investment Going Mainstream*, UBS Real Estate Research, London

60 B. Case, W.N. Goetzmann, & K.G. Rouwenhorst (1999) Global Real Estate Markets: Cycles and Fundamentals, *Yale ICF Working Paper No. 99–03*

61 C. Jackson, S. Stevenson, & C. Watkins (2008) NY-LON: Does a Single Cross-continental Office Market Exist? *Journal of Property Investment and Finance*, Vol 14, No 2, pp79–92

62 S. Stevenson, A. Akimov, E. Hutson, & A. Krystalogianni (2011) Concordance in Global Office Market Cycles, *Working Papers in Real Estate & Planning*, Vol 18, No 11 Henley Business School, University of Reading

63 Valuation Office Agency (2019) *Commercial & Industrial Floorspace Statistics*; Hillier Parker (1986) *Commercial and Industrial Floorspace Statistics, England, 1983–1986*, Valuation Office Agency, London

64 Ramidus Consulting (2009) *London Office Policy Review*, Greater London Authority, London

65 P. Jenkins & E. Hammond (2013) Canary Wharf Claims High Ground on City, *Financial Times*, 13 May

66 R. Preston & P. Stephens (1992) Banks Face $3bn Losses on Their O&Y, Loans *Financial Times*, 16 May, p1

67 R. Preston (1992) Banks Abandon Canary Wharf, *Financial Times*, 29 May, p1

68 P. Rodgers (1995) City Outrage at Being Sold Down the River, *Independent on Sunday*, 9 April, p1

69 M. Breheny, P. Cheshire, & R. Langridge (1983) The Anatomy of Job Creation? Industrial Change in Britain's M4 Corridor, *Built Environment*, Vol 9, No 1, pp61–71

70 M. Nathan, E. Vandore, & R. Whitehead (2012) *A Tale of Tech City: The Future of Inner East London's Digital Economy*, Centre for London, London

71 London & Partners (2019) *London and UK Top European Tech Investment Tables*, Press Release, 8 January

72 Colliers (2016) *London Tech Market Monitor*, Colliers, London, Spring/Summer

73 Harry de Quetteville (2018) The Silicon Joke? From Roundabout to Revolution, *The Telegraph*, 19 March, www.telegraph.co.uk/technology/the-silicon-joke/

7 Mediating

From advice to service

In Chapters 5 and 6 we met the principals of the property process, the developers and investors who speculate, who risk. But of course, they are surrounded by myriad mediators, or professional advisors who, in composite, make up the built environment professions: the architect, the builder, the contractor, the engineer and the surveyor. These are populated by myriad specialisations. The early years of the office economy saw a professional structure evolve that was shaped by social and cultural forces as much as by functional and technical ones. And we live with the consequences of this today.

It was not until the 1980s that the supply industry grew more aware of the needs of its customers. The emergence of the digital workplace (Chapter 8) encouraged new approaches to building design generally, and there was a growth in research activity that mediated the needs of occupiers and the priorities of owners and developers.

Also since the 1980s, but to a great extent since the turn of this century, there has been a rapid growth and evolution of the flexible space market. This sector offers space to occupiers on much less onerous terms than are available with traditional leases, and it offers these easier terms along with service and quality management. The operators in this sector mediate the traditional priorities of the owner with the rapidly changing needs of the twenty-first-century office economy occupiers.

7.1 Professions and professional jealousies

Between the late eighteenth century and the start of the twentieth century, the real estate development industry underwent several phases of profound change, as the economy and society, and building materials and techniques, evolved rapidly. For much of this time supplying buildings for the office economy was a relatively peripheral role: dedicated office buildings did not become commonplace until the latter quarter of the nineteenth century. Before this time, building activity was dominated by residential buildings, and then by industrial and other commercial buildings, built mostly for their occupier owners.

However, it was during this earlier period that the component parts of the real estate development process were formally organised into a structure that is not dissimilar to that with which we are familiar today. The commercial property supply industry is, in some ways, the antithesis of the fast-changing, unpredictable world of work described in Chapters 8 and 9. It is relatively slow moving, rigid, reactive and long-term. Kipling's *A Truthful Song* from 1922 was not so very wide of the mark:

> I tell this tale, which is strictly true,
> Just by way of convincing you
> How very little, since things were made,
> Things have altered in the building trade.

By the time that Kipling scribed his poem, architects, builders, engineers, planners and surveyors, for example, had all become distinct professions. But what is apparent from this early organisation is that the key drivers were more social and cultural than functional. The early professions were searching not only for functional identity but also social position and respect. This is critically important because today's built environment activities continue to suffer from some of these socio-cultural origins and constraints. Chapter 6 illustrated well the role of the developer (an activity that has never professionalised, curiously). Here we examine the development of the other principal roles.

Many of today's professions have their origins in the mid-to-late nineteenth century. Among the earliest professions to organise were the medical and legal professions. Solicitors, barristers, surgeons and general practitioners all had national organisations by 1832. From 1834 to 1860, similar organisations were formed for architects (1834), dispensing pharmacists (1841), veterinary surgeons (1844), actuaries (1848) and dentists (1856). These were then followed by surveyors (1868), accountants (1870), optometrists (1895), insurers (1897), psychologists (1901), structural engineers (1908) and town planners (1914).

From the very beginning, 'professionalisation' was imbued with an aspirational quality: it was deemed to reflect the social standing of those who refer to themselves as having a profession: something different to or, more crudely, better than the run-of-the-mill job. In many countries, professional status has endowed upon the individual higher social status, more secure employment and even political power, or at least, influence. But there was another, equally important factor: depending upon your point of view, professions either helped to maintain a certain level of standard in the service they provided, or they were a closed shop for those with a vested interest in maintaining prices and reducing competition.

No doubt professionalisation reflected a blend of these factors, but it is undeniable that they carried social and cultural importance as well as functional identity. Neal and Morgan observed a staged process to the establishment of the early professions:[1]

• the occupation became full-time rather than practiced as part of another activity;

- a training regimen was established, formalised by the articles system, in which articled assistants were trained by established professionals;
- a professional association was formed in order to obtain greater status;
- qualifying exams were introduced to stress high levels of competence;
- a Royal Charter was sought to secure legal protection of the profession, protecting members from competition from non-members, and
- routes to academic qualifications were formed to enhance standards and status.

In the built environment, the specific origins and early development of each profession varied widely, and the number of professions multiplied as buildings and construction generally became more complex and demanding of narrow technical skills. But they were clearly territorial in nature: placing boundaries around certain activities that could not be undertaken by unqualified tradesmen. Nowhere was this more the case than in the formation of the architectural profession.

Architect

The building process was once the realm of the architect. All powerful and central, the architect held a close relationship with his (for it was such) client, designing and delivering buildings, using where required the services of others to fill specialist roles. From these early times, the architect perceived himself as having 'social parity' with the client, unlike the tradesfolk who placed one brick upon another. Powell described the role in the later nineteenth century:

> After the person who paid for the building, the central figure in this procedure was the architect. The profession developed significantly in nature and size, largely in consequence of heavy building activity and increasing numbers of commercial clients.[2]

One such architect was the neo-Classical practitioner Sir John Soane (1753–1837), who was perhaps most renowned for his work on the Bank of England, where he worked for 45 years. The son of a bricklayer, Soane rose to the peak of his profession, becoming professor of architecture at the Royal Academy and an official architect to the Office of Works, before being knighted in 1831. In 1788 he defined the role of the architect as follows:

> The business of the architect is to make the designs and estimates, to direct the works and to measure and value the different parts; he is the intermediate agent between the employer, whose honour and interest he is to study, and the mechanic, whose rights he is to defend. His situation implies great trust; he is responsible for the mistakes, negligences, and ignorances of those he employs; and above all, he is to take care that the workmen's bills do not exceed his own estimates.

Soane finished this description of the role of the architect by drawing a clear line between architecture and building:

> If these are the duties of an architect, with what propriety can this situation and that of the builder, or the contractor be united?[3]

This was a bold position to take, and one which attracted a good deal of opposition. Soane was arguing for the separation of profession and trade. He was bringing social stratification into the property process, setting in train the first step in the distancing of the architect from the building process itself. Soane sought to define what he saw as the clear blue water between the professional and skilled architect who acted on behalf of the client, and the 'workmen' who laid one brick atop another. Soane's notion was one of social stratification as much as one based on functional separation. Bowley captured the mood of the time well:

> Since the wealthy classes demanded a unified aesthetic setting for their lives, architects could be fully occupied with design and supervision only. Organisation of the actual building process could be viewed as a chore, a waste of artistic talent; it could be delegated to others.[4]

Bowley further suggested that the profession was "moulded on the social and aesthetic pattern of the eighteenth century, when architecture was one of the

Figure 7.1 Sir John Soane, architect

arts which contributed to the aesthetic milieu of the well-to-do". This issue was compounded by the prevailing view that "buildings were of interest to an architect primarily as the vehicles of his art, the purpose of the building being secondary".[5] Bowley was forthright in her analysis:

> It can be regarded as one of the accidental tragedies of history that the architectural profession crystallised at this time, just as a new era of building for an industrial society started.

Despite Soane's wishes, the boundaries between design, development and building remained very fluid well into the nineteenth century. For example, the great architect and contemporary of Soane, John Nash (1752–1835), acted as both speculator and contractor during much of his career, including during his development of Regent Street in the second decade of the nineteenth century.

Before Soane's wider aims were achieved, there remained a lack of formal education and certification for architects which meant that virtually anyone could describe themselves as an architect, leading to the inevitable spread of bad practice and incompetence. So much so, in fact, that in 1849, the architect John Burley Waring (1823–1875) declared that: "In no occupation are there more quacks than in architecture".[6] Similarly, Micklethwaite described the professional circumstances of the 1870s:

> Any man worth a brass plate and a door to put it on may dub himself an architect, and a very large number of surveyors, auctioneers, house-agents, upholsterers, etc, with a sprinkling of bankrupt builders and retired clerks of works, find it in their interest to do so.[7]

But change did begin. Powell notes that while initially architects were hardly distinguishable from builders and measurers, commonly combining their skills with those of others, from around the 1820s "they gradually began to divorce themselves from direct involvement in building". Egged on by Sir John Soane,

> they sought to represent and protect promoters' interests (an influential position in the hierarchy). The emerging building procurement method required firm cost estimates (and hence firm design) at the outset. At the same time closer alliance with promoters was achieved at cost of greater distance from practical building production.[8]

Soane was influential in replacing the Architectural Society (founded in 1831) with the Institute of British Architects in London in 1834 (which received its Royal Charter in 1837), with the aim of advancing interest in and spreading knowledge of architecture. However, his wider vision of separation was not to be fulfilled until a century after his death when, in 1938, the passing of the

Architect's Registration Act saw the practice of architecture finally separated from that of contracting.

In the meantime, the latter half of the nineteenth century was a period of dramatic and irrevocable change for architects. The architectural historian J Mordaunt Crook describes this period as one which saw the 'fragmentation' of the architect into the surveyor, designer, builder and engineer largely in response to the spread of new building techniques, materials (such as iron, mass-produced bricks and glass) and innovations such as plumbing and gas. For example, the role of the engineer was becoming more prominent as building complexity increased. Then there was the spread of general contracting and tendering which tested architects in terms of the drawings and specifications they needed to produce.[9]

Architects were faced with the problem of defining what they were responsible for in the building process and their status on the job site. Many architects sought a professional status equal to lawyers and doctors, which meant changes in the way they learned and practised. Pupillage, the apprenticeship to a master for seven years, was the primary means of education at the beginning of the century, but this was gradually being replaced by new university degree programmes.

The principal role of the architect, in terms of the patronage of a client, on whose behalf the architect organised the various trades, was also under threat from a new player – the developer. While it was not always possible to control (own) the land for development, the growing complexity of the process pointed to an increasingly important role for those who could manage the process in a way that the more aloof architects did not desire.

In 1887 the principle of separation between architects and builders was incorporated in the supplementary charter of the RIBA. This laid down *inter alia* that no member of the Institute could hold a profit-making position in the building industry and retain his membership.[10]

The outcome of this early fragmentation of roles in the development process, and the role of patronage between wealthy private clients and their chosen architects, resulted in a linear structure (now referred as the traditional procurement system), where clients in need of a building would seek an architect, produce a brief outlining requirements and provide an estimate of the capital to be invested in the project. Based on this information, the architect would develop a proposal for the client, and then a bill of quantities would be drawn up by a quantity surveyor. The contract would be put out to competitive tender to builders who would seek to undercut each other with the lowest possible prices. This new arrangement stratified responsibilities and led to social and professional niches.

> By the outbreak of the First World War the industry had assumed its modern pattern of contractors and subcontractors, estate developers and speculative builders, and of distinct and separated designing professions. This was

part of a process of adaptation and resistance to economic expansion and technical change.[11]

The long-term effects of the emergent structure were profound: the replication of English social stratification in the structure of the property process. The client and architect formed the uppermost tier, propped up by the other professions, with the lowly builders at the base of the pile; an arrangement that has worked to the detriment of the industry and the consumers of its products ever since. The mass provision of speculative offices and other buildings during the 1960s and 1970s that lacked any architectural merit or client-side interest in functionality were a symptom of this inadequate process.

Builder

The organisation of the building industry into general building firms, contracting or speculative, and specialist firms is comparatively modern. The emergence of the general contractor is less well documented than that of either the architect or the developer. As we saw earlier, at least until the middle of the eighteenth century, the building owner or his architect employed crafts directly, with the former frequently supplying the materials, while individual workers were engaged by master craftsmen who were the employers in each craft. The work of co-ordination of the various stages, general contracting, was not a separate enterprise; and it was not unusual for the designer or architect to combine the functions of contractor into their remit.

One of the clearest histories of the early, industrial age building trade was provided by JH Clapham in his great work *An Economic History of Modern Britain*. Clapham begins by observing that that building trades had avoided any revolution in technique before 1825, thereby establishing a point at which, it might be said, modern contracting began. The entrepreneurial nature of building at this time was captured in the phrase

> in the full flood of capitalism, there was no industry in which the handicraftsmen more frequently rose to be a small jobbing employer and perhaps, eventually, a builder on a larger scale.[12]

Clapham was identifying not only the mechanism for an expanding industry, but also the socio-cultural dimension in which the building trade was a route to professional success, particularly once it began to take on its modern contracting character. Clearly, rapid industrialisation and urbanisation were creating plentiful opportunity for builders in providing the means through which expansion could occur. They were also organising into distinctive groupings.

> By 1830, the 'respectable builders' of London were already specialised into definite groups. There was a small group, contractors as we should say, who did little but erect public buildings. A second, larger group, devoted

themselves to the building of shops and business premises. Third, perhaps not at all respectable, came . . . those who took risks with private houses, 'speculative builders' as they were already called.[13]

Clapham suggests that while "deservedly criticised as the jerry-builder has been, no one has even suggested how the ever-swelling British urban population could have been housed without him". The result, however, was not dissimilar to the situation that architects found themselves in, leading to John Burley Waring's aforementioned reference to the prevalence of "quacks" in the occupation. There was clearly a need – socially as much as functionally – for 'respectable builders' to distinguish themselves from their less reputable competitors.

The Builders' Society was formed in 1834, with 18 founding members, following a series of meetings in a coffee house on Ludgate Hill. The Builder's Society was incorporated under the Companies Act and became the Institute of Builders, with the stated aim of promoting excellence in the construction of buildings and just and honourable practice in the conduct of business. The London Master Builders' Association was formed in 1839.

Honour and conduct were central tenets of these early professional bodies. As Clapham observed, "All the large builders had been organised into the very respectable London Master Builders Association since 1839".[14] In 1867 it accepted through its Secretary the description of 'a club of gentlemen belonging to your profession' and, probably like the National Association of Master Builders of 1877–1878, it had the rule 'no speculative builders admitted'.[15]

But social structures run deep. Powell suggests that while the "pioneering big building contractors" were overlapping "in the hierarchy with the emerging professionals", and while many were better off than their professional peers, "their status was limited by their being engaged in trade".[16] In this important respect, they were distinguished from their architect colleagues. One way for builders to enhance their status was to extend their influence across the development process, and this was achieved by becoming a general builder.

Bowley points out that the general builder

> grew out of and was dependent on the traditional craft organisation of the building industry. Since the amount of work belonging to each craft varies with the type of building, general builders require the different craftsmen in varying proportions according to the work in hand.[17]

Often, general builders specialised in some aspect of construction themselves, subcontracting the work of other trades as and when needed. As Bowley notes, while the emergence of general building faced bitter opposition from the small master craftsmen, the system prevailed, and this "is the practice of the building trade today", in which the trades are held together by the general builder with a mixture of direct employment and subcontracting in a highly elastic fashion.

As we saw in Chapter 4, Thomas Cubitt emerged in the early nineteenth century with his innovative approach of directly employing craftsmen under craft foremen on a full-time, permanent basis to undertake whatever jobs he had at a particular time. While he later engaged in speculative building to keep his employees busy, he was a general contractor. So successful was this new approach that by the 1830s, "there were already very substantial building firms in London, for by this time a London contractor of not quite the largest type employ 170–235 men".[18] Nevertheless, the great majority of firms were, as they are today, small family businesses and small private companies employing small numbers of workers.

Despite organising themselves into a more professional set-up, throughout the nineteenth century and into the twentieth, no matter how wealthy or successful they were, the builders continued to be seen as tradespeople rather than professionals, and "small builders were little more than glorified craftsmen, 'cap-in-hand builders' as they are sometimes termed". The system remained riven with social status: development was "hierarchical both socially and in terms of working organisation. The builders and their workmen formed the lowest ranks, the architects the highest; the other professions came in between".[19] A century later, the industry remains deeply divided by such notions.

Surveyor

Surveying has a long history, and as such it has gone through many transformations as the economy has changed to present new challenges and areas of work. The original growth of the surveying profession can be traced back to the land reforms of the sixteenth century. Changes to agricultural tenure and management led to the demand for measuring, surveying and valuation.[20] Population growth, together with economic changes and a more intense use of land, led to a demand for more sophisticated approaches to the measurement and valuation of land. Ownership rights, for example, which were often bitterly contested over many years, could be resolved or avoided with accurate, measured records.

As the land market matured during the seventeenth century, surveyors became more involved in transactional activity. Land supply was being bolstered as monastic, crown and church estates were being broken up[21] and demand for measuring and valuing grew as a result. However, the role of the surveyor really got underway during the eighteenth century, when landlords began to appoint estate managers to improve the efficiency and productivity of their land.[22] Surveyors were also instrumental in establishing the lines of canals and supervising their construction.

Although recent mergers and acquisitions, combined with globalisation of the profession, have changed the structure of the industry, it is possible to trace the origins of many of the firms in existence today back to this time. For example, Cluttons was founded in 1765; Richard Ellis (now part of CBRE) in 1773, and Jones Lang Wootton (now Jones Lang LaSalle) in 1783. Figure 7.2

Historic name	Year founded
RH & RW Clutton	1765
Richard Ellis	1773
Jones Lang Wootton	1783
Farebrother	1799
Chestertons	1805
Healey & Baker	1820
Grimley & Son	1830
St Quintin	1831
Cluttons	1851
Savills	1855
Weatherall Green and Smith	1860
Fuller Peiser	1883
Strutt & Parker	1885

Figure 7.2 Some of the oldest names in surveying

shows a selection of the oldest firms, many bearing the names of their founding partners. Until relatively recently they were all partnerships, in which the equity partners shared the profits of the business and took an equal share of responsibility for the performance of their firms.

During the nineteenth century, as the Industrial Revolution reached its zenith, the demand for surveyors grew stronger, and their roles diversified further. The huge expansion of the railway network, the digging of canals, the building of sewers and drainage, and the expansion of factories all demanded surveying skills. While on the one hand, surveyors responded by providing the by now traditional skills of measuring and valuing, they also developed into an advisory capacity by acting for principals involved in the acquisition of land by public utilities.[23] This latter activity set the tone for the development of surveying practices, because it was those firms that "steered clear of the measuring side and concentrated on the negotiating and valuing who made good, and established themselves as the commanding figures of the mid-Victorian years".[24]

Solicitors started the process in 1739 with the formation of the Society of Gentlemen Practisers of Law, which evolved into the Law Society in about 1800, before incorporating by charter in 1831. Similarly, in medicine, the College of Surgeons was founded in 1745 with the foundation of the College of Surgeons, later becoming the Royal College of Surgeons, in 1800.

We saw earlier how the newly founded Institute of British Architects in 1834 sought to distance the practice of architecture from other aspects of the building process, including contracting and surveying. As Thompson observed, as the architects sought to dissociate themselves from both building, and surveying and measuring, they chose to "smear the measurers and surveyors, who were widely held to be a set of rogues, at best cantankerous, and usually dishonest". This was a sign, according to Thompson, that architects

> were becoming desperate to rescue themselves from the low repute into which their profession had fallen, and in order to do so were obliged to root out the so-called architects who practiced measuring, were involved in the building trades, or extorted commissions from builders for works done under their direction. The tactic was to exalt the one essential function of the architect, design, and to denigrate all the non-architectural functions which any particular architect might also perform.[25]

Moreover, while their tactic was understandable as they sought to "clean up their own underworld", they "failed to notice that at the same time surveyors and measurers were successfully differentiating themselves from builders".

Activities that we now associate with surveyors were reasonably well developed by the late eighteenth century. But they had failed to 'professionalise'. Whilst reasonably well developed, the activities of surveyors were still not performed by specialist, full-time surveyors.[26] Indeed, part of the effect of their tactic was to make 'homeless' measurers and quantity surveyors which led indirectly to the coalescing of various surveying activities into a cohesive whole. In the same year as the Institute of British Architects was formed,

> It was a group of old-established land surveyors, however, not the new-fangled quantity surveyors, who got together in 1834 and took the first step towards professional association. In the same months as the RIBA was forming, six London surveyors . . . decided it would be useful and agreeable to form a club [and] on 27 June 1834 at the Freemason's Tavern, they held the first meeting of the Land Surveyors Club.[27]

The Land Surveyors intended to advance the dignity and status of their profession by using their association for active professional promotion, although progress towards professionalisation was slow. On 23 March 1868, some 20 surveyors met at the Westminster Palace Hotel under the chairmanship of John Clutton and resolved to establish the Institution of Surveyors.[28] While evidence is minimal, Thompson suggests that the Institution of Surveyors grew out of the Land Surveyors Club, based on the fact that 11 of the original 20 that met on 23 March were members of the latter.

John Clutton was highly influential in the early years of surveying. His grandfather, William Clutton, inherited his father-in-law's surveying business in Sussex in 1765 and soon began trading under his own name. The business was successful, and William was succeeded by his son, also William, who had three sons. Two sons remained in Sussex to continue the family business, trading to this day as the independent RH and RW Clutton. The third son, John (1809–1896), moved to London in 1837 to found a new business.

John was very successful, building up a large multi-department office. By 1848 his office earned a net income of £9,000 (c£750,000 today) from a turnover of £13,000. He lived "an upper middle-class life in Sussex Square", and sent his sons to Harrow. By his own example, he was creating "an image of the surveyor as an independent, forthright professional man of undoubted integrity, and of considerable consequence in the great world".[29]

This latter point highlights a common theme with the professionalisation of architects and builders: the search for social relevance, or standing. Not only had Clutton's "great material success and eminence helped to mark off the surveyor from the civil engineer, architect and untrained land steward" (notice that builders are not even mentioned on this list), but he had also helped "tip the scales in favour of regarding the term 'surveyor' as the designation of a profession rather than as the description of a type of appointment".[30]

The Institution of Surveyors' first Associate Member, John Horatio Lloyd, defined the objects of its formation as intellectual advancement, social elevation and moral improvement. He went on to observe that all such associations "tend to raise the members to a higher level in the social scale". So, once again, we have a search for recognition by wider society as a primary motivation for professionalisation: "the prospect of the Institution leading to a rise in their standing in society was a potent argument to use in advertising the attractions of the Institution to surveyors".[31]

On 9 November 1868, the Institution was ready to open its first annual session, in its own home at 12 Great George Street, where it resides today. It received its Royal Charter as the Surveyors' Institution on 26 August 1881 and became the Chartered Surveyors' Institution in 1930. In 1946, George VI granted the title "Royal", and in 1947 the professional body became the Royal Institution of Chartered Surveyors.

While the surveying profession has been multi-disciplinary since the start (including at the outset auctioneers, valuers and land agents), adding more and more disciplines over time, there is one sub-group which has had an uneasy relationship with its 'parent profession'. Although activities described as 'measurer', 'custom surveyor' and simply 'surveyor' had been used for decades to describe the activities, it was when larger, more complex buildings became common that quantity surveying was formalised into a sub-discipline: "accurate estimating, project cost control, informed choice between alternative building materials, control of sub-contracting, and measurement of development cost against site potential . . . can only be gained in large-scale individual objects".[32]

Through the latter decades of the nineteenth century, quantity surveyors began to lobby for their recognition as a distinct profession, amidst a growing sense that the Institution of Surveyors was doing nothing to promote their distinct role in the development process. By 1889 the quantity surveyors "felt strong enough to ask for increased representation on the [Institution's]

Council"[33] and were duly ignored, leading many to doubt whether the Institution was the best means of advancing their particular professional interests. This sense of belonging to a distinct profession led to the founding of the Quantity Surveyors Association in 1903. Despite the "atmosphere of acrimony and recrimination" at the time, relations soon improved, so much so that the arrangement lasted only until 1922 when the two organisations joined together once more.

Surveyors provide a bewildering array of specialisms grouped into areas including broking, building, surveying, construction, facilities management, investment, minerals, planning and development, project management, quantity surveying, rural land management, valuation and waste management. Their most high-profile role with respect to the story being told in this book is their transactional, or agency, role. This encompasses either buying or selling office buildings as standing investments, primarily for property companies and investment institutions, or transacting leasing deals, working either on behalf of the lessor (owner) or lessee (occupier). For both these roles, surveyors are widely referred to in the market as either 'agents' or 'brokers'.

Other professions

The foregoing has traced the early professionalisation of architects, builders and surveyors, but there were of course other professions (Figure 7.3). The Town

Architects	Architectural Society	1831
	Institute of British Architects in London	1834
	Royal Charter	1837
Builders	The Builders' Society	1834
	London Master Builders' Association	1867
	Institute of Builders	1884
	Institute of Building	1965
	Chartered Institute of Building	1980
Surveyors	Land Surveyors' Club	1834
	Institution of Surveyors	1868
	Royal Charter	1881
	Quantity Surveyors' Association split from the Institution	1903
	Re-amalgamation of QSA and the Institution	1922
	The Chartered Surveyors' Institution	1930
	Royal Charter	1947
Civil Engineers	Institute of Civil Engineers	1818
	Royal Charter	1828
Structural Engineers	Concrete Institute	1908
	Institute of Structural Engineers	1922
	Royal Charter	1934
Town Planners	Town Planning Institute	1914
	Royal Charter	1976

Figure 7.3 The built environment professions

Planning Institute was founded in 1914 and received its Royal Charter in 1976. The term 'town planning' gained common currency only in the early years of that century, becoming a statutory practice following the Housing, Town Planning, etc Act 1909. In these early years, town planning formed a distinctive element in the already-established professions of architecture, engineering, law and surveying.

The notion of town planning as a distinct profession took hold following the appointment, in 1910, of Thomas Adams as the first Town Planning Inspector at the Local Government Board. He convened meetings with a small group of practitioners before an invitation was sent to a wider constituency to join a Town Planning Institute. Adams chaired the inaugural meeting in November 1913; a Council was elected in December, and Adams was elected President on 13 March 1914. An inaugural dinner on 30 January 1914 marked the public launching of the Institute, and the Articles of Association were signed on 4 September 1914, which is designated the date of the Institute's founding.

The Concrete Institute was founded in 1908 as the representative body for professions related to concrete. Architect Edwin Sachs (1870–1919) was instrumental in the move, and the membership attracted architects, engineers, chemists, manufacturers and surveyors. The first meeting of the council took place in the smoking room at the Ritz Hotel on 21 July 1908. Then, in 1922, the Institute changed its name to the Institution of Structural Engineers, which was granted its Royal Charter in 1934.

The Institute of Civil Engineers has a much longer heritage, having been founded in 1818. In this year, a group of engineers met in a London coffee shop and founded the Institution of Civil Engineers (ICE), the world's first professional engineering body. After two years of struggling to attract new members, ICE asked Thomas Telford to become its first President. His was a very successful appointment, leading to a Royal Charter in 1828.

Social and functional stratification

We saw in Chapter 5 the emergence of the developer with reference to Nicholas Barbon, James Burton and Thomas Cubitt. We saw how their emergence mirrored the growing complexity and scale of buildings required to house a rapidly industrialising economy. And we have seen in this chapter how, for similar reasons, the built environment professions evolved, to provide their developer clients with ever-growing knowledge specialisations.

By the early twentieth century, the main built environment professions – architects, builders, surveyors and engineers – were all firmly established, and the "industry had assumed its modern pattern of contractors and subcontractors, estate developers and speculative builders, and of distinct and separated designing professions".[34] There was a clear understanding of their social standing both with each other and with wider society. In this context, Marion Bowley is worth quoting at length.

> The architects were members of a profession concerned with a major art, they were the confidants of gentlemen and to a considerable extent arbiters of taste, Successful architects at least were members of the upper middle-class elite. They were gentlemen or regarded themselves as such. Engineers on the other hand were closely associated with trade and industry; their training had no cultural significance; they were not artists; they might be industrialists . . . Surveyors of all sorts were even more definitely socially inferior. Builders, however wealthy or successful, were in trade and small builders were little more than glorified craftsmen, 'cap-in-hand builders' as they are sometimes termed.[35]

The long-term effects of the emergent structure and its rationale were profound: the replication of English social stratification in the structure of the property process, with the client and architect forming the uppermost tier, propped up by the professions with the lowly builders at the base of the pile is a structure that has been maintained to the present day.

The mass provision of speculative offices and other buildings during the 1960s and 1970s that lacked any architectural merit or client-side interest in functionality were a direct outcome of this structure. So too is the on-going separation of interests between the ownership and management of buildings where the client-side roles of ownership and building design are distant from the on-going occupation of buildings. The lack of feedback between the two sides has been largely responsible for the supply-led culture of the industry and almost complete disconnect between supplier and customer.

The shortcomings of the traditional stratification of the property process were thrown into particularly sharp relief with the arrival of the digital revolution, which fundamentally changed, first, office work and, secondly, office management. The white-collar factories we met earlier in this book have been replaced by environments that were fundamentally different: their occupation and management have been revolutionised with the emergence of the knowledge economy and office technology. The traditional professional structure was simply not prepared to anticipate, analyse and respond to the rapid changes in occupier demand.

7.2 Research and market intelligence

Chapter 6 emphasised the supply-led nature of the property industry – creating assets for owners, rather than products for customers. Up to the mid-1980s, there was very little market research; no feedback into the design process, and no co-operation in the procurement chain. There was some technical research into materials and construction technologies; and the larger surveying firms had research departments, although these were largely focused on transactional data such as rents and yields, take-up and vacancy levels.

For much of the post-war period, design was driven primarily by consideration of how cheaply buildings could be constructed. Most speculative office buildings were designed and built with an inflexible floor plan, low-capacity service provision and a low-quality internal specification. Small, shallow floors were standard with a structural floor-to-ceiling height of 2.4–3.0 m (8–9 feet). Floors were generally finished in woodblock or tiles, with no provision for under-floor cabling; while suspended ceilings were largely unheard of – light fittings were fixed directly to the plastered ceiling. Single glazing, coupled with large window areas, gave rise to problems of heat gain in many buildings. To let accommodation, the minimum of tenant facilities was required by the developer, and toilet areas and entrance halls were generally extremely basic.

The buildings were designed to accommodate highly structured and unchanging organisations with staff in small or individual offices, and little attention was paid to the environmental or ergonomic conditions on the occupied floor. Centre Point in London's West End is a classic of this building genre. The industry was simply not organised to address the needs of the companies who occupied its buildings. There was no perceived need for intelligence, or research: the industry had a standard product, available on standard terms.

However, this began to change as buildings became more complex and as the demands placed upon them by occupiers began to increase. Research began to change the general understanding of the relationship between building form and building occupation, and how the former could have such a major influence over the latter. For example, the relationship between buildings services (air handling, lighting, power and so on) and the layout of space became more critical to manage. The growing use of partitions, for example, required that air handling was distributed not only to an open floor but also to enclosed areas. Thus it was that the relationship between the basic building structure and its various 'grids' became critical to the effective planning of office space.

One of the pioneering texts in this respect was *Planning Office Space,* written in 1977 by Francis Duffy, Colin Cave and John Worthington.[36] Figure 7.4 is a key diagram which shows structural and planning grids and their interaction with planning and services. The columns around the perimeter set the basic geometry of the building. These are divided by window mullions, which determine where partitions can be placed. This is then overlain by services provision such that air, lighting and power are delivered everywhere in a uniform fashion.

Few industry practitioners today will even be aware of *Planning Office Space,* yet its influence was profound. Most of the post-Big Bang buildings in London reflected the thinking of the work in their physical configuration.

The authors note the relationship between structural and planning grids, pointing out that many difficulties in using a layout planning grid stem from earlier choices of structural and servicing grids. This thinking is taken more

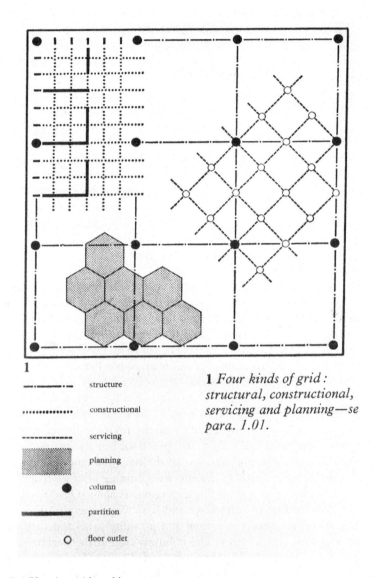

structure
constructional
servicing
planning
column
partition
floor outlet

1 *Four kinds of grid :
structural, constructional,
servicing and planning—se
para. 1.01.*

Figure 7.4 Planning grids and layouts

Source: Image by kind permission of AECOM/University of Reading, Special Collections.

or less for granted in today's development process, but it was the first explicit articulation of the relationship between the efficient and effective planning of floorspace and its underlying design parameters. Its practical application was to gain strong currency during the early stages of planning the Broadgate development in the City of London.

Planning Office Space proved to be the progenitor of a raft of research and market intelligence that collectively, fundamentally, changed the nature of London's office buildings and assisted the office economy in maturing to a new level.

Product development

Number One Finsbury Avenue in the City of London (Figure 7.5) was completed in 1984. The 300,000 sq ft (28,000 sq m) building was developed by Rosehaugh Greycoat Estates and designed by Arup Associates. The building was a magnificent addition to the City's office stock and remains one of the most elegant commercial buildings in London. But its significance is much greater than its aesthetic: it fundamentally changed approaches to the design and development of office buildings by integrating analysis of occupier needs.

This was the first joint project of Godfrey Bradman and Stuart Lipton, then of Rosehaugh and Greycoat, respectively. Sometimes referred to in the trade as 'Phase One Broadgate', the project was indeed the test bed for the development

Figure 7.5 Number One Finsbury Avenue

Source: © Ramidus Consulting Ltd.

duo's more famous project on the adjacent site, the most significant development in the post-war office economy.

Finsbury Avenue was pioneering, although not entirely experimental. Designer Arup Associates had tried the striking, modernist approach on projects of a clearly similar pedigree, notably the Wiggins Teape building in Basingstoke, originally known as Gateway 1, and later for IBM in Portsmouth. For Lipton also, the watchword was evolution rather than revolution, after experimenting with Cutlers Gardens in the City and Victoria Plaza in Victoria. But Finsbury Avenue was infinitely more refined than anything constructed before. Something had changed; but what exactly? Essentially, the building challenged three fundamental principles of the development industry.

The first challenge to accepted wisdom was on location. Sited as it was just beyond the sharply defined City boundary, it was one of the first City buildings to be built outside the Square Mile, in the London Borough of Hackney. While the building now sits comfortably among a number of other City buildings, all providing for the needs of City companies, back in 1982 when the project got underway, 'bank walks' (that defined the acceptable walking distance that banks could physically locate from the Stock Exchange) dictated that Finsbury Avenue was well beyond the pale. Yet, the building was fully let four months after completion. Finsbury had successfully challenged the age-old monopoly of the Square Mile and shook traditional thinking about location.

Secondly, Finsbury Avenue demonstrated that speculative buildings did not have to be cheap, built to a maximum density and to a minimum quality. By contrast, Finsbury Avenue was bold, exciting and of an exceptionally high quality. The building has bronze-coloured façades with horizontal metal grille sun screens and exposed cross-bracing, a stepped section, planted roof terraces and, what became a Lipton trademark, a cavernous atrium with sculpture and large-scale planting. The finishes are beautifully detailed, and components such as balustrades and light fittings are made of fine-metal sections, demonstrating craftsmanship in manufacture and attention to detail in specification.

It is relatively low rise (at eight floors), with large rectangular 46 ft (14 m) deep floor plates, and structural floor heights of 11.5 ft (3,5 m). More importantly, it was specifically designed so that the internal environment could be repeatedly redesigned without compromising or being constrained by the longer life span structural elements. It was delivered to a 'Category A' finish, meaning that while suspended ceilings and raised floors were provided, internal fittings such as carpets and partitions were not, thereby allowing in-coming occupiers to install an interior to suit their needs.

Thirdly, Finsbury Avenue challenged the notion that speculative development was, by definition, an entirely supply-led process. Indeed, what really set the building apart was the effort that was invested in understanding the needs of the companies that might be expected to occupy the completed space. This was the first time that a commercial developer had commissioned a speculative building specifically to meet the needs of targeted occupiers. It signalled the beginnings of a new approach to development in which the product was subjected to major advances in specification and design as the result of research into occupier needs.

One of the earliest signals that Finsbury Avenue was going to be different was the fact that Greycoat was a co-sponsor of the first study on the impact of information technology on the design of office space, the ORBIT study, a multi-client study carried out in 1982 by DEGW with Building Use Studies and EOSYS.[37] This work is discussed in Chapter 8, suffice to say here that it was the first research to link the physical characteristics of buildings (such as floor size, depth, sectional heights and servicing capacity) to their ability to accommodate the emerging information technologies – both desktop services such as personal computers but also, critically, building management systems.

Academia also produced some notable work around this time. In 1983 the Centre for Advanced Land Use Studies (CALUS) published a very high quality piece of research entitled *Property and Information Technology – the Future for the Offices Market*.[38] The report demonstrated how technology would call into question many of the traditional beliefs upon which investment criteria were based. CALUS found that investors would need to be concerned with providing the flexibility to arrange office areas in different layouts, and in providing tenants with flexibility through the provision of a basic office shell. Also, they would have to pay increased attention to the quality of their offices because the standard of development was likely to become increasingly dictated by occupiers.

Anticipating the demand for business park space, the report found that office parks were one form of development for which there was likely to be increased demand, due to an increasing demand for property that is interchangeable between office and non-traditional office activities, including light assembly, research and development, and product testing.[39] Also from the same stable, in 1983, appeared a book by Anthony Salata entitled *Offices Today and Tomorrow*. Referring to the growing numbers of occupiers dealing with 'sophisticated products', Salata rather quaintly observed that they will "increasingly demand at least a modest degree of sophistication from the buildings that they are to occupy" and that those "in the property professions must be mindful of their needs".[40]

Customer studies

Despite the focus on technology in much of this work, a host of non-technological or technologically enabled changes were also growing in importance. For example, the City was becoming more international and was beginning to confront the prospect of deregulation (see Chapter 6). Competition was driving a reorganisation of businesses through mergers and acquisitions; and the economy as a whole was being transformed as traditional industries were eclipsed in their importance by the burgeoning office economy.

From these big issues, it was apparent that the businesses that occupied office buildings would begin to make radical demands on their premises, and this justified a closer look through customer-focused research. Following the success of Finsbury Avenue, Stanhope led the way, this time through its joint venture development at Broadgate (Figure 7.6) with Rosehaugh, and largely through a unique collaboration with design and research practice DEGW. Together they undertook work that helped transform property development into a more customer-focused and innovative process.

The first stream of work was a series of focus groups involving individuals from City companies. The discussions provided a forum to explore the pressures that were being faced by City businesses, how they were changing as a result, and how this change was having an impact on their use of buildings. The findings of the research were fed directly into the architectural briefing process for the buildings being developed by Rosehaugh Stanhope at Broadgate. A number of interesting facts emerged from the groups. First, it became quickly apparent that the individuals attending the workshops were unaccustomed to talking to each other about their premises. Despite the fact that they worked in a tightly knit community in the City, there was no evidence of a shared property experience, and the idea of gathering a group of people together for this purpose was an entirely novel concept.

Secondly, once they were gathered together, it also emerged that many of the participants actually had common concerns. Those involved were specifically

Figure 7.6 Exchange House, Broadgate

Source: © Ramidus Consulting Ltd.

selected because they were involved in some way with the business of the City, so there is no great surprise in this. But a consensus emerged around a set of building issues that would have to be addressed and resolved if City businesses were to avoid being hamstrung in their ability to respond to a changing financial services industry by inappropriate buildings.

Thirdly, it became abundantly clear that the scale of change that the participants' businesses were beginning to confront was enormous. Many of the participants were not specialist facilities managers. A good number were directly involved in their companies' core businesses and were therefore able through the focus groups to bridge the gap between what was happening to their businesses and the issues that the focus groups were set up to explore. The focus groups revealed a deep-seated feeling among the participants that the buildings they occupied were inadequate to meet their operational needs in a rapidly changing business environment.

The results of the work were summarised in *Accommodating the Changing City*.[41] The report pointed to three trends changing the way business was done in the City: internationalisation of financial trade; increasing competition from large US and Japanese integrated merchant banking and securities houses, and proposed changes in the Stock Exchange rules and in the Bank of England. The report argued that these trends would have a profound effect on the organisational structure of the City institutions,[42] and that autumn 1986 would be critical because this was to be the time of Big Bang, when fixed commissions would end and single capacity would begin.[43] The key conclusions of the report suggested the emergence of profound changes in the demands of occupier businesses:

- shifts in the business structure of the City, compounded by advances in technology;
- growth and restructuring, making many existing buildings obsolete;
- need for lower occupancy costs, larger floors and flexibility to cope with change;
- getting the right space becoming more critical than location and image, and
- shortage of space, forcing companies to consider locating outside the City.

A number of specific building requirements also emerged:

- floors of at least 10,000 sq ft (929 sq m) and up to 30,000 sq ft (2,800 sq m);
- uninterrupted space with generous structural floor heights;
- finished floor to ceiling heights of at least 9 ft (2.7 m);
- highly serviced buildings with adequate vertical ducts, raised floors and zoned lighting, power and air handling, and
- dealing floors of 60 ft (18 m) depth, very deep raised floors; support space to cope with data/telecom needs, and boosted air conditioning.

In 1985, these were radical demands that lay well outside the institutional specifications. They were also, as we saw in Chapter 4, completely missed by the
Corporation of London as it prepared to release its *Draft Local Plan* in November 1984. The focus groups were an enormous success, with a direct causal link
being established between a changing business context and novel demands on
buildings.

Such was the success of the work that Rosehaugh Stanhope and DEGW
investigated further sectors of demand. Typical of the work was *The Space
Requirements of Professional Firms in the City of London*.[44] Semi-structured inter-
views supported by industry analysis were used to investigate the business
context of the leading firms, before implications for premises requirements
were elucidated. Following deregulation of the financial services industry,
professional firms were also undergoing radical change, although this was
barely recognised within the property industry. The report began with some
observations on generally held perceptions on the premises needs of professional firms:

- they generally share the same premises requirements;
- they are not demanding space users;
- they do not use much information technology, and
- the typical professional firm is slow to change.

The research found that professional firms had their own subset of requirements within the occupational market; that they were growing rapidly – often
through merger, and that their needs were changing rapidly, especially in terms
of location criteria, image and the uptake of information technology. More
specifically, the work found that

- they were being squeezed by the demand for office space from financial
  services, and they would consider more fringe locations;
- they were growing at an annual average of 13%; diversifying and
  internationalising;
- the amount of space they required had grown by around 40% since 1981;
- they needed larger buildings and deeper floors;
- they needed more sophisticated provision of support space, and
- they were beginning to adopt much more intensive use of technology,
  both desktop and centralised, with implications for mechanical and electrical servicing.

The research painted a picture of change and upheaval within a sector that
had previously exuded an image of stability and conservatism. For example,
the report found that the use of technology by fee earners in legal firms was
projected to increase markedly, and that their offices would be networked for
electronic mail and to allow access to common data. The main implication
of such a development was that cable management and raised floors would

become more important to deal with highly cellular space and a high take-up of information and communications technology. And this would mean significant changes to the institutional specification.

Some of the mechanics and findings of the research might seem a little naïve in today's market. For example, it is difficult now to imagine a City legal practice without electronic mail and "access to common data". But at the time it represented a mind shift. The shift that was taking place was a fundamental one: from a position where for three decades buildings had been designed and built to institutional standards in a kind of 'one size fits all' regime, to a position where product innovation through understanding and responding to customer needs became an acceptable means of developing buildings. The importance of the research also lay in the fact that it sought to discover what was happening within the businesses that occupied buildings and worked outwards to the implications for their occupancy needs and, ultimately, to the changing morphology of the City.

As well as pioneering product development through customer research, the Stanhope and DEGW collaboration also resulted in the first rigorous approaches to evaluating the qualities of buildings in a consistent and comparative manner. The earliest work, *Four Properties Compared*,[45] was also undertaken for the Finsbury Avenue project, and compared its physical and design attributes with three other projects. The report covered areas such as floor efficiency, servicing (electrical and mechanical), parking and amenities, fitting out costs and service charges.

The report was followed in 1986 by an altogether more ambitious project and perhaps the most influential report of the genre: *Eleven Contemporary Office Buildings: A Comparative Study*.[46] The work built upon the Planning Office Space in recognising the importance of integrating the planning of the services backbones of buildings with their efficient space planning to creating buildings that could cope with increasing organisational complexity; an emergent need for flexibility in terms of, for example, occasional re-stacking of space to accommodate corporate changes, and the growing influence of office technology. The work systematised the comparison of different building layouts and objectively assessing their relative merits. Figure 7.7 is an extract from *Eleven Contemporary Office Buildings*.[47] This illustration examines the depth of space, which is a key element of layout planning.

The style and format of *Eleven Contemporary Office Buildings* were simple to use and visually pleasing, and the method gained widespread acceptance within the London property market in a very short time. The report itself made very clear:

> The purpose of this comparison of eleven contemporary office buildings in the City of London is to remedy the damaging lack of information about building performance. The data presented here are not about how buildings are constructed but what they can do – [and] what capacity they have to accommodate the new kinds of City organisations.[48]

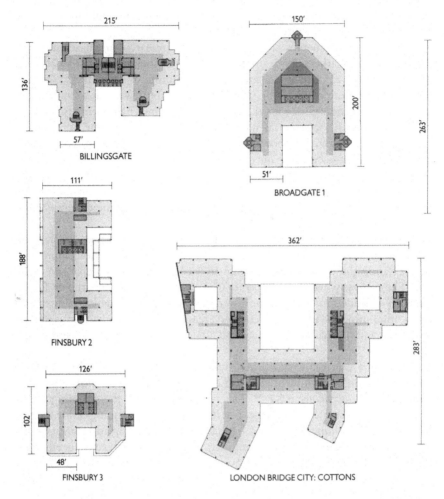

215'

150'

136'

200'

263'

57'

BILLINGSGATE

51'

BROADGATE 1

111'

362'

188'

283'

FINSBURY 2

126'

102'

48'

FINSBURY 3

LONDON BRIDGE CITY: COTTONS

Figure 7.7 Planning office space at Broadgate

Source: Image by kind permission of AECOM/University of Reading, Special Collections.

The report suggested that the design community needed this comparative information in order to

> put together building blocks based on user requirements in ways which are inventive and which will perform far better than the fragmented, underserviced, poorly engineered, ugly structures which were typical of developers' architecture in the City of London.[49]

Broadgate was at the time the largest development in the City, but soon most developers were having their projects evaluated. The effect of the building appraisal techniques was to generally raise the quality of building specification in the City, as competitive pressures between developers led to a widespread adoption of comparative studies. Combined with the customer research described earlier, there emerged a general consensus that product development was a positive force in the market. Perhaps more significantly still, for the first time in commercial property development, there was feedback from users to the development and design process. Market gossip and the received wisdom of the agency community were dramatically relegated to irrelevance by a user-informed, research-led and systematic approach to design and development by a new kind of developer with a new kind of product.

This new approach to design and development coincided with the emergence of the digital office, where technology assumed a far more fundamental role in the work and workstyles of occupier businesses. It was also coincident with the rise of the knowledge economy (Chapter 3). Herein laid the seeds of another fundamental challenge to the traditional, professional structure of the real estate industry. This was the rise of the 'flexible space' market with its focus on simplified tenure arrangements and the provision of service. Flexible space providers, or operators, in effect disintermediated traditional professions, by assuming overall control of the delivery process, and presenting to the client a finished working product, on flexible terms and with services.

7.3 Towards service

Three decades ago, the property industry offered, essentially, two contractual options to occupier businesses: freehold or long (25-year) leasehold, complete with onerous obligations. However, in response to the recession of the early 1990s, when vacancy levels rocketed, average lease lengths shortened, and serviced offices arrived from the USA as an alternative to long-term commitments. Since the turn of the century average lease lengths have shortened further and other forms of occupation have emerged. Together, these various products form the flexible space market (FSM).

In 1989 a new company appeared in the property sector. This was not another 'PropCo', nor was it a new form of investor. It was not even an advisory company offering any of the professional skills described earlier in this chapter. Rather, it was an entirely new kind of company, an *intermediary* in the property process. Mark Dixon – with no previous connection to the commercial property sector – took an idea from America and launched Regus in Europe, specifically in Brussels. Regus was a serviced office provider. For the first time, there was an actor in the supply process between the traditional landlord and the occupier.

Dixon was, as far as the property sector was concerned, unconventional. He was not a Chartered Surveyor, nor did he have a track record in the acquisition and management of buildings. He was what later became known as a 'disruptor' – he aimed to turn the traditional landlord-tenant relationship on its head. While in the end he succeeded, it was by no means an easy journey.

Born in 1959, the son of a Ford motor car engineer and a 'housewife', he left the fairly unremarkable Rainsford Comprehensive school in Essex at the age of 16 to fulfil the life of an entrepreneur. His first business in 1976, called 'Dial-a-Snack', was a telephone ordering business in Chelmsford. He delivered sandwiches on a butcher's bicycle. He was moderately successful, but then he sold the business to a local sandwich shop. He then went on an extended break, travelling the world, funded by a series of short-term jobs, arriving back in the UK in 1979.

In 1983 he set up his second business, investing £600 on a van from which to sell burgers and hot dogs from a pitch on the North Circular Road in London. Before long, he had a fleet of seven vans selling burgers across Essex. When Dixon encountered problems with the supply of buns for his burgers, he set up a full-time bakery in St Albans and ended up selling to businesses nation-wide. A decade after returning to the UK, he secured a life-changing return of £800,000 by selling The Bread Roll Company; whereupon he move with his new wife to Brussels. While pondering what to do next, he decided to set up an office. This was not as simple as he had expected, and therein he saw his opportunity. He quickly established a serviced office business.

Dixon set out to offer occupiers flexible terms, rather than the more tra-ditional institutional lease with its onerous and long-term obligations. The basis of the business model was to take traditional leases from landlords and then break the space up into smaller units, offered to occupiers on licence for months rather than years and on 'easy-in, easy-out' terms.

Dixon capitalised on the recession of the early 1990s by taking space from distressed landlords, who were suffering record vacancy levels and were happy to offload empty property to the upstart entrepreneur. By 1999, Dixon had grown Regus to over 230 centres operating in over 40 countries, with revenues of c£200 million. He was yet to have his 40th birthday. It was at this point that a degree of hubris led to Regus's first stumble.

Dixon decided to float the Regus business, and when the pathfinder pro-spectus for the business was unveiled, it suggested a value somewhere between £1 billion and £1.5 billion, with Dixon's personal stake worth up to £800 mil-lion. But then Dixon cancelled the float at the last possible moment in Septem-ber 1999. Question marks had been raised over the valuation of the company (which was around ten times higher than the previous years' sales of £112 mil-lion, together with an operating loss of £9 million) and its growth forecasts (sales of £1.2 billion and profits of £184 million by 2005).

Despite the severe setback, within a year, Dixon was planning a fresh attempt at flotation, this time, successfully. The company floated in October 2000 at 260p, valuing the company at £1.5 billion. At this point the company oper-ated over 300 centres in 47 countries. Despite various economic shocks, the business survived. In 2013 Regus opened in its 100th country, Nepal, and its 1,500th centre in Pune, India, and paid £65 million for control of MWB Business Exchange, the UK's second largest serviced office provider. In 2015, the business acquired Dutch coworking brand Spaces; in 2016 it established a new holding company called IWG plc, and in 2017 it acquired the rights to Stockholm-based members club No18 under a franchise agreement. It was firmly established as the world's largest provider of flexible workspace.

What Dixon and Regus had done was to establish 'flexible space' as a viable alternative to the traditional landlord lease. He did this on the back of the dramatic expansion of internet-related technology companies, and the multitude of related and spin-off businesses, together with countless knowledge economy micro-businesses. While the dot.com crash sent Regus shares into freefall, it proved to be a temporary adjustment in the face of fierce speculation. Once the technology markets had settled, demand rose once again. And as more and more companies adopted emerging technologies, so demand for flexible space surged. Dixon had tapped into one of the key symptoms of the 'knowledge economy' or the 'weightless economy' – the desire for flexibility, including a reduced commitment to real estate.

Drivers of change

The flexible market evolved during the 1990s and 2000s but remained relatively small in terms of the number of providers and the number of centres. A paper from 2001 discussed the growth of Regus while highlighting the relatively small-scale nature of the flexible space sector. Nevertheless, the article underlined the growing importance of flexibility to occupier businesses, and the importance therefore that the industry "allows its customers to adapt to rapidly changing business environments with space designed, procured and managed for fluidity, connectivity and support".[50]

And growing market acceptance of the new model was eventually forthcoming. From around the Global Financial Crisis in 2008 rapid growth took hold. By 2014, an industry report found almost 600 centres nationally, accommodating over 21,000 firms and 104,000 workers, occupying c17 million sq ft (1.6 million sq m).[51] There then followed a ramping up in activity, and by 2019 there were reported tb be around 6,000 centres nationally.[52] While there might be some data compatibility issues between the two sources, the scale of growth is very significant. London's flexible space market accounts for between 5% and 7% of the capital's total office stock of 240 million sq ft (22.3 million sq m), or between 12 million sq ft (1.1 million sq m) and 16 million sq ft (1.5 million sq m).[53]

As shown in Figure 7.8, the rapid growth in the sector was underpinned by the maturing knowledge economy, evermore ubiquitous and mobile technology, rapid corporate change and increasingly agile workstyles.[54]

The diagram shows a range of different drivers (which might vary according to the business sector being analysed). The companies that respond to the drivers cover a broad spectrum of types (from self-employed, start-ups and SMEs to larger firms). They will be influenced in their locational and premises choices by a range of factors (attractors), and they have available to them a range of 'property products'.

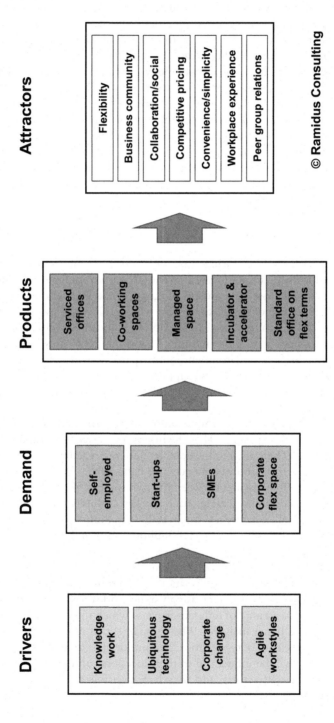

Figure 7.8 The dynamics of the flexible space market

This shift in the relationship between demand and supply represents the 'commoditisation of space' – no longer a long-term commitment with onerous obligations, but available as-and-when needed, on flexible terms. This is a direct response to the disruptive impact of the technology-driven, knowledge economy.

For larger occupier businesses, the flexible space market provides agility and flexibility. It offers 'easy-in, easy-out' terms and allows businesses to avoid the capital costs normally associated with establishing a new office, including fit out, furniture and fixings. Over the next decade, many larger, corporate organisations will migrate towards some form of networked, 'hub and spoke' model. The largest occupiers will maintain at least one corporate headquarters in a key city to provide a 'corporate hub', to bring employees and clients together, nurture the corporate culture, encourage collaboration and learning, and instil a sense of 'shared purpose'. Elsewhere, they will occupy flexible space which will afford them the ability to change their occupational footprint speedily and cost effectively.

For smaller businesses, the model means that they do not need to pay for expansion space which might be needed at a future date, nor for space that is used only occasionally such as large meeting rooms or conference facilities. Businesses in flexible space are able to combine their buying power for services such as a reception, connectivity, technology, security, telephony and meeting rooms. It gives them flexibility and the opportunity to have a presence at the heart of their market cluster, on terms that suit their business models.

The Centre for Cities & Cambridge Econometrics (2015) noted that many SMEs operating in the creative and digital industries are sole traders or employ a small number of staff, that they "often prefer smaller, more flexible premises" and that increasingly flexible work spaces "are allowing for co-location, lower overheads and the capacity for businesses to grow quickly".

There remains within the real estate supply industry a level of scepticism over the flexible space market. It is sometimes described as a 'fad' or 'fashion' that will re-adjust in harsher economic conditions. It is also viewed with some suspicion by the investment community, which has a problem with valuing buildings which receive unpredictable income rather than a gold-plated institutional rental income. Part of the continuing scepticism is related to the recent history of provider, and market leader, WeWork.

The growth of WeWork in London and the UK was unprecedented. The business leased its first building in February 2014: 40,400 sq ft (3,750 sq m) in Sea Containers House on the South Bank. A further three leases followed in 2014, giving a year total of 136,300 sq ft (12,660 sq m). The following year, WeWork leased almost 640,000 sq ft (60,000 sq m). By the end of 2019, WeWork had leased 4,516,000 sq ft (420,000 sq m) in 63 separate leases in London (Figure 7.9). Six of the deals had ranged between 150,000 sq ft (13,900 sq m) and 350,000 sq ft (32,500 sq m). In just six years, WeWork had grown from nothing to become London's largest private sector occupier, by a wide margin.

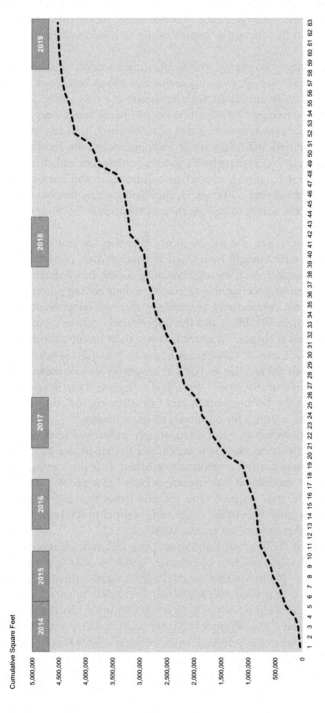

Figure 7.9 Cumulative lettings (sq ft) by WeWork, 2014–2019

In 2017, WeWork took its first lease in a UK regional city (at Spinningfields in Manchester), and through to the end of 2019, took a total of one million square feet (100,000 sq m) in Birmingham, Cambridge, Dublin, Edinburgh and Manchester.

But then, towards the end of 2019, WeWork's expansion halted; an October stock market listing was pulled; recently negotiated rental deals fell through, and discussions started with its main creditor, Softbank. The very existence of the company was being questioned. But the company survived, only then to be confronted, like the rest of the industry, with the Covid-19 crisis. Obviously, shorter-term licences are easier to exit than longer-term leases.

Time will tell how the flexible space market recovers from recent events, but the fundamental drivers summarised in Figure 7.6 above remain firmly in place to suggest that recovery will at least follow the industry-wide trajectory. It might even outpace the wider market as occupiers recognise the benefits of flexibility.

According to recent research from JLL, upwards of 30% of all commercial office space might be consumed as flexible space by 2030;[55] while Simonetti & Braseth predicted that the largest commercial landlord of the future will own no physical space and that, by 2030, most companies' real estate requirements will be outsourced and consumed on demand, with only 20% taken via traditional, long-term obligations.[56]

Products and emerging market structure

Fundamentally, flexible space has two defining characteristics: shorter, less onerous contractual terms and a greater or lesser degree of service provision.[57] With a combination of simpler contracts and a level of service provision, the FSM responds to the highly uncertain business context within which most firms now operate, when business planning horizons are short term and when occupiers demand more service and more choice. Figure 7.10 compares the characteristics of a conventional lease with terms in flexible space.

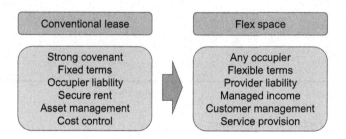

Figure 7.10 Conventional lease compared with flex space

Source: Image by kind permission of Investment Property Forum.

The conventional market relies upon strong covenants, fixed terms and secure rent, in a bond-style arrangement. 'Assets' are managed (lightly) and cost containment is a principal driver.[58] By contrast, the FSM pays very little attention to an occupier's covenant. Flexible terms ('easy-in, easy-out') are paid for with a fixed, 'unitary charge' which is all-inclusive (rent, rates, service, technology, etc.), without onerous exit terms (such as dilapidations). Another key attractor is the ability to 'flex', or grow/shrink in terms of space occupied. Whether the customer is a company in early growth phase and looking to scale up, or a larger corporate looking for project/temp space, or space to establish a foothold in a new city, the appeal is strong.

Broadly speaking, the market breaks down into managed, serviced, coworking and member products (Figure 7.11). The diagram shows these products in relation to conventional FRI leases and the recently introduced 'CAT A+' solution which is, essentially a CAT B fitout with 'plug and play' capability. Hybrid space can also be offered with more than one offer in the same centre or building, for example managed and serviced, or managed and coworking.

In addition to these main components of the flexible space market, there are more specialist offerings. Some are sector-specific, for example catering for 'digital' firms or 'fintech' firms; while others cater for those in music and

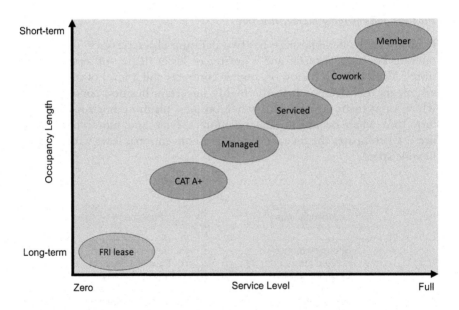

Figure 7.11 Flexible space market products

Source: Image by kind permission of Investment Property Forum.

media. There are also incubators and accelerators which are based around business support and mentoring, where occupiers enter into time-limited, formal start-up programmes.

The emergence of the flexible space market reflects a fundamental shift that has taken place in our understanding of the role of real estate. Where once it was seen as a fixed, leaden backdrop to work, suddenly it was coming to be seen as a corporate resource to be managed like other corporate resources (Chapter 9). The flexible space represented a structural shift in the market, calling into question traditional supply processes. Longer, more onerous leases will reduce significantly in coming years as a great proportion of the market moves to flexible space (Figure 7.12).

Today, there is more activity providing shorter-term space with a low level of service (blue box), and much more activity providing high-quality experience on short-term contracts (yellow box). Good-quality space on relatively long terms with moderate service provision (pink box) also remains important – particularly for corporate occupiers.[59]

7.4 The industry reforms

The organisation of the property supply industry into its constituent professions largely took place almost two centuries ago; with the boundaries between activities ossifying nineteenth-century socio-economic structures into a 'property supply process'. The supply process has become ever more sclerotic as professional jealousies have prevailed over market requirements and as each major division – architecture, surveying, construction – has spawned countless sub-disciplines, many of which overlap, to create a labyrinthine process that is all but impenetrable to the customer.

The first genuine attempts to understand the nature of demand and to allow this to influence the nature of the product took place in the City of London in the run-up to Big Bang in 1986. A collaboration between an unconventional developer and an unconventional advisor produced some highly innovative thinking – and an altogether better product that became a new standard. The quality of office buildings made great strides, responding at last, to the rapidly changing demands of the customer base.

But, while a huge step forward, a better product continued to hide the inadequacies and inefficiencies of a nineteenth century supply process. Changes to the latter only began to emerge in the 2000s, when the disruptive impact of technology and a new breed of entrepreneurs combined to change the supply process. The rise of the flexible space market was a critical response offering, for the first time, a genuine alternative to the restrictive and onerous landlord-tenant contract, with a service based, experience-based, easy in - easy out customer-operator relationship.

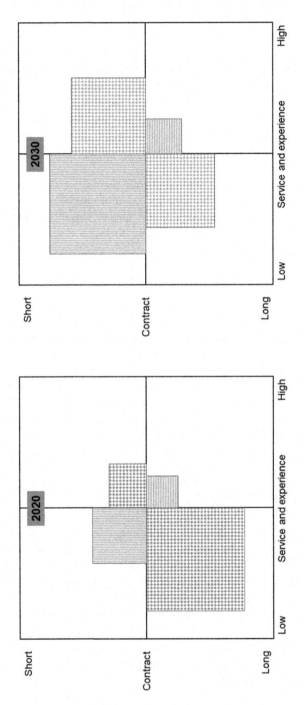

Figure 7.12 The growing flexible market in 2020 and in 2030

Source: Image by kind permission of Investment Property Forum.

Notes

1 M. Neal & J. Morgan (2000) The Professionalization of Everyone? A Comparative Study of the Development of the Professions in the United Kingdom and Germany, *European Sociological Review,* Vol 16, No 1, pp9–26

2 Powell (1996) *Op cit,* p76

3 Cited in: O. Bradbury (2018) *The Overlooked Influence of John Soane on Architecture from 1791 to 1980: Critical Appraisal,* Doctoral Thesis, Sheffield Hallam University

4 Bowley (1966) *Op cit,* p341

5 *Ibid,* p348

6 J.B. Waring (1849) The Diploma Question, *Builder,* 24 November, pp559–560

7 J.T. Micklethwaite (1874) *Modern Parish Churches,* King, p236 Cited in Powell (1996) *Op cit*

8 Powell (1996) *Op cit,* p30

9 J. Mordaunt Crook (1969) The Pre-Victorian Architect: Professionalism and Patronage, *Architectural History,* Vol 12, pp62–78

10 Bowley (1966) *Op cit,* p342

11 *Ibid,* pp348–349

12 J.H. Clapham (1950) *An Economic History of Modern Britain: The Early Railway Age 1820–1850* (2nd edition), Cambridge University Press, Cambridge, p162

13 *Ibid,* p164

14 J.H. Clapham (1932) *An Economic History of Modern Britain: Free Trade and Steel 1850– 1886,* Cambridge University Press, Cambridge, p146

15 *Ibid*

16 Powell (1996) *Op cit,* p31

17 Bowley (1966) *Op cit,* p336

18 *Ibid,* pp335–336

19 *Ibid,* p350

20 F.M.L. Thompson (1968) *Chartered Surveyors: The Growth of a Profession,* Routledge & Kegan Paul, London

21 *Ibid,* p25

22 A. Leyshon, N.J. Thrift, & P.W. Daniels (1987) *Large Commercial Property Firms in the UK: The Operational Development and Spatial Expansion of General Practice Firms of Chartered Surveyors,* Working Papers on Producer Services Series No 5, University of Bristol and University of Liverpool

23 *Ibid,* p5

24 Thompson (1968) *Op cit,* p109

25 *Ibid,* pp92–93

26 *Ibid,* p65

27 *Ibid,* p94

28 *Ibid,* p128

29 *Ibid,* p133

30 *Ibid*

31 *Ibid,* p152

32 Powell (1996) *Op cit,* p280

33 *Ibid,* p285

34 Bowley (1966) *Op cit,* p349

35 *Ibid,* p350

36 F. Duffy, C. Cave, & J. Worthington (Editors) (1977) *Planning Office Space,* The Architectural Press, London

37 F. Duffy (1983) *The Orbit Study: Information Technology and Office Design,* DEGW/ EOSYS, London

38 Centre for Advanced Land Use Studies (1983) *Property and Information Technology,* CALUS Research Report, College of Estate Management, University of Reading

39 *Ibid*, pp1–2
40 A. Salata (1983) *Offices Today and Tomorrow,* College of Estate Management, Reading, p85
41 DEGW (1985) *Accommodating the Changing City,* Unpublished Report DEGW, London
42 *Ibid*, p6
43 *Ibid*, p8
44 DEGW (1987) *The Space Requirements of Professional Firms in the City of London,* Unpublished Report DEGW, London
45 DEGW (1983) *Four Properties Compared,* Unpublished Report DEGW, London
46 DEGW (1986) *Eleven Contemporary Office Buildings: A Comparative Study,* Unpublished Report DEGW, London
47 *Ibid*
48 *Ibid*, p4
49 *Ibid*
50 Harris (2001) From Fiefdom to Service: The Evolution of Flexible Occupation, *Journal of Corporate Real Estate,* Vol 3, No 1, pp7–16
51 IPD (2014) *The UK Business Centre Market,* Business Centre Association, London
52 Instant Group (2019) *UK Market Summary: Flex is Leading the Way,* Instant Group, London
53 Ramidus Consulting (2020) *Property Ownership in a Flexible World,* Investment Property Forum, London, p6
54 R. Harris (2019) Real Estate in the Digital Era, *Journal of General Management,* Vol 44, No 3, pp119–127
55 JLL (2017) *Workspace, Reworked: Ride the Wave of Tech Driven Change* JLL, London
56 R. Simonetti & H. Braseth (2017) *Your Workplace, On Demand: Five Predictions for the Future of Work,* Convene, London
57 Ramidus Consulting (2020) *Op cit*, p3
58 *Ibid*, p4
59 *Ibid*, p9

8 Working

From corporatism to individualism

In this chapter we turn to the office as a place of work – from the clerical factories of the nineteenth century, through the corporate offices of the twentieth, to the digital workplaces of modern times. Across this span of time, the design of the workplace, its servicing, furniture and equipment and its management have changed radically. Chapter 9 examines the management of the workplace, while this chapter deals with the world of work.

Early on in the modern era, the office performed an ancillary function to manufacturing and trading activities, largely recording and accounting for transactional activity. Before the end of the nineteenth century, purpose-built office buildings were increasingly common, housing the burgeoning banks and insurance companies. As the economy generally evolved, so the office economy grew, eventually providing accommodation for the corporate behemoths of the twentieth century. The past three decades have seen the arrival of the knowledge economy and the digital workplace – transforming again how office work is undertaken.

Organisation of work

Frederick Winslow Taylor (1856–1915) was born in Germantown, Philadelphia, into a wealthy, highly educated and liberal-minded Quaker family. His father was a lawyer and graduate of Princeton, and his mother an abolitionist and feminist. Following a three-year 'Grand Tour' of Europe, Frederick enrolled into the highly respected Phillips Exeter Academy in New Hampshire. However, his health was poor, and despite passing the Harvard entrance examination, he left Exeter prematurely. After several months of recuperation, and in his 18th year, he chose a vocation in engineering, joining Ferrell and Jones (later the Enterprise Hydraulic Works). Despite his social background and obvious intellectual strengths, Taylor followed the pure and principled nature of his upbringing and started his new vocation as a humble pattern maker's apprentice. On completing his apprenticeship in 1878, Frederick moved to the Midvale Steel Company to take a post as a machinist.

His inquisitive and innovative mind soon marked him out for greater things, and he scaled the corporate ladder rapidly, rising to research director and then

Figure 8.1 Frederick Winslow Taylor

chief engineer. His dedication was to be demonstrated by his ability to earn an engineering degree from the Stevens Institute of Technology in New Jersey, at the age of 25, while working full-time. Taylor was an innovator and a rationalist, and he was imbued with a belief in the role of science and technology in the advancement of mankind.

While at Midvale, Taylor studied industrial productivity and methods for measuring output. He introduced piecework, developing detailed systems aimed at maximising the efficiency of men and machines in the factory. His work involved time and motion studies to determine the best methods for performing tasks in the least amount of time. His basic thesis was that incentive-based wages were ineffective unless they were used alongside carefully planned and managed tasks, backed up by supportive and co-operative management. He argued for the careful selection and development of people, identifying clear goals and tasks, organising feedback, providing training and encouraging group support. His methods were successful and output rose significantly. Scientific management was born.

Taylor became a consulting engineer at the age of 37, often making enemies in the companies where his methods resulted in job losses aimed at increasing efficiency. His most important client became the Bethlehem Iron Company, but following various disputes with the management, he was fired in 1901. This was to be his last job. His most famous published work, *The Principles of Scientific Management*,[1] was based largely on transcripts of talks given long after leaving employment. He contracted influenza in 1915 while on a speaking tour and died the day after his 59th birthday. Taylor's legacy was to influence management thinking, both in the factory and the office, for nearly a century, as offices came to resemble factories stuffed full of white-collar workers, processing paper rather than widgets.

Taylor also left a string of scientific management disciples to continue his work: Henry L Gantt (critical path scheduling), Frank B Gilbreth (hospitals), and William H Leffingwell (secretarial work) were all prominent. The last of these published *Scientific Office Management* in 1917 and the mammoth 800-page *Office Management* in 1925.

Taylor's formative years in industry coincided with the vertical integration of the US economy, when American business switched structural models. Up to the 1880s, the American economy comprised mainly local, entrepreneurial, owner-managed businesses and was characterised by many layers of production and distribution. The new model was to be dominated by large, vertically integrated businesses, which gained significant economies of scale and stripped out the add-on costs of intermediaries, brokers and distributors. This time was the dawn of the 'corporation' – the large, complex, multi-layered, multi-departmental organisations that were to dominate the twentieth-century office economy. Europe and the UK were to follow similar models of corporate organisation, although never on such a scale in such numbers.

The new businesses practised centralised command-and-control management, and there was a consequential rise in white-collar work as the emerging behemoths generated a demand for process management, administration, sales, accounting and so on. The demand for new office buildings rose accordingly. In just one example, during the latter half of the 1880s, 400 office buildings were completed in central Pittsburgh.[2]

Largely in the shadow of Taylorism, offices retained their rigid, command-and-control layouts through the 1920s and 1930s. Constant supervision, no privacy, sedentary work and uncomfortable furniture were commonplace. There was also growing separation between 'management' and the 'workers', and a growing divide in the quality of furnishings and perks. In many cases, management would retreat behind glass-walled private offices to ensure that they could continue to monitor the work schedule, whilst at the same time enjoy privacy and more comfort.

8.1 The tools of the trade

Enormous economic change during the latter half of the nineteenth century led to industrialisation and mass production processes, with goods being manufactured on a scale unimaginable in the early part of the century. Driven by world trade – in both goods and commerce – industries expanded rapidly, and in doing so, they mechanised and automated. The growing availability of cheap power in the form of electricity served as a reinforcing agent. The office economy was not immune from this tumultuous change. We saw in Chapter 3, for example, how buildings evolved rapidly during this time, with the introduction of the elevator and reliable sanitation allowing buildings to increase in size and height.

Within the office building, the workplace itself also saw dramatic changes during the later years of the nineteenth century, with new forms of communication, machines and equipment replacing manual labour and allowing companies to scale up their operations: "administrative spaces were transformed within a few decades from places that resembled medieval scriptoria into the machine-rich environments we associate with the modern office".[3]

Many of the innovations from this time have passed firmly into history (the telegraph and mechanical adding machines); the vestiges of some remain with us (the QWERTY typewriter keyboard and telephone); while others remain with us almost complete (the ballpoint pen and the paper clip). The picture of innovation also reflects changes in social relations as well as economic relations, with a growing female workforce and an increasingly educated workforce.

Telegraph, telephone and typewriter

The telegraph was first used commercially on the railway, specifically between London's Euston Station and Camden Town. Here, in 1837, William Fothergill Cooke and Charles Wheatstone demonstrated to the management of the London & Birmingham Railway company how the technology could contribute to railway safety. Cooke and Wheatstone went on to set up the Electric Telegraph Company which would ultimately morph into BT. Telegraph cables were then laid across the English Channel in 1851 and, after a few failed attempts, across the Atlantic in 1866. Just four years later, in 1870, a telegraph link was completed between Europe and India. A year later, a cable reached Australia.

Significantly, a method of printing the telegraph message was invented in 1854, and was developed in the USA as Hoase's printing telegraph. In 1867 the New York Stock Exchange installed the electric stock ticker. Telegraph lines multiplied in Europe and the United States in the latter part of the nineteenth century, greatly increasing the volume of work done by the office.

> As messages came in over the telegraph they would have to be recorded and processed. And the very rapidity and ease of this new form of communication would increase the volume of messages. In order to keep up with competitors, or with the news of world events which now flowed in with increasing rapidity, an administrator must issue orders and directives more quickly.[4]

At the same time, the telegraph was critical in making the detachment of administration and production more practical. The resulting spatial and organisational impacts reinforced the growth of the office economy as a distinct entity, particularly in the city centre.

It was not long before the telegraph had a competitor in wireless technology. Guglielmo Giovanni Maria Marconi (1874–1937) was born in Bologna into Italian nobility. Home-educated and without higher education, the young Marconi became an inventor and engineer, undertaking world-changing work on long-distance radio transmission; he is credited with inventing the radio. He found success as an entrepreneur in London, where he founded the Wireless Telegraph & Signal Company in 1897, the forerunner to the Marconi Company. In May 1897, Marconi sent his first message over the open sea – a full 6 km between Flat Holm Island in the Bristol Channel and Lavernock Point in Penarth, Wales; in 1901, he despatched the first transatlantic message.

Effective wireless communication, whether by telegraphy or telephony, was firmly established in the early years of the twentieth century, linking far-distant trading posts with a head office. The two technologies co-existed for many years: "wireless communication complemented and did not replace submarine telegraphy; when Britain's international telegraphy companies were consolidated into a single unit in 1928 it embraced both and from 1934 was known as Cable & Wireless Ltd".[5]

The telephone was the third major office workplace innovation from this era. American-born Thomas Edison and Scottish-born Alexander Graham Bell introduced their telephone systems in 1879 and 1880, respectively. The success was immediate: there were more than one million telephones in the USA by the end of the century.[6]

The telephone changed office activities in a similar fashion to the telegraph a few decades earlier. First, it provided people with instant contact with others located at a distance – colleagues in distant offices, customers, supply chain, factories, warehouses and so on. The main impact of distance – length of communication – had been overcome, ultimately allowing the spatial reorganisation of business. Secondly, within the office itself, the telephone helped change processes and the internal organisation of space, as well as increase the workload.[7] Confirmation of orders, instructions, appointments and so on became one of the major components of office activity.

The first submarine telephone cable, laid in 1891, linked Paris with London. But this technology was slow to develop, and it was many years before long distance, radio-based services became available. Transatlantic telephone calls between New York and London began in 1927, quickly spreading tentacle-like across the North American continent and mainland Europe. Shortly afterwards Britain and Australia were linked, but all such services were for many years prohibitively expensive, and telegraphy remained the clear leader.[8]

The telephone and other contemporary innovations led to new forms of organisational structure and, as Cowan and colleagues observed, helped create an office economy that was no longer simply part of other facets of urban life but a facet in its own right.[9]

> But if we look back to the end of the nineteenth century we can see that the office function had become well established as a separate entity in the city: new buildings were being constructed to house it; more and more of the labour force were working in offices, and women especially were beginning to go into office work in large numbers. The outlines of the office as we know it were laid down.[10]

But the overriding point was that the means had been created to link trading networks across regions and the globe, and communications were now possible on the same day, when before a written communication could take days or weeks. The telegraph and telephone symbolised a new era of global trading, and intensification of activities in trading capitals, including London. At one and the same time, the new communications overcame the friction of distance between production and administration, while encouraging the physical concentration of the growing number of specialist roles in the latter.

While perhaps lacking the glamour of the new 'tele-communications', the typewriter was in many ways equally world-changing, becoming perhaps *the* defining icon of the office landscape, until the arrival of the computer. It symbolised the growing scale and complexity of organisations by formalising the means of written communication and, with copying and printing, *mass* written communication. The typewriter was first patented in the USA in 1868 by Sholes, Glidden & Soule; and sales figures for the follow-up Remington No2 grew from "146 in 1879 and rose to an annual figure of 65,000 in 1890".[11] But the adoption of the typewriter was not immediate.

> At first, the typewriter was not seen as indispensable to business. The personal touch of the hand-penned letter was considered so important that the typewriter was ruled out of commercial use by the pundits.

Soon, no business

> could be indifferent to the greatly increased pace set by the typewriter. It was the telephone paradoxically that sped the commercial adoption of the typewriter. The phrase, 'send me a memo on that', repeated into millions of phones daily, helped to create the huge expansion of the typist function.[12]

The operators of these early machines typed 'blind', unable to see the emerging text. The Remington No2 set the standard for the universal key layout, the 'QWERTY system'. The Remington brand became established in Europe in the mid-1880s and 1890s.

Figure 8.2 Remington Standard Typewriter, 1890

Mumford summed up the impact of the evolving office technology on organisational structure – internally and spatially – and the emergent office economy.

> With the manufacture of the typewriter in the seventies, and the coincident spread of high speed stenography, more and more business could be conducted on paper. Mechanical means of communication and mechanical means of making and manifolding the permanent record, mechanical systems for audit and control – all these devices aided the rise of a vast commercial bureaucracy, capable of selling in ever remoter territories.[13]

Suddenly, the means of communication was at the heart of the growing office economy, and the typewriter was a fundamental part of it. Right through to the mid-1980s, almost every business employed a raft of secretaries and it helped to influence the design and layout of offices because, in larger organisations, workflow often involved typists being centralised in specialised typing pools, replacing copy clerks. For eight decades and more, typewriters were used to produce letters, memoranda and reports and documents, often on headed notepaper.

Between 1870 and 1914, many offices were transformed by the increased use of female clerks using the new technology: female clerks grew nearly tenfold between 1871 and 1911.[14] In this way, the typewriter had an enormous impact on social relations, drawing into the office a whole new class of worker-women, whose labour had been previously confined to the factory or to domestic service. Cowan *et al* noted that "the number of female workers rose quickly, so that they very soon began to outnumber men in the more routine sections of the office".

The telegraph, typewriter and telephone revolutionised the way in which office-based businesses communicated with one another, but one of the biggest by-products of these innovations was paper – vast amounts of paper on which to specify, order, verify, promise, record and log countless contracts and transactions from around the world.

Paper, filing, storing and copying

Offices are often referred to as 'paper factories', reflecting the fact that while the production of paper had evolved over several centuries, during the nineteenth century and the Industrial Revolution, the process advanced rapidly through mechanisation. In 1797, French engineer Louis-Nicolas Robert is credited with having invented the first machine to produce continuous paper, a forerunner to the Fourdrinier machine. Following a military career, Robert became a clerk at a Parisian publisher, the family business of Didot. In his book *Papermaking: the History and Technique of an Ancient Craft*, Dard Hunter noted that Robert had declared that "constant strife and quarrelling among the workers of the handmade papermakers' guild" drove him to invent a machine to replace labour.

The prevailing method involved paper being made one sheet at a time, by dipping a rectangular mould into a vat of pulp, which was then drained, leaving a pulp which was set aside to dry. When this was complete, the frame was re-used to repeat the process all over. Robert's construction, by contrast, had a moving screen belt that received a continuous flow of stock and delivered an unbroken sheet of wet paper to a pair of squeeze rolls.

Following a failed prototype, a successful model was produced in 1798. However, plans for patenting and production were frustrated by the small matter of the French Revolution. Plans were further complicated by disagreement between Robert and Didot. Not to be deterred, Robert brought the patent to England, where he was introduced via his brother-in-law John Gamble to Henry and Sealey Fourdrinier. The brothers owned a stationery house, and they agreed to finance the detailed design and manufacture of the machine.

The first Fourdrinier papermaking machine was built and commissioned in Frogmore, Hertfordshire, in 1803. A patent was not granted until 1806. The machine could make a continuous web of paper, which not only transformed productivity but also led to production in roll form, for applications such as wallpaper printing.

Despite its technical success, both the Sealey brothers and Robert failed to benefit. The patent proved difficult to protect, and the system became widely copied. Having sunk a fortune into its development, the brothers became bankrupt. The personal impact on Robert was greater. Having lost control of his invention, and having failed to make his deserved fortune from it, in 1812 Robert returned to France a broken man. He opened a school in the chaos of post-Napoleonic France, earning a living, just, until his demise in 1828.

As is so often the case, invention creates demand. So it was with paper. The potential of mechanised production revolutionised the role of paper in the office. But one of the side-effects was a growing shortage of cotton rags, a key material in the process. A step forward was made in 1843, when German machinist and inventor Friedrich Gottlob Keller invented a wood-grinding machine that produced groundwood pulp ideal for papermaking. Machine-made paper became the norm and, by 1850, 90% of paper in Britain was machine-made.[15]

The next big step occurred in 1854, when Hugh Burgers and Charles Watts used chemicals to convert wood into pulp. The steps from rags to pulp, and from mechanical to chemical pulping, allowed paper to be mass produced. Together with the advent of the practical fountain pen, ink wells and pencils, cheap paper not only transformed the office economy but society more generally: novels, children's books and reference texts, magazines, newspapers and school books, all were now more widely available and affordable, along with paper for writing letters, diaries and sketching. Having found the means to mass produce, new means had to be found to store the product.

As organisations in the office economy grew in size and complexity, so they began to find novel ways to store documents. Early filing 'systems' were very basic, often as simple as metal spikes onto which orders and receipts, for example, were impaled. However, such "primitive systems would only be effective when record volumes were relatively small, as was the case with the vast majority of businesses before the nineteenth century".[16] Also common until the late nineteenth century were pigeonholes and cabinet drawers, which were widely used to store rolled or folded papers. Loose-leaf ring binders were available by the end of the century, as was the 'manila' file in which letters were

> placed in a strong paper or linen cover (which can be had in various colours to distinguish departments, subjects, geographical division) on the back of which are placed metal strips, which pass through corresponding holes pierced in the document, the ends of the strips being bent down over the top letter and held in place with a clip.[17]

With more and more paper being produced, and the growing size and complexity of businesses, especially in banking and insurance in the City of London, so there was a growing need

> not just for relatively sophisticated file stationery, but also for complex filing systems subject to greater intellectual control. This was especially so for 'strategic' information, say about a particular project, transaction or policy, as opposed to 'routine' information such as invoices and remittance advices or papers relating to, say, an insurance policy or bank account.[18]

Filing was revolutionised in 1898 by the invention of vertical filing systems. The inventor of this simple system was Edwin G Seibels (1866–1954). While raised on a cotton plantation at Mount Willing in Edgefield County, South Carolina, the young Seibels spurned an agricultural career and went to work aged 16 in the family's insurance business in Columbia, E. W. Seibels and Son (now the Seibels Bruce Group). While there he attended South Carolina College, from which he graduated with a degree in mechanics and engineering in 1885. In his day job, Seibels was involved in exporting and insuring cotton, particularly to Russia, until the outbreak of World War One. However, it was not for his prowess in insurance, let alone his knowledge of cotton, for which Seibels is remembered today. His legacy to the world was the art of vertical filing – the humble filing cabinet.

Seibels' central insight, that papers could be kept in large envelopes standing on end vertically in a drawer, obviated the need for folding and rolling. Not only that, records could be arranged according to subject, contract, customer, supplier, or whatever instead of, as was widely practiced, by date of receipt or despatch. Cabinets and boxes could replace the older, bulkier bound ledgers and bundles of paper. Greatly excited, Seibels convinced the Globe–Wernicke Company of Cincinnati to make five wooden filing boxes to his specifications. Confident in his invention and with an eye to personal wealth, he applied for a patent, only to be brusquely informed that his invention was not a device, merely an idea, and ideas were not patentable!

Seibels eventually took over the helm at the family firm, where he spent much of the rest of his working life. In 1909 he was elected to the South Carolina House of Representatives. He died on 21 December 1954, aged 88. Whilst financially comfortable, his 'idea' never made him the fortune that would have flowed from a patent.

While letter copying presses were used in the final decades of the eighteenth century, the mass production of cheap paper for all, together with vastly enhanced means of physical storage, fed the urge to copy more widely. As early as the 1870s, it had become commonplace to insert carbon paper or tissue between sheets of paper, while using the pressure of a pen or a pencil to produce a number of copies, up to four or five, and sometimes more. It became routine to indicate to the recipient of the 'top copy' to everyone else who had received the same document, or a carbon copy. Hence the origin of 'CC', now a ubiquitous element of business emails.

According to Orbell, a favoured solution for situations demanding less high-quality finish was found in duplication using stencils initially cut by handheld pens. The pen was used to make patterns of fine holes in impermeable paper through which ink was subsequently forced to create an image of handwriting on sheets of paper below. The earliest commercially viable stencil copying product, the papyrograph, was patented in 1874 and was commonplace by the 1880s.[19]

Of course, producing limited numbers of copies became ever more constraining as organisations grew in scale and complexity, as well as geographically dispersed, not to mention their external relationships, such as supply chains. The number of people needing to be informed of 'corporate decisions' and 'corporate deliberations' grew rapidly, and new means had to be found to produce paper copies in great numbers. It was therefore not long before mechanised copying arrived. The Neostyle Duplicating machine was one such machine. It could make 1,000 to 2,000 copies from one original document, with the operator sitting at the machine, overseeing the automated supply of ink, automatic impression and automatic copy production.

Scribing and addressing

The quill pen and ink bottle were familiar staples of the office through the transition from the eighteenth to the nineteenth century. Awkward to use and in need of constant sharpening and filling, quills remained in common usage well into the 1830s. The development of steel nibs began in the 1820s and were commonplace by mid-century. While more reliable in legibility terms, metal nibs suffered the same need as quills to be constantly re-filled. It was not until the 1880s that pens with an integral ink reservoir were invented. Lewis Edson Waterman (1837–1901) patented his first fountain pen in New York in 1884. Mass production followed and, of course, Waterman pens remain with us to this day.

The first ballpoint pen was designed by American inventor John Jacob Loud (1844–1916), who worked as a cashier at the Union National Bank in Weymouth, Massachusetts. Loud obtained a patent for the first ballpoint pen in 1888, but it had a flaw: while suitable for use on rough surfaces such as leather and wood, it was not commercially viable. The patent eventually lapsed, and it was to be another 50 years before a commercial product emerged.

László József Bíró (1899–1985) was a Hungarian-Argentine inventor whose major gift to society was the ballpoint pen – the eponymous biro. Bíró's innovation was a new form of tip, a free-turning ball, for which his chemist brother, György, helped devise a new type of viscous ink, stored in a cartridge. When the ball turned, it collected ink from the cartridge and deposited it onto paper. Bíró showed the first production of his product at the Budapest International Fair in 1931, and patented the invention in Paris in 1938.

The brothers fled to Argentina during the war, where they continued to develop and produce the ballpoint pen. In 1945, the patent for Bíró's pen was bought by Marcel Bich, whose company in Paris made parts for fountain pens and mechanical lead pencils. Bich paid US $2 million for the patent, which soon became his company's biggest seller. The company continues to trade today as Société BIC Group, selling millions of BIC pens each year.

As American corporations became larger and larger, spreading over ever greater distances, and as their customer bases broadened to include not only other businesses but the individual customer, it was necessary to find means of communicating *en masse*. One means of achieving this was to automate the addressing of envelopes, invoices, ledgers and so on. Orbell notes that there were two early forms, one based on stencils (offering flexibility as they could more easily be made in-house) and the other on embossed metal plates (giving the clearest print). Addressographs developed greater capacity and flexibility, and by the turn of the twentieth century, treadle-operated machines were available; by the 1930s, automatic selection of address plates and selection of only certain fields for printing was possible.[20]

Adding and accessories

While crude adding machines date back to the early seventeenth century, they were large and had very limited functionality; they were not suitable for office work, using levers rather than keyboards. As the office economy was expanding, and as trading volumes in many firms were growing and internationalising, there was a growing need for a mechanised form of calculation. However, the first mechanical calculator to enter mainstream use was not created for offices, but for military deployment. It was invented by the wonderfully named Frenchman, Charles Xavier Thomas de Colmar (1785–1870).

Born in Colmar, capital of the Alsace wine region, Colmar joined the French army in 1809. He rose in seniority rapidly, until he became Inspector of Supply for the whole of Napoleon's army. It was in this early logistics role, with the vast number of calculations that had to be made, that he conceived the idea of a mechanical calculator. Colmar created his Arithmometer in 1820, for which he was made Chevalier of the Legion of Honor. The Arithmometer was very successful and sold well through to the First World War.

However, almost inevitably, corporate America eventually caught up. In 1877 George B Grant of Boston began producing the hand-cranked Grant mechanical calculating machine, capable of addition, subtraction, multiplication and division; while in 1886, William Seward Burroughs introduced the P100 Burroughs Adding Machine. The Burroughs family business, which became the Burroughs Corporation, was enormously successful, releasing grandson William S from the drudgery of business life to write novels including *The Naked Lunch*.

Perhaps the first office calculator was the Comptometer, first produced by Felt & Tarrant Manufacturing Company in 1887. The Comptometer took calculating into the push button age. Two years later, the same maker introduced the Comptograph, which came complete with the printer.

In an echo of earlier industrial innovations, Turck stated that adding and calculating machines "met with very strong opposition for the first few years", and that "It was strongly evident that the efforts of book-keepers and counting-house clerks to prevent these machines entering their department were inspired by the fear that it would displace their services".[21]

In addition to the new telecommunication technology and the arrival of machines in the office, many more modest inventions in the nineteenth century became indispensable accessories for the office worker. These included rubber bands (introduced by British firm Perry & Co in 1844), brass paper fasteners with two 'legs' inserted through paper and then bent in opposite directions (US patent 1866) and envelopes which, when mass production methods were created alongside the falling cost of paper, got going in the 1850s and 1860s.

In 1867 inventor George McGill received a patent for a machine that would press the fastener through paper. His stapler reached the mainstream market for the first time in 1869. In 1895, the first stapler to use a long strip of bendable staples, wired together, was introduced by the E.H. Hotchkiss Company.

All this before one of the humblest office accessories – the paper clip. The Gem-paperclip, the double-oval-shaped paperclip with which we are so familiar today, was designed and manufactured by British company Gem Manufacturing Co from the early 1870s. But it was not patented. In 1899 William Middlebrook, of Waterbury, Connecticut, patented a machine for making paper clips of the Gem design.

The tools and equipment of the office evolved quickly in the final decades of the nineteenth century. As businesses grew in scale, geographical spread and complexity, so a symbiotic relationship with their tools and equipment allowed them to mature further still. Never before, and not for another 100 years, would the technology of the workplace have such a profound impact on the businesses which it underpinned.

Mechanisation was labour saving but also allowed mass production; while the diversification of tools led to increased job specialisation and new social relations with a growing female workforce. From the typewriter and telegraph, through filing and printing, through adding and addressing machines and

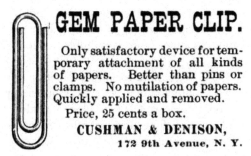

Figure 8.3 Advertisement for the Gem Paper Clip, 1893

through paper, pens and paper clips, the tools of the emerging office enabled scale and complexity in business structures, and they led to ever more intricate bureaucracies.

Controlling the environment

Whether in Chicago's tall skyscrapers or Europe's lower rise buildings, floor sizes generally were getting larger. As corporate organisations grew in size, so they sought to bring more and more people together into the head office. This meant that the size of individual floors was growing, which meant in turn that novel ways had to be found provide 'services' – air handling, heating/cooling, lighting and power.

Two key technologies allowed the office floor to get larger and deeper, and thereby accommodate growing numbers of sedentary workers: 'man-made weather' and lighting. Both air handling and fluorescent lighting promised greater flexibility in office design, but what they meant in practice was that managers could lay claim to the external façades for their offices, while office interiors became deeper and deeper, allowing ever denser, production-style environments.

Air conditioning first became available to cool industrial processes rather than provide personal comfort to office workers. Indeed, the term 'air conditioning' was in fact coined by Stuart W Cramer (1868–1940), founder of Cramerton. Cramer was deeply involved in the textile industry and secured many patents for humidity control and ventilating equipment in cotton mills. He first used the term in an address to the American Cotton Manufacturers Association in 1906.

However, New Yorker Willis Haviland Carrier (1876–1950) is known as the father of air conditioning. He graduated from Cornell University in 1901. He had created cooling systems in the early 1900s, particularly for printing and publishing works, but scaled new heights in 1911 when, at the annual meeting of the American Society of Mechanical Engineers, he presented his *Rational Psychrometric Formulae* – the cornerstone of the air conditioning industry to follow. Carrier's paper explained how relative humidity, absolute humidity and the dew point temperature could be harnessed to build air-handling systems that could provide constant, manged temperatures in buildings – most significantly here, in office buildings. Carrier's patent solved the problem of humidity removal by condensing the water vapour on droplets of cold water sprayed into an airstream.

Carrier founded the Carrier Engineering Corporation with a group of colleagues in 1915, and he continued to develop his industrial applications. But it was clear that other applications were being called for, and in 1922 Carrier created a smaller and more powerful centrifugal refrigeration compressor, which became the precursor to modern air conditioning systems.

Carrier installed the first air conditioning system in an office building at the 21-story Milam Building in San Antonio, Texas, designed by George Willis, in 1928. It had a central refrigeration plant in the basement that supplied cold water to small air-handling units on every other floor; these supplied conditioned air to each office space through ducts in the ceiling; the air was returned through grills in doors to the corridors and then back to the air-handling units.

The first air conditioned office building in London was the sprawling Broad-casting House, for the BBC, designed by G Val Myers. Completed in 1931, the building set particular challenges for studios which needed acoustic treatment and air conditioning, and administrative offices which needed to maximise daylight. The air conditioning system was designed and installed by Carrier Engineering.

Along with the Taylorist thinking of the time, it was widely believed that better lighting would improve productivity. Big improvements were brought by the widespread availability of the incandescent light bulb, which changed the way offices were lit. As early as 1860, in England, Sir Joseph Wilson Swan (1828–1914) developed a primitive electric light which used a filament of car-bonised paper in an evacuated glass bulb. But the lack of a vacuum helped reduce its reliability and lifetime. Twenty years later, Edison patented an incan-descent bulb – largely based around Swan's design – with a filament that gave off light when heated to incandescence by an electric current.

Initially, Edison used a carbonised thread for the filament before experi-menting with card paper and then carbonised bamboo. Extruded cellulose fila-ments were introduced by Swan in 1883. The next significant innovation came with the invention of the tungsten filament in Europe in 1904. The new fila-ments lasted longer and were brighter. But it was the introduction of the fluo-rescent lamp in the 1930s that revolutionised office lighting: early fluorescents provided more than twice as much illumination per watt as incandescent lamps, without the heat output.

The first credible fluorescent lamp is usually credited to Edmund Germer (1901–1987), born in Berlin, Germany, and educated at the University of Ber-lin, where he wrote a doctorate in lighting technology. Together with Frie-drich Meyer and Hans Spanner, Germer patented an experimental fluorescent lamp in 1927. However, Germer's invention was not successfully commercial-ised. This was not to occur until scientists at General Electric in the USA, led by George Inman, created an improved and practical fluorescent lamp, first sold in 1938.

The physical organisation of work and our ability to control the working environment are intrinsically linked. People need lighting, air handling and furniture, all of which need to be laid out efficiently and effectively. No matter where a desk is situated – in the centre of a floor, on the façade or somewhere else – it needs to be heated, cooled and lit in the same way as every other desk in the building. This requires co-ordination between the organisation of work and the services needed to support it.

Thus it was that office building technologies such as air conditioning and lighting combined with office equipment and improvements in office management and processes to enable the growth of very large, corporate organisations. However, the two World Wars, the Spanish Flu, the Great Depression following 1929 and a worldwide economic malaise combined to mean that the office economy changed relatively little during the first half of the twentieth century. Certainly compared to the tumultuous years of the later nineteenth century, it was a quiet period for the office economy. This all changed after the Second World War.

8.2 Company man to agile worker

Chapter 3 provided some insights into the office workplace of the later nineteenth century, not least through Charles Lamb's experience at the East India Company's headquarters in the City of London, when he described how "wearisome it is to breathe the air of four pent walls, without relief, day after day, all the golden hours of the day between ten and four, without ease or interposition". Work conditions in offices of this period were poor, reflecting their ancillary function to the main activities of businesses. Conditions were often cramped, poorly lit and poorly ventilated; and the work was often repetitive and physically demanding.

As offices grew larger and more complex in the later nineteenth century, so obviously did their workforces which had increasingly to be 'organised', physically, in space. And as work became more 'organised', the office workplace came to reflect two key drivers: status and function. Those in more senior positions (both within the office and socially) were afforded more space and greater comfort; office workforces resembled military regimentation, with the other ranks, NCOs and officers, each respecting of the others' rank and duties. The other ranks were assembled in large platoons, undertaking tiny, abstract, aspects of the overall operation and without sight of the greater plan. They were corralled and directed by NCOs who received daily operational orders from the officer class, some distance away on another, usually higher, floor of the building.

> On the shop floor, a corps of supervisors acted as an information conduit between management and problems arising in the production process. . . . Information was starkly centralised in a bureaucratised hierarchy, workers' informational input to the production process having been brutally reduced by de-skilling and a minute division of labour.[22]

The organisation of the office and the organisation of office-based enterprises, reflected Taylorist management thinking; physical arrangements mirrored production processes; and the tools and machinery were heavy and fixed.

While *building innovations* enabled changes in building form (principally larger and higher), so *workplace innovations* shaped the organisation of office work. And as organisations (and buildings) grew in size, and as workplace innovations increasingly demanded regiments of relatively unskilled clerks, so the role of information assumed ever greater import. Consequently, new approaches to information management appeared, including:

> statistical analysis, graphic representation, the internal memo, the staff magazine, the management meeting, schemes for classifying documents, the procedural manual and written protocols.[23]

Social status also became more pronounced as offices became more uniform – ways had to be found to differentiate space to reflect social or corporate standing. The most obvious step was the divide between open plan and cellular space. Increasingly, private offices claimed the outward-looking parts of office floors, pushing open-plan clerical staff into the less well-lit, central areas of the floors. Private offices proliferated as 'middle management' expanded rapidly in the post-war years. Even in the open plan there was hierarchy, expressed through desk size and pedestals – even carpeting was used to designate areas and status.

The parallel development of the hardware and the software led to ever more subtle division of labour and ever more subtle layers of seniority and responsibility. Those who copied documents were less skilled/lowlier than those who typed them, who in turn were less skilled/lowlier than those who dictated them.

The rise of company man

In the early years of the twentieth century, the US economy continued to vertically integrate, creating vast companies that controlled supply chains from production to sales, and the national economy also evolved from a "society of island communities into a homogenous national community".[24] The implications for the office economy were profound – both in the USA and worldwide as other national economies mimicked the American corporate model.

Large, inflexible, departmentalised, hierarchical and slowly moving businesses dominated by clerical, process-based work and with clear demarcations between grades and functions of work grew and dominated the office economy in the early twentieth century. The headquarters of large manufacturers, banks and insurance companies, advertising and media businesses, lawyers and accountants and so on all concentrated into large buildings. Layer upon layer of management evolved, and companies divided like amoebae into complex departmental structures. This was the age of corporatism, and 'Company Man'.

Micklethwait and Wooldridge suggested that Company Man's character had been formed as early as 1920, to reflect two characteristics: professional standards and corporate loyalty.

> He was part of a professional caste and adopted Frederick Taylor's motto that there was 'the one best way' for organising work and sneered at rough-hewn entrepreneurs for not knowing it.[25]

The same authors refer to the "triumph of managerial corporatism" in which the "multidivisional firm was an important innovation . . . because it professionalised the big company and set its dominant structure". While referring to largely industrial conglomerates, including King Gillette, William Wrigley, HJ Heinz and John D Rockefeller, the authors were targeting their office activities. They "hired hordes of black-coated managers to bring order to their chaotic empires", and American cities were "redesigned to provide these managers with a home – the new vertical filing cabinets knows as skyscrapers".[26]

William Hollingsworth Whyte (1917–1999) was an American journalist, urbanist and organisational analyst. Born in West Chester, Pennsylvania, Whyte graduated from Princeton University in 1939 and served in the US Marine Corps between 1941 and 1945, including time as an intelligence officer in the Guadalcanal campaign. In 1946 he joined *Fortune* magazine.

In 1952, Whyte coined the term "Groupthink" by which he meant not so much the politically loaded definition familiar today, but more a form of rational thinking leading to conformity with a set of group values that work in a positive manner for the greater good. His best-known work was *The Organisation Man*,[27] which was a polemic on conformity, specifically, conformity to employing organisations by a generation of risk-averse middle managers. Whyte's Organisation Men are those who 'belong' to the organisation; they are "the ones of our middle class who have left home, spiritually as well as physically, to take the vows of economic life, and it is they who are the mind and soul of our great self-perpetuating institutions".[28]

Whyte argued that, before the rise of corporatism towards the end of the nineteenth century, a key assumption of the Protestant Ethic was that success was due to neither luck nor the environment "but only to one's natural qualities". By contrast "the big organization became a standing taunt to this dream of individual success", and it became clear that in corporate organisations, those who survived and succeeded "were not necessarily the fittest but, in more cases than not, those who by birth and personal connections had the breaks".[29] In Whyte's description of individualism versus collectivism, organisation man surrenders his individuality (through for example, decision-making) to the organisation in return for stability and security.

It was this sacrificing of flawed individuality on the altar of more perfect decisions of the organisation that came to symbolise the huge American corporations rising up across the USA, and also Europe during the twentieth century. Anthony Sampson captured the essence of the phenomenon in his book *Company Man: the Rise and Fall of Corporate Life*.[30] He suggested that the corporate organisations that emerged late in the nineteenth century in the USA, with their massive structures, rigid hierarchies and lifetime employment, created the professional manager, or company man, thereby allowing the middle class to grow immensely.

By the 1920s and 1930s, Sampson argued, British corporations were emerging along similar lines with their own armies of loyal company men (for they were largely such) occupying managerial roles in corporate organisations typified by the likes of BP, Shell and Unilever: corporate leviathans that bestrode the economic landscape for much of the post-war period.

The 1950s and 1960s, it now transpires, were the golden years for Company Man, because this stable, almost paternal corporate environment began to change in the mid-1980s when, as Sampson argued, the security of Company Man began to be undermined, and a changing economic landscape sounded the death knell for large, slumbering, multi-layered and complacent organisations.[31]

Despite the success of 'corporatism', as measured in its own terms, the work environment had failed to modernise. The economic, social and physical damage wrought by the Second World War was, of course, immense, and there ensued a period of deep, prolonged austerity which was reflected in the workplace. Drab, grey environments proliferated in cost-conscious organisations. Office furniture was basic; equipment (typewriters, copiers and so on) was noisy, and regimentation in organisation resumed. Chairs were often made of wood; they were non-adjustable. As paper-based systems and processes grew rapidly, so an army of drab grey filing cabinets mustered on every floor, often being used as surrogate walls or partitions to separate teams and departments.

Indeed, in 1977, Frank Duffy and colleagues felt able to write

> Older architectural magazines reveal with cruel clarity the prevailing standard of office design in the late 50s and early 60s. Advertisements display more vividly than editorial what seems to us now to be an astonishingly uncomfortable internal physical environment – hard floor finishes, lots of glazed partitioning, surface mounted light fittings, bulky steel desks.[32]

The authors went on to suggest that, in the UK, technical problems rather than user requirements or experiments in plan form were what obsessed architects. However, change was on its way, to strike the first blows in a long campaign to fundamentally change approaches to the layout and management of the workplace.

The birth of the modern workplace

The year 1958 was very significant for the office economy, and specifically for the office workplace. Two coincident events, on each side of the Atlantic, took on the prevailing Taylorist thinking in the command-and-control approach to office management. The first of these was the appointment of Robert Propst at Herman Miller in Colorado; the second was the establishment of the Quickborner consulting group in Hamburg, Germany.

Propst (1921–2000) was already a prolific creator and intellectual when Herman Miller president Hugh DePree enticed him from his part-time role as professor of art at the University of Colorado in 1958. The firm was keen to diversify beyond its core market of office furniture design, into the health and education sectors. Propst was to head Herman Miller's new Research

Corporation in Ann Arbor, Michigan. Tellingly, Propst read widely in areas including anthropology and sociology – areas of considerable influence over his work at Herman Miller, which lead him to such observations as:

> The caveman was undoubtedly very pleased to find a good cave but he also undoubtedly positioned himself at the entrance looking out. Protect your back but know what is going on outside is a very good rule for survival. It is also a good survival rule for life in offices.[33]

Propst also regarded himself as an antidote to the Taylorist thinking that was still prevalent in office design and layout. Rather than seeing the office floor as a white-collar version of the factory, in which the objective was to increase the output of widgets with minimal input of additional resource, Propst's vision was one led by an understanding of office work as a mental activity.

> Today's office is a wasteland. It saps vitality, blocks talent, frustrates accomplishment. It is the daily scene of unfulfilled intentions and failed efforts. A place of fantasy and conjecture rather than accomplishment. It fosters physical and mental decline and depresses capacity to perform. It is the equivalent of doing business on clay tablets in an age of the computer and instant communication.[34]

Just as his anthropological reading taught him about the relationship between man and his environment, so it informed his thinking about the worker and the office. It was but a short leap to addressing what features of the work environment could be enhanced in order to improve the productivity of the worker. Clearly, layout and furniture were fundamental to such considerations.

In 1964, as corporate America was expanding at a phenomenal rate, Herman Miller unveiled the fruits of Propst's labours – the Action Office. The name was no accident: the system that Propst had come up with was the antithesis of the sedentary furniture of the past. Instead, workers were offered lightweight, colourful furniture that enhanced both privacy and movement. It was, however, a commercial failure. Several reasons have been advanced for this, but the most likely reason seems to be that it was 'over-designed', with elegant lines and generous material, and was also not sufficiently flexible. It was also deemed to be expensive for mass deployment. Some have argued that it was simply ahead of its time. Among its positive points was its system approach and the use of vertical elements for shelving and display, as well as to provide the occupant with an element of privacy.

So, Propst went back to his drawing board and started work on Action Office II (AO-II), in search of a more egalitarian solution that would introduce real flexibility into rigid office environments. The results appeared in 1967. The revised furniture system comprised two- and three-walled, fabric-covered hinged dividers, built for movement and flexibility. The kit was modular, allowing many arrangements; it was standardised and elements were interchangeable: ultimately flexible (Figures 8.4 and 8.5).

As teams changed, or the business changed, so the kit could be simply and cheaply disassembled and re-built elsewhere. It afforded individuals a level of privacy and personalisation whilst also providing for shared space and collaboration. Screens of varying height doubled as storage units, shelving and display space, whilst also allowing workers vistas across the office floor, as per his cave analogy.

The simple, system approach combined space delineation (the partition), with storage shelves and racks with desks and chairs to produce a 'workstation' fit for the emerging knowledge worker. Instead of tying the worker to

Figure 8.4 Herman Millar Action Office

Source: Image by kind permission of Herman Miller.

Figure 8.5 Image from *A Facility Based on Change*

Source: Image by kind permission of Herman Miller

an immovable desk to undertake repetitive and routine tasks on a factory line principle, the new system approach provided for flexibility, agility and differential tasks.

In 1968 Propst published *The Office: A Facility Based on Change*. Part academic paper; part brochure and part operating manual, the document described change as a new constant force – 'the new master' – which "the office as an institution will no longer be able to ignore". He described the "lack of mobility in our physical facilities as the most stubborn laggard in offices".[35] Having set out the conflicting needs of privacy and collaboration, the book then set out the principles of the Action Office in terms of office layout. Figure 8.6 shows an image from the document.

Propst continued to enhance and innovate on the basic AO design. Eventually, this included floor-to-ceiling panels with windows and doors – full height partitions. Space could be easily and cost-effectively subdivided to create discrete departmental and functional areas within the office.

Little wonder than that AO-II was a commercial success. However, not all applications were true to Propst's vision: many adopted a more 'box-like' approach to space planning, leading to the commonly used cubicle, immortalised in Scott Adams's *Dilbert* cartoons.

Propst was 77 years old when he was interviewed by *Metropolis* magazine. In the profile, *The Man behind the Cubicle*, Propst was very frank with his interviewer Yvonne Abraham in his reflections on the longer-term success of his creation.

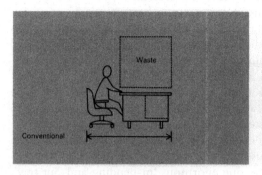 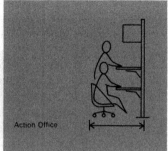

Figure 8.6 Herman Millar Action Office
Source: Image by kind permission of Herman Miller.

> The dark side of this is that not all organizations are intelligent and pro-gressive. Lots are run by crass people who can take the same kind of equip-ment and create hellholes. They make little bitty cubicles and stuff people in them. Barren, rathole places . . . I never had any illusions that this is a perfect world.[36]

Indeed, whatever Propst's original vision, the deployment of Action Office and derivatives thereof across the global office economy were used to perpetuate the search for efficiency in office layout. While superficially rejecting the principles of Taylorism, the new system furniture somehow reinforced command-and-control management in the workplace. This is less a criticism of the furniture itself than a recognition of the stultifying impact of corporate gigantism, as described by William Whyte and Anthony Sampson, and the art of packing thousands of people in one building, all geared to pursing the business objectives of a small managerial elite. Propst died two years after the *Metropolis* interview.

While expected to improve internal communications and interaction and to enable the faster and cheaper reconfiguration of space and people, in real-ity Action Office allowed offices to be planned at higher occupancy densities, and certainly did nothing to address hierarchy and isolationism; for most, the dull thud of office routine continued daily. In an echo of Propst's "daily scene of unfulfilled intentions and failed effort", sociologist Studs Terkel opened his book *Work* with the memorable "This book, being about work, is, by its very nature about violence – to the spirit as well as the body", followed by "To survive the day is triumph enough for the walking wounded among the great many of us".[37] Terkel then gave us his take on what work should be:

> about a search for daily meaning as well as daily bread, for recognition as well as cash, for astonishment rather than torpor, in short, for a sort of life rather than a Monday through Friday sort of dying.[38]

In the same year that Robert Propst moved to Herman Miller, on the other side of the Atlantic in Hamburg, Germany, two brothers were setting out on a very similar path. Wolfgang and Eberhard Schnelle had previously been working as assistants in their father's furniture studio. Like Propst, they could see that contemporary office furniture and layout solutions ossified activities and suffocated innovation with their serried rows of desks, rigid corridors, production line management and reinforcement of hierarchy. The brothers totally rejected the Taylorist scientific management approach to the workplace and focused instead on the flow of knowledge, or information, between individual workers. The Schnelle brothers seem to have been amongst the earliest to recognise that the emerging office economy workplace was not a production line, one enormous 'in-pending-and-out tray', but a complex set of relationships, where interaction was more important than process.

In 1958 the brothers left the family business and set up a consulting group – Quickborner Team. Their unique approach to space planning quickly became known as Bürolandschaft, a German term that translates to "office landscape". The basic concept was for a more natural, or organic, layout. The change involved stripping out uniform rows of desks, internal walls and private offices. Desks were arranged according to need, and the floor was softly divided by plants and low-level screens. The result was an open plan, free flowing and conducive to better communications.

Duffy and colleagues summarise three key features of Bürolandschaft. First, the approach addressed managerial and physical arguments in parallel. "The apparently idiosyncratic layout was not an arbitrary design whim but, so it was argued, an inevitable physical concomitant of a certain managerial style". Second, the arguments for Bürolandschaft "had a distinct moral edge. Not only must you have office landscaping but you ought to adopt this managerial style". Finally, approach complete: "Once the decision had been made, the techniques for applying office landscaping were already codified (in 68 rules) and ready to be put into effect".[39]

While modest today, these were bold steps at the time. Previously, there was a binary choice: the inefficiency and anonymity of cellular offices, or the efficient sterility of the American open plan. With office landscaping, the choice was much richer. Whilst at first glance Bürolandschaft appears chaotic and random, the layout is in fact a carefully considered reflection of the needs of individual teams for interaction with others. There are break-out and informal meeting areas; there is planting and carpets, and there are no private areas. The layout, or rather the approach to layout, had many advantages, including its ultimate flexibility and the speed and economy with which change could be effected.

But there were also some downsides. Not least of these is one of the most commonly identified problems of the open plan office to the current day – noise. The open plan allowed noise to travel and amplify, forming a significant distraction. The open plan also negated one of Propst's big achievements – a sense of privacy for the individual and the ability to focus on solitary work.

In an attempt to balance the strengths and weaknesses of the Action Office and Bürolandschaft, the variant 'combi-office' was adopted in some offices. As suggested, the combi-office consists of a mix of enclosed offices and open-plan space. The enclosed offices are generally either for a single person or a small group of people. They are typically arranged around a common area that includes break-out space, filing and storage, vending areas and so on.

Figures 8.7–8.10 show a series of images capturing the shifting styles of office layouts during the twentieth century. Figure 8.7 shows the 'clerical factory' of the early years, with its production line layout. The next three images show variations of the corporate environment. Figure 8.8 shows the more flexible low-level partitioning of the corporate office in the United States in the 1950s–1970s. Figure 8.9 shows the European response, with Bürolandschaft. Finally, Figure 8.10 shows the combi-office which sought to balance open plan and enclosed space in a palette of settings for different activities.

It seems that neither the Action Office nor Bürolandschaft pulled off a 'killer blow' in terms of radically changing approaches to office management and layout. As late as 1993, Frank Duffy and Jack Tanis felt able to observe:

> Taylor's ideas changed the face of the workplace 100 years ago and still dominate office design to this day. Compare such leading offices of the turn of the century as Frank Lloyd Wright's Larking Building of 1904 with those of corporate America of the last few years. The physical similarities are striking, the conservatism colossal.

Figure 8.7 Office layout styles: clerical factory

Source: Image by kind permission of Michael Bedford.

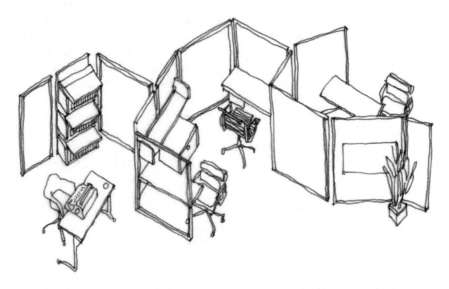

Figure 8.8 Office layout styles: system

Source: Image by kind permission of Michael Bedford.

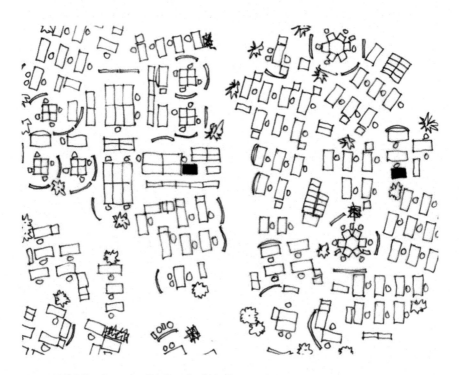

Figure 8.9 Office layout styles: Bürolandschaft

Source: Image by kind permission of Michael Bedford.

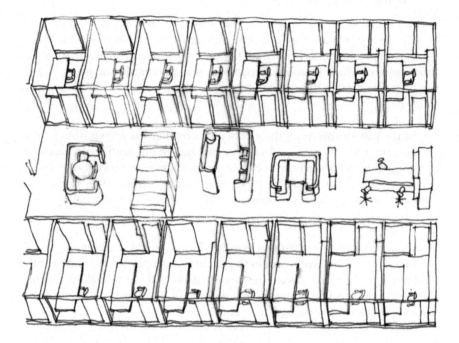

Figure 8.10 Office layout styles: combi-office

Source: Image by kind permission of Michael Bedford.

Noting the resemblance between the Larkin Building and a factory, the authors go on:

> Taylor's ideas of the mechanical, top-down, inhuman, status-rich, invention-poor, alienated workplace live on, manifested in every workstation, ceiling tile and light fitting.[40]

Duffy and Tanis put this down to the fact that the process by which office buildings, and especially office interiors, were procured had hardly changed since the first textbooks on office administration and office layout were compiled in the first decade of the twentieth century. Undoubtedly there is some truth in this. But what is equally true is that the *modus operandi* of most office work itself had not been shaken fundamentally. Work itself had not changed greatly: in several respects, the workplace contained many of the rigidities of earlier times: hierarchical, spatially divided according to strata of seniority, process-based and uniformly dull. Yet, the office economy was on the cusp of fundamental change: it was rapidly approaching the digital age; when new office technologies would help re-write the rule book of office workplace design, layout and management.

The transition to a digital workplace

William Goodwin was a geography professor at the University of Wisconsin. In 1965 he published a paper, *The Management Center in the United States*, in which (and picking up a theme from Chapter 3) he identified a knowledge gap about cities in which business management is important. In the paper, Goodwin drew attention to the emerging digital workplace in terms that, today, look antiquated.

> In the past twenty years the electronic computer has grown from a curiosity to a much-used tool of management. Many routine decisions are programmed for electronic computers; many data are processed by punch cards rather than by pencil, paper, and desk calculators. This 'revolution' has reduced the number of clerks needed to prepare raw materials for decision making.

Goodwin went on to suggest that this 'revolution' would result in one of two outcomes: either the "reduction in the number of employees needed to staff a headquarters office may hasten the concentration of decision making in a few locations", or "it may well mean the dispersion of headquarters because of the flexibility of data flows through a computer".[41]

Ten years later, in 1975, Peter Daniels was defining the office function in terms of five activities: receiving information, recording information, arranging information, giving information and safeguarding assets. Further, he defined the office environment as a place where "people sit at desks or tables, equipped or supported in their work by paper and pens, typewriters, dictating machines, telephones, telex, calculating machines and more advanced data-processing machines".[42]

Six years later International Business Machines (IBM) released its first PC, and just three years after that the Apple Mac was launched. The dawn of the digital office was, at one and the same time, both a very long time ago and also quite a recent event! The digital era has transformed office work. The spread of office technology – replacing all of Daniels's machinery – has defined a new era, in which technology-enabled work has utterly changed how office workers engage with the office place. Yet all this has been achieved easily within the lifespan of a single working generation.

The early stages of the transition saw small, tentative steps. Office equipment was becoming leaner and more efficient. For example, electric typewriters were improving the quality of office documents with, for example, proportional character spacing to mimic printed documents. While IBM was producing these in the early 1940s, the 'golf ball' typewriter was not to appear until the early 1960s (Figure 8.11). This machine, the IBM Selectric, featured an interchangeable typing head, permitting a variety of fonts.

Gestetner started his own duplicating business in the City of London as long ago as 1861. But copying took a great leap forward in the 1960s through the

Figure 8.11 The IBM Selectric typewriter 'Golf Ball', 1961

Xerox machine. The 1960s also saw the replacement of duplicating machines by the photocopier. As with the electric typewriter, early versions were being developed in the 1940s, but it was Xerox Corporation's 914 model, the first automatic, plain-paper commercial copier, launched in 1959, that set the pace of change. The machine could produce seven copies per minute and was one of the most successful Xerox products ever produced.[43] In 1963, Xerox unveiled its first desktop copier, the Xerox 813.

The Xerox Corporation was also responsible for the first modern fax machine. The word 'fax' is a contraction of facsimile and stands for the transmission of scanned printed material, via telephony, to a distant printer. In 1964 Xerox launched the LDX (Long Distance Xerography), followed two years later by the Magnafax Telecopier. The fax later integrated copying and scanning capabilities. By 1988, there were almost 400,000 fax machines in the UK, reflecting their dispersal from central stations to departmental and, in some cases, to individual desktop level.[44]

As well as these tentative steps, more fundamental progress was being made. For example, by the mid-1950s and early 1960s, one of the USA's most successful corporations to date had established itself as the uncontested leader in the development of computer technology. IBM was a sprawling corporation with manufacturing, R&D, distribution and sales facilities across the USA. IBM specialised in selling mainframe computers to big businesses, and in 1965 it introduced one of its most successful products – the System 360, with a wide range of scientific and commercial applications, that allowed customers to upgrade without the expensive cost of reprogramming applications.

By the 1970s, IBM and other manufacturers were making major advances in the manufacture of computers, resulting in the introduction of what would come to be known as the personal computer (PC). In 1981, IBM launched its PC in a move that was a catalyst for a revolution in software, spawning countless new firms developing products for word processing and spreadsheets, as well as applications with the prefix 'computer aided', as in computer aided design, or CAD.

The 1981 IBM PC spelt the death knell not only for the large corporate mainframe, with its desktop terminal linkages, but also for a range of other office tools, including the typewriter. IBM was also quickly joined by a plethora of competitors in PC production; for example, the Commodore 64 and Apple Macintosh. Still, compared to the computers we know and use today, the PCs of the 1980s were somewhat primitive. As an example, the Apple Macintosh 128K had a 6 MHz processor, 128 kilobytes of RAM and a black-and-white screen with 512 by 342 pixels, all dressed in a beige cover and weighing 7.5 kg (16.5 lbs).

But specification quickly improved. During the 1980s, Intel released a series of processors, the 80286, the 80386 and the 80486, the last of which had a 32-bit processor which had more than a million transistors on a single chip, a clock speed of 25 MHz and a 4-gigabyte memory space. And it was not just under the bonnet improvements. Apple's revolutionary 1984 Macintosh computer came with a Graphical User Interface (GUI). The screen of the 'Mac' was clear, mimicking paper with its black characters on a white background, and came with the iconic 'mouse' for clicking and dragging. It is difficult to conceive, over three decades later, how novel and exciting such developments were. Even so, the GUI did not gain universal application. By the end of the 1980s, PCs were everywhere in the office economy.

During the 1980s, desktop PCs with their large CPUs and CRT monitors spread rapidly, leading to a new generation of workstations capable of taking the PC footprint and leaving room for other work. Cable management become a major issue.

One of the most iconic developments of the 1980s was, the mobile phone. The 1980s saw the first commercial cell phones brought to market. Motorola launched the DynaTAC 8000X in 1983, complete with a $4,000 price tag, and in 1986, the 4500X which, together with its large battery pack weighed in at c3.5 kg. But the technology improved rapidly. In 1989 Motorola released the MicroTAC (the flip phone), and in 1996 Nokia released its 8110 (the slider phone); which it followed up in 2000 with the 3310 which gave Nokia market domination, not least due to its introduction of SMS (Short Message Service) a kind of precursor to tweets.

The first phone to call itself 'smart' was the Ericsson R380 in 2000. It was the first device to use the new Symbian OS. In the same year, Samsung launched its Uproar – the first with MP3 music capability (for up to one hour). The Palm Treo (2002) could dial from a list of contacts and send emails; while the Sanyo SCP-5300 (2002) was a camera phone in which photos could be viewed on screen. The Blackberry and the iconic Apple iPhone were launched in 2003 and 2007, respectively.

In 1990, the internet arrived, quickly followed by the release of MS Windows 3.1 in 1992. The latter came with major user interface improvements, including a True Type font system, thereby introducing desktop publishing capability for the first time. It also came with a new system allowing icons to be dragged and dropped.

While email was invented in 1971 by Ray Tomlinson, using the internet's predecessor, Arpanet, the more familiar formats of AOL and Outlook were released in 1993, with free email becoming available in 1996. Email gave companies instant communication, allowing people to share information, messages and files in a truly revolutionary manner. It also transformed how employees interacted with clients and colleagues. Connectivity was also greatly improved in 1999 with the release of WiFi.

This cocktail of technological innovations fundamentally questioned how office work was undertaken. Large, inflexible, hierarchical and slowly moving businesses dominated by clerical, process-based work and with clear demarcations between grades and functions of work dominated economic life through to the late 1980s. But the stable, paternalistic corporate environment began to change dramatically with the revolution in technologies available to every office worker, sounding the death knell for large, slow-moving, multi-layered organisations.

Organisations were evolving rapidly, and they were looking less like 'corporate islands' and were more extensively involved in complex supply chain relationships. The power of networks, involving collaborative production and multi-disciplinary skills, was coming to be realised, and there was a rapidly growing 'contingent' workforce: those who are self-employed or within small businesses, who provide specialist input to the business activities of larger firms.

Agile working

One of the most profound impacts of all this technology was to enable a wholesale change to the way in which we actually worked. In 1985, an article appeared in the *Harvard Business Review* that was particularly prescient in its claims for the unfolding impact of technology on the nature of work. It was written by Philip Stone and Robert Luchetti, from Harvard's Faculty of Arts and Sciences, and was titled *Your Office Is Where You Are*. The authors set out to "challenge the customary ways of thinking about offices" and to show how managers could "integrate physical layout, design, and communications to support organizational objectives".[45] These objectives were to emphasise informal exchange, reassign people to different work teams and study groups, provide employees with access to specialised equipment, value individual initiative and mobility, derive payoffs from serendipity, attract talented employees and increase productivity while reducing office costs.

Stone and Luchetti proposed that managers should rethink how both information and people flow in an office, and adopt 'activity settings' to provide a richer office experience with appropriate environments to suit the work in hand. Such thinking was a world away from the command-and-control systems that had been so dominant for so long. It was holistic, it was purposeful and it recognised technology as an enabler, not a driver. The newly emerging corporate real estate management profession (see Chapter 9) started to recognise the growing implications of organisational and technological change for the delivery of support services to enable new organisational structures and new patterns of work.

The notion that 'your office is where you are' encapsulated as well as any phrase the essence of the knowledge worker – someone whose skills and value to their employer related to their cognitive ability rather than their mastery of the processes of the white-collar factory. It presaged the oncoming 'new economy' or 'knowledge economy' (set out in Chapter 2), typified by knowledge and technology-intensive jobs and economic activity; and investment in knowledge-based assets or 'intangibles' that are enabled by "powerful and cheap computers and the 'general purpose' information and communication technologies".[46] Moreover, the knowledge worker was the 'agile worker', no longer shackled to a single desk, but released to work in a variety of settings, both within the office and beyond – at home, clients' premises, coffee shops, in transit via rail and air nodes, and so on.

As workers became more mobile, or agile, with greater choice over where and when to work, the office became less a place to go to work, largely alone, on a set of prescribed tasks, and more a place to visit and interact with colleagues and use support services. Cairncross argued that the office would become a place for the social aspects of work, such as celebrating, networking, lunching and gossiping.[47] And she was right. The ubiquitous impact of mobile phones, laptops, the internet and email presaged an era in which work itself has been transformed, conducted in ways entirely different to even the recent past.

As agile working spread, a new language in office design emerged. Group areas, encouraging interaction, co-operation, innovation and a higher density of use, replaced enclosed offices, which were seen to reinforce hierarchies and status, as well as consume valuable and expensive space. New approaches to space use such as, for example, hot desking, hotelling and virtual offices became commonplace, giving rise to touch-down areas, lagoons, oases and docking stations. Some reception areas became 'concierge' services where desks could be booked on an as-needed basis, and where personal belongings could be collected for the duration of the stay in the office.

One of the earliest examples of agile working in the corporate sector was that introduced by IBM at Bedfont Lakes, near Heathrow in 1992. On moving in, IBM had 600 'SMART' workers who used the office as a base but had no fixed, personal space. The sales staff were expected to be away from the office for large parts of their time, and were allocated a pool of available desks when in the office. Each desk could be personalised to accept an individual's telephone calls and to allow access to personal files. Each worker was equipped with a laptop for use at home or on clients' premises.

BT was one of the strongest advocates of agile working, with a blueprint known as 'Workstyle 2000' in 1995. The concept was first tried at Westside, near Hemel Hempstead. Around 1,500 people were accommodated in the 180,000 sq ft (16,500 sq m) building, with 1,100 using permanent desks; 200 hot desking and 200 using the building as a base. The workspace was provided with touchdown centres, hot desks, project rooms and quiet rooms. The ground floor provided restaurant, café, shop, conferencing facilities and social space.[48]

One of the most celebrated examples of new ways of working was that provided by British Airways in its purpose-built headquarters at Waterside, near Heathrow. The building was officially opened in the summer of 1998 and was designed specifically to help transform the way that people worked, to allow the business to streamline its processes and to increase efficiency. The architect for the building was Niels Torp, designer of the famous SAS Airways building in Stockholm. Waterside was formed of 6 four-storey buildings, all radiating from a central, enclosed 'street'. The street is 175 m (575 ft) long and includes cafes, a bank, supermarket, travel centre, newsagent and so on. The project involved the consolidation of nearly 3,000 people from disparate sites and was designed to accommodate a highly mobile workforce, with widespread hot desking and over 650 of the staff having no fixed desk.

Design practice DEGW (Chapter 7) was one of the first to categorise different workstyles and the work settings needed to support them.[49] DEGW identified four generic work environments – den, club, hive and cell – to express both the physical and behavioural characteristics of different kinds of work (Figure 8.12). The den environment supported group work where teamwork was high and autonomy low.

The club environment supported creative and knowledge-based work where concentration was important and where interaction and autonomy were important. The hive supported process work where both interaction and autonomy were low. And the cell described individual work where interaction was low and autonomy high. This categorisation has stood the test of time, and remains a useful way of gaining a rapid understanding of how an organisation's work processes translate into types of workspace.

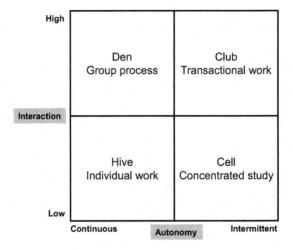

Figure 8.12 Workstyles and work settings

Source: Image by kind permission of AECOM/University of Reading, Special Collections.

DEGW had captured the essence of the strategy of IT-enabled agile work-ing, as had been outlined by Stone and Luchetti, and provided an intellectual underpinning for those organisations seeking to implement their own, specific projects. As the twenty-first century dawned, organisations were coming to be managed less as huge corporate machines and more as systems or networks in which work was increasingly about short-term assignments, communication, mobility and connectivity. In such organisations, knowledge becomes the cur-rency of exchange, and traditional approaches to command-and-control struc-tures break down.

Despite the enormous potential of technological advancement, our ability to absorb its implications and adapt rapidly is often limited by behavioural traits that anchor us in the present. Indeed, the full potential of agile working took time to settle in; so much so that as late as the mid-1990s, Micklethwait and Wooldridge[50] were able to make the rather sobering comment that the most horrifying thing about the future of work may be just how similar it will be. Indeed, through the 1990s, for the great majority of people, the office was not actually very different to that of 20 or so years before. While in a physical sense, offices were generally of a higher quality than those of the 1960s and 1970s, the activity of office work in the 1990s remained quite familiar.

Many people had yet to be liberated from a fixed workstation; many were neither able nor permitted to structure their own work pattern; many remained task oriented and felt oppressed by the layers of seniority above them. Some employers did not allow or encourage internet access and home-working for fear that such 'privileges' would be abused. And for large sections of the office workforce, the grim reality remained a daily commute to an office building and a workstation that, on a good day, just about allowed them to do their job, in only minor discomfort. Nevertheless, the genie was out of the bottle: the digital office had arrived.

8.3 The digital office

The first two decades of the twenty-first century have seen enormous changes in work and workstyles, driven by a spectrum of forces. Figure 8.13 summarises some of the change factors within which more specific drivers have helped shape the occupier agenda, which, in turn, feeds through to workplace implications.

For example, as businesses became leaner, cost conscious, and focused on productivity and access to skilled labour, so they sought to reduce their overall footprints, create a more effective workforce and shift their locational priorities, and generally become more 'flexible'. In turn, these had workplace implications in terms of the provision of commodity space, more active management of the workplace experience and an adaptable fitout.

Collectively, these features of corporate change altered the traditional bedrock of demand which once comprised large, relatively unchanging and predictable 'corporate islands' that were mainly process-based and which could plan ahead with a comparatively high degree of certainty. 'Company man' flourished in this highly divisionalised, hierarchical, command-and-control, relatively secure environment. *The Economist* magazine described an 'old model' in which workers tended to receive security, benefits and a regular salary 'for life', while employers had a stable workforce in which they could invest.[52]

But as technology transformed work, so the operational environments of businesses had to respond. This happened in many ways, but, in short, organisations became less hierarchical and more agile; they were less predictable, and there were now fewer places where a 'job for life' could be secured with ease. Equally, organisations were becoming more 'permeable', working in complex supply chain relationships; there was more team-based and collaborative work, and there was a growing focus on wellbeing and productivity.

One symptom of the increasing adaptability and flexibility of organisations is the growth of a 'contingent' workforce. In many corporate organisations, an increasing proportion of the workforce is not directly employed; they are consultants, interims, contractors, part-timers and supply chain partners. How these staff are housed and managed raises important questions for demand planning within buildings.

The critical nature of connectivity, changing corporate structures, the priorities of knowledge workers and the reduced importance of the 'corporate island' in favour of a more complex web of supply chain relationships have all altered the nature of demand for space. Organisational change is directly related to the nature of work and skills required. This is developed in Figure 8.14, which describes four key aspects of the digital office.

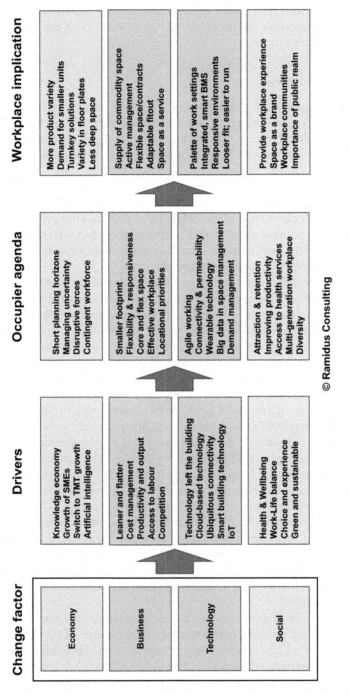

Figure 8.13 Change factors, drivers and workplace implications

Source: Ramidus Consulting (2018).[51]

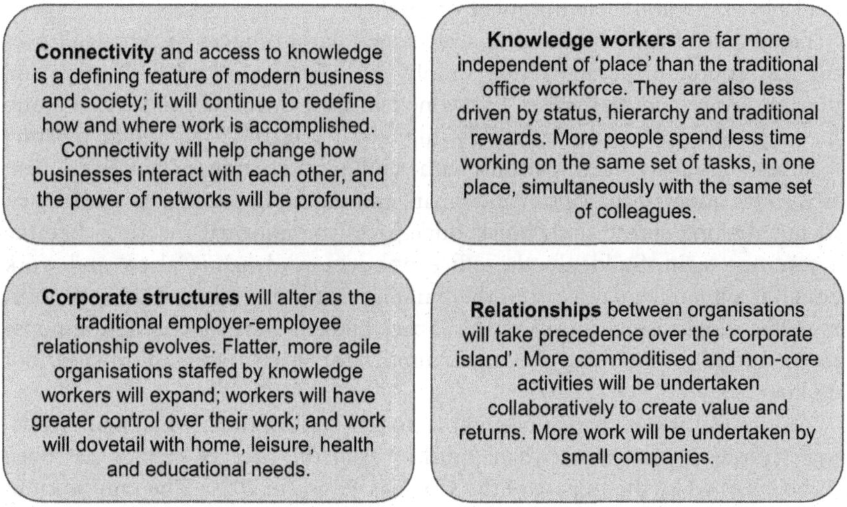

Connectivity and access to knowledge is a defining feature of modern business and society; it will continue to redefine how and where work is accomplished. Connectivity will help change how businesses interact with each other, and the power of networks will be profound.

Knowledge workers are far more independent of 'place' than the traditional office workforce. They are also less driven by status, hierarchy and traditional rewards. More people spend less time working on the same set of tasks, in one place, simultaneously with the same set of colleagues.

Corporate structures will alter as the traditional employer-employee relationship evolves. Flatter, more agile organisations staffed by knowledge workers will expand; workers will have greater control over their work; and work will dovetail with home, leisure, health and educational needs.

Relationships between organisations will take precedence over the 'corporate island'. More commoditised and non-core activities will be undertaken collaboratively to create value and returns. More work will be undertaken by small companies.

Figure 8.14 The digital office landscape

Source: Adapted from Ramidus Consulting (2015).

Occupiers today operate within short-term planning horizons, responding to an ever-changing economic landscape and seeking to maximise their flexibility to adapt. The power of networks, involving collaborative production and multi-disciplinary skills, is coming to be realised. More commoditised and non-core activities are being undertaken by specialists; more work is being undertaken collaboratively, and more work is being undertaken by small companies.

All of this means a different approach to real estate, and to the role of the workplace and its design and management. In a re-run of Cairncross' *Death of Distance* in the mid-1990s, there has been widespread debate in more recent times about the future role of the office, much of which has focused on its demise. The premise for this thinking, broadly and as before, is that as technology dispenses with the need to be tied to a place, knowledge workers will become nomadic, moving from place to place, connected only by wireless data and smart devices. However, the draw of the office remains strong: the office is not disappearing; but its role is evolving, once more.

While in the past people commuted to the office to access the paraphernalia for work – phone, personal computer, filing and stationery – increasingly they do so for different reasons. These include socialising, mentoring, training, collaboration, corporate opportunity, meeting and so on. Workers like to socialise, and employers wish to create a common purpose. Similarly, the home is often not set up for work for one person, let alone two. For many, the quality of public realm, access to shops, leisure and services, and office support in the form of nutrition and wellbeing, are often far superior in the office to that which is available in the home environment or locale.

Figure 8.15 shows how the 'space budget' within offices has changed over the past couple of decades. Very simply, the office once had desk space and meeting space, and that was it. Everyone had their own desk, and when group meetings took place, the attendees migrated to, usually, centralised meeting facilities – grouped meeting rooms with vending and stationary facilities. Then, as workers adopted the laptop and email, and started to become less sedentary, so hot desking spread and 'break out' spaces for informal meetings became a common sight. Suddenly, the office was a more dynamic place; and work could be undertaken away from the building – at home when writing reports, in coffee shops or in airport lounges. Space budgets were squeezed in response to economic pressure and the opportunities for space saving afforded by hot desking.

This trajectory of change brought us to the current time, when agile working is more normal rather than 'quirky'. And the pace of change has been turbo-charged by the impact of the Covid-19 crisis in 2020. The ramifications of Covid-19 are discussed further in the final chapter, suffice to suggest, at this point, that the office workplace is likely to continue to have an important role, albeit one that will change significantly. It is rapidly moving away from

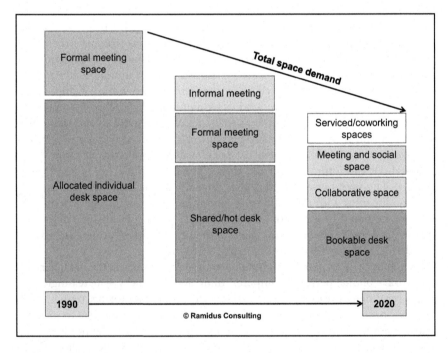

Figure 8.15 The changing office space budget

Source: Ramidus Consulting (2018).

being a static backdrop for process-based, largely routine and solitary work, to an increasingly actively curated environment, managed more like a hotel than a traditional office, with a high level of service and experience for 'guests'.[53]

As noted in recent work for the Corporation of London: "the role of the office is increasingly acknowledged as enabling people to interact and collaborate"; it is "expected to provide a wider range of settings in which individuals and groups [can] work in more dynamic ways compared with much of the more solitary work of the past", and the office is "becoming less a place to go to work on a set of prescribed tasks, and more somewhere to visit and interact with colleagues".[54]

The office is becoming more of a social hub for an increasingly agile workforce that utilises a range of work settings. The emerging priority for tomorrow's workplace will be to energise and motivate people, and make them productive and effective; it will provide for social interaction and knowledge transfer; it will provide workers with more personal control and it will provide a wide range of support services. It will form part of a more distributed work pattern in which the home, local offices, public transport, client premises – indeed, just about anywhere is used to work.

The underlying, and most profound aspect of the latest stage in the evolution of the office is the potential switch from corporatism to individualism. The criticisms levelled at the corporate office in this chapter, notably from Frank Duffy and Jack Tanis, Robert Propst, Studs Terkel and William Whyte all, in their own ways, focused on the crushing impact of corporatism and made pleas for individualism and individual expression. Now, in the early twenty-first century, there is a real opportunity to transform work and the workplace from a transaction to a social activity.

The four ages of the office economy discussed thus far have traced the development of the office workplace from the original markets and 'coffee houses', to the 'clerical factory', to the 'corporate office', to the 'digital office'. In this chapter we have traced the evolution of modern office work, from 'Company Man' who dedicated his life to the corporation, to the knowledge worker who is 'self-sufficient' and 'self-determining'; from 'corporatism' to 'individualism'. The next phase in the evolution of the office economy is likely to be characterised by the 'network office'.

Notes

1 F.W. Taylor (1911) *The Principles of Scientific Management,* Dover Publications, New York
2 J. Shlaes & M.A. Weiss (1993) Evolution of the Office Building. In: J.R. White (Editor), *The Office Building: From Concept to Investment Reality,* Counsellors of Real Estate, Chicago, p9
3 Black *et al* (2007) *Op cit,* p15
4 Cowan *et al* (1969) *Op cit,* p31
5 J. Orbell (2017) Changes in Office Technology. In: A. Turton (Editor), *The International Business Archives Handbook: Understanding and managing the Historical Records of Business,* Routledge, London, Chapter 3, pp54–86, 68

6   Daniels (1975) *Op cit*, p11
7   Cowan *et al* (1969) *Op cit*
8   Orbell (2017) *Op cit*, p70
9   Cowan *et al* (1969) *Op cit*
10 *Ibid*, p34
11 Daniels (1975) *Op cit*, p10
12 M. McLuhan (1964) *Understanding Media,* McGraw-Hill, New York, Cited in Cowan *et al* (1969) *Op cit*, p32
13 Mumford (1938) *Op cit*, p226
14 J.E. Lewis (1988) Women Clerical Workers in the Late Nineteenth and Early Twentieth Centuries. In: G. Anderson (Editor), *The White-Blouse Revolution – Female Office Workers Since 1870,* Manchester University Press, Manchester
15 Orbell (2017) *Op cit*, p56
16 *Ibid*, p74
17 L.R. Dicksee & H.E. Blain (1906) *Office Organisation and Management,* Pitman, London, p43
18 Orbell (2017) *Op cit*, p75
19 *Ibid*, p62
20 *Ibid*, pp59–60
21 J.A.V. Turck (1921) *Origin of Modern Calculating Machines,* Forgotten Books, London, pp144–145
22 Black *et al* (2007) *Op cit*, p25
23 *Ibid*, p26
24 J. Micklethwait & A. Wooldridge (2003) *The Company: A Short History of a Revolutionary Idea,* Orion Publishing, London, p102
25 *Ibid*, p109
26 *Ibid*, p103
27 W.H. Whyte (1956) *The Organization Man,* Simon & Schuster, New York
28 *Ibid*, p3
29 Whyte (1956) *Op cit*, pp16-17
30 A. Sampson (1995) *Company Man: The Rise and Fall of Corporate Life,* HarperCollins, London
31 *Ibid*
32 F. Duffy, C. Cave, & J. Worthington (1977) *Planning Office Space,* The Architectural Press Ltd, London, p68
33 R. Propst (1968) *The Office: A Facility Based on Change,* Herman Miller Inc., Zeeland, MI, p25
34 R. Propst (1964) A New World to Work. In: *The Action Office Brochure,* produced by Herman Miller Inc., Zeeland, MI
35 Propst (1968) *Op cit*, p29
36 Y. Abraham (1998) The Man behind the Cubicle, *Metropolis,* February, pp76–115
37 S. Terkel (1972) *Working: People Talk about What They do All Day and How They Feel about What They Do* Pantheon Books, New York, pXI
38 *Ibid*
39 Duffy *et al* (1977) *Op cit*, pp68–69
40 F. Duffy & J. Tanis (1993) A Vision of the New Workplace, *Site Selection and Industrial Development,* Vol 162, No 2, March/April, pp1–6
41 Goodwin (1965) *Op cit*, p3
42 P.W. Daniels (1975) *Office Location: An Urban and Regional Study,* G Bell, London, pp4–5
43 https://xeroxnostalgia.com/2015/02/08/xerox-914/. Accessed 24th July 2020
44 Orbell (2017) *Op cit*, p71
45 P.J. Stone & R. Luchetti (1985) Your Office is Where You Are, *Harvard Business Review,* March–April, pp102–117

46 Brinkley (2010) *Op cit*, p4
47 F. Cairncross (1997) *The Death of Distance*, Orion Business Books, London
48 Cited in: Workstyle 2000: A New Flexibility, *Flexible Working*, February 1996, pp28–29, 32
49 DEGW (1998) *Design for Change: The Architecture of DEGW*, Watermark Publications, Haslemere, p98
50 J. Micklethwait & A. Wooldridge (1996) *The Witch Doctors*, William Heinemann, London
51 Ramidus Consulting (2018) *Op cit*, p7
52 The Economist (2015) Briefing: The Future of Work, *The Economist*, 3rd January, pp13–16
53 S. Brunia, I. De Been, & T. van der Voordt (2016) Accommodating New Ways of Working: Lessons from Best Practices and Worst Cases, *Journal of Corporate Real Estate*, Vol 18, No 1, pp30–47
54 Ramidus Consulting (2015) *Future Workstyles and Future Workplaces in the City of London*, Corporation of London, London, p6

9 Managing

From liability to corporate resource

At one level, an office building provides the physical environment within which businesses undertake their work. Whether the occupier is a bank, an advertising agency, local government or a university, the building provides the envelope within which to accommodate the activities that define the organisation's *raison d'être*. In addition to the envelope, the building is also expected to provide comfort and safety in the form of clean air, fresh water and sanitation, heating and lighting. Then it is expected to provide the equipment necessary to undertake the work: furniture, computers, telecoms, photocopiers, printers and all the other paraphernalia that makes the modern office work. Finally it provides services such as vending, catering, meeting facilities and social space.

All this comes together, every day, in the planned, seamless provision of environment and services. Until it goes wrong: the hot summer's day when the air conditioning fails; the morning rush when the lift fails; the day when the server goes down and there is no means of communicating with the outside world; or simply the on-going niggle of a back pain caused by a badly designed chair or a poorly configured workstation. Welcome to the world of real estate management.

The management of office space (and real estate generally) has evolved rapidly over recent years and has been referred to variously as asset management, building management, facilities management, property management, real estate management and workplace management. The variety of titles reflects the professional jealousies and sclerotic division of built environment activities outlined in Chapter 7. For example, surveyors prefer either asset or property management because these are 'professional ' activities, while they see facilities management as a manual activity. Despite this, in practice, the terms are broadly interchangeable, and for reasons of simplicity, this chapter adopts real estate management (REM) as an umbrella term, except where quoting secondary sources.

This chapter describes the emergence and maturing of REM, largely through its transition over the past three decades of the digital office, from a routine and reactive janitorial function, into an integrated workplace resource management role. Over this time, the office has evolved from being a static backdrop to office activities (a liability in cost terms), into a corporate resource which must be professionally managed in order to ensure that it has a direct and positive impact on organisational performance and the wellbeing of workers.

9.1 Growing building complexity

Until quite late in the twentieth century, office buildings remained relatively simple in terms of their technical specification. As late as the mid-1970s, only 15% of office units in the City of London were larger than 10,000 sq ft (929 sq m), and over 60% of units were smaller than 3,000 sq ft (c279 sq m); the buildings themselves were uncomplicated, and the demands placed upon them by their occupiers were relatively light. Even at this time, the City comprised predominantly small-scale offices, often in buildings constructed for different purposes (including residential and light industrial). As we saw in Chapter 3, London had to wait until 1967 for the 365 ft (111 m) Centre Point; while the eponymous 600 ft (183 m) Nat West Tower (later re-branded Tower 42) was not built until 1981.

As described in Chapter 5, the 1950s, 1960s and 1970s saw some of the most brutal and ugly commercial office buildings ever built in the UK. The era of the 'lowest price wins' bidding process had arrived. Architectural firms found innumerable ways to cut the costs of construction, with consequent impacts on the quality and durability of both the façade and the interior work environment. There was no meaningful understanding of how the use of the building and its management would interface.

Marriott estimated in 1967 that between half and three-quarters of post-war speculative office blocks in Greater London had been designed by just ten architectural practices.[1] Marriott explained that "What the developers wanted from their architects was a commercial service. They needed functional buildings designed to a certain price, usually the lowest possible"; furthermore, they "were just not concerned with design". To emphasise the point, Marriott presented some comparative costs. "Whereas developers put up office blocks for £4 10s 0d to £7 per square foot . . . the owner occupier perhaps spent £8 to £10 a square foot on average".[2] When speculative developers were spending half as much as owner occupiers, the impact on the quality of the resulting work environment needs little further illustration.

The situation began to change in the 1960s when furniture and layout were systematically scrutinised and recognised as having an impact on the quality and effectiveness of the work environment (Chapter 8). Even then, the approach was a largely spatial matter – the physical arrangement of space – rather than one that encompassed on-going management.

It was not until the mid-1980s, with the sudden spread of digital office technology, initially to support the trading demands of global investment banks, and very quickly thereafter spreading to all office occupiers, that the management of real estate underwent any meaningful change. Almost overnight, demands on office buildings were transformed. There were fundamental changes in building form, with the sudden imperative to accommodate enormous amounts

Figure 9.1 Tower 42 (The Nat West Tower), 1981

Source: Erik Brown, 2020.

of technology hardware – physical space and voids to cope with cables and kit; together with mechanical and electrical systems to cope with high power demands and heat output. Buildings had to be designed with 'deeper' space to accommodate large trading floors and higher concentrations of workers, along with raised floors and drop ceilings, and generally highly engineered systems and infrastructure.

Technological capability and technology adoption proceeded in a kind of space race in which more capability served only to increase the appetite of occupiers for more and more technological capability. As the bulk of occupiers caught up on one innovation, so there seemed to be a dozen others to succeed it. The knock-on effect for buildings was profound. The amount of cabling, the capacity and demands on heating and cooling, power loadings and the proliferation of kit (leading to server rooms, communications cabinets, computer rooms) all increased greatly.

The market shifted almost overnight from no raised floors, to floors sometimes a metre deep. Air conditioning became *de rigueur*. Building depth, typically 12–15 m previously, grew until 20 m plus became quite normal. Such depth could accommodate much larger numbers of workers, as well as the new dealing floors, some the size of football pitches, accompanying Big Bang in the City.

As part of the space race, the 'groundscraper' was born at Broadgate in the City (Figure 9.2), to cope with the new demands. Respecting Europe's uneasy relationship with skyscrapers (see Chapter 2), these large (by historic standards) buildings were relatively low rise at around 8 to 10 storeys. And it was not just the financial services companies that got involved: the traditionally staid environments of lawyers, accountants and consultants were changed forever

Figure 9.2 One Exchange Square, Broadgate, City of London
Source: © Ramidus Consulting Ltd.

when the leading firms began to intensify the use of their buildings and adopt desktop technology throughout. And as this revolution in building design was taking place, the buildings themselves began to assume a less passive role in the lives of their occupiers.

The pivotal role of ORBIT

Frank Duffy, co-founder of DEGW in London, produced some of the first research exploring the potential impact of organisational change and emerging technologies on buildings.[3] This research, *Organisations, Buildings and Information Technology* (ORBIT), was amongst the first empirical evidence to link the physical characteristics of buildings (such as floor size, depth, sectional heights and servicing capacity) with their ability to accommodate the emerging information technologies – both desktop services such as personal computers and, critically, building management systems.

ORBIT was the result of a multi-client study, carried out in 1982 by DEGW with Building Use Studies and EOSYS. Privately published in 1983, under the authorship of Frank Duffy, it examined the growing impact that information technology was having on buildings. The fact that the study was a multi-client study, carried out with a number of sponsoring companies involved in the office supply process, meant that the results got into the bloodstream of the industry very quickly.

The main findings of the research focused on the conclusion that information technology would have a profound effect on the design, management and occupation of buildings in coming years. It sounds simple more than three decades later, but the conclusion illustrates how much our relationship with office buildings has changed. In 1883, the lift, incandescent lamp and typewriter were changing our relationships with the office; 100 years later it was information technology and organisational change.

ORBIT recognised both the direct and indirect effects of technology on design. Direct effects included, for example, that technology equipment would require novel forms of servicing, with an obvious impact on design in areas such as heating, cooling, ducting and so on. Indirect effects concerned the ways in which technology would change organisations by, for example, changing the mix of skills required; introducing different forms of decision-making, and changing work patterns and locational imperatives. The report's key conclusions included the following.[4]

- most technology expected to have the greatest impact already existed;
- IT was not easy to assimilate into office buildings;
- demands of technology on new buildings were more stringent than assumed;
- problems in accommodating computer equipment were not short term;
- IT would change the way in which organisations used buildings;
- IT changes patterns of space use in buildings;
- IT was changing the rules of office choice;

- buildings vary greatly in their ability to absorb technology, and
- many existing buildings were in danger of premature obsolescence.

ORBIT was followed in 1985 by ORBIT 2,[5] which was placed in the North American context and was conducted by Harbinger Group Inc (a subsidiary of the Xerox Corporation) and Facilities Research Associates, Inc. The report was published by Harbinger under the joint authorship of Gerald Davis, Franklin Becker, Frank Duffy and William Sims. The focus of ORBIT 2 was on the growing interface between people, places and technology, and it provided guidance on how their interactions should be managed. The breakthrough of the report was that it developed a systematic and comprehensive building appraisal methodology that either consultants or occupier organisations themselves could use to rate the suitability of different buildings for specific organisations, or groups within organisations.

The overriding importance of the ORBIT research was that it set out consciously to address those technological forces of change that were acting on occupier businesses, and to understand these changes for their implications on building and workplace design. There were, however, a host of non-technological or technologically enabled changes that were also growing in importance. For example, the City was becoming more international and was beginning to confront the prospect of deregulation. Competition was driving a reorganisation of businesses through mergers and acquisitions. And the economy as a whole was being transformed, as traditional industries were eclipsed in their importance by the burgeoning office economy.

It was clear in the mid-1980s that white-collar factories were in the process of yielding to a fundamentally different approach to office design, management and occupation. By the end of the decade, signature developments in London, such as Broadgate and Canary Wharf, were beginning to emerge, setting in train a new phase in the physical expression of the office economy. Large buildings with deep floorplates and large sectional heights were appearing, with the capacity to cope with not only larger organisations, but also intensive provision of servicing to cope with new building management systems and the distribution of desktop technology.

The emergent real estate management role

Chapter 7 described the professionalisation of activities within the built environment. The mid-nineteenth century was a particularly busy period in this respect: the structure of the professions, and the motivations of many of those involved, reflected British social structure at the time as much as essential functional separation. In the early years, professional affiliation afforded status, respect and security. The deliberate separation by architects of their activities from building 'trades' was a classic case in point. The professions quickly became rigid and exclusive organisations that were more interested in protecting the *status quo* and the self-interests of members than responding to a rapidly changing context.

Building management first became a recognisable stream of business management in the USA in the late 1970s, where it emerged in the industrial sector, before becoming established in the office sector during the early years of the 1980s. It became recognised and practised in the UK in the mid-1980s, where it adopted the US terminology of Facilities Management.

Various elements of REM existed before the collective label arrived, but its novelty and achievement lay in a growing recognition that an integrated and co-ordinated approach to the management of occupied space could add significant value to the operational performance of businesses. From the outset, REM was functionally, "concerned with the integration of property management, with the management of its utilization and with the full range of services provided to support a business operation".[6]

The 1980s was a period of rapid growth and change for business generally, particularly following the deregulation of financial services in October 1986. The corporatism of the 1960s and 1970s was given a further fillip with technology-enabled globalisation of services. Expansion and diversification spread as the 'bigger is better' corporate philosophy grew; planning and decision-making were short term in the scramble for expansion and stock market performance, rather than focused on long-term strategy and performance. John Renzas observed that during this period of economic growth,

> companies often entered new business areas that were poor strategic fits with their core businesses . . . The financial terms under which these new acquisitions were made were often predicated upon continuation of aggressive rates of growth, which may seem unrealistic in today's environment.[7]

More specifically, rapid growth meant that businesses could subsidise the property industry; they could afford the escalations in rents demanded by landlords, and the mistakes in the construction process that meant too much money was being spent unnecessarily on the product; few questions were asked of the long-term utility of the buildings (either by providers or occupiers). It is extraordinary, with the benefit of so much hindsight, that until the mid-1980s there was no recognised role for managing workplaces. There was very little recognition that real estate required any particular management time, and generally, real estate was regarded as an expensive necessity of business.

It was often the role of the company secretary or finance director to pay the rent and deal with the managing agents on issues such as repairs, maintenance and security. In large organisations such as the clearing banks and insurance companies, it often fell to line managers who had reached the extent of their potential in the core business, to move across and look after property. Sometimes there would be a 'building manager' – someone whose responsibilities were largely janitorial: managing the cleaners, ensuring that catering and vending were delivered, and ordering stores such as stationery and the other paraphernalia of office life.

The genesis of a new way of thinking can be traced back to the USA, where a Harvard University study in 1981 highlighted the low priority afforded to property in the public and private sectors, and the continuing failure to manage real estate as a resource.[8] These findings were confirmed and expanded on in a later study at MIT.[9] The public sector in the UK showed, not for the first time, that its thinking was ahead of the corporate sector. A report from the Department of Health and Social Security in 1982 expressed concern over the management of property within the NHS.[10] Further public sector reports were to follow from the Audit Commission[11] and the National Audit Office,[12] all addressing the role of property in operational efficiency within the public sector.

In the first work of its kind in the UK, Martin Avis, Virginia Gibson and Jenny Watts[13] made an extensive study of the way in which major corporate occupiers managed property. They uncovered a profound lack of appreciation of the role of property and could find little evidence of understanding of the contribution it made to organisational performance. They found that the majority of occupiers did not clearly establish property objectives as part of their overall corporate objectives. This was despite the fact that more than half the organisations interviewed for the study claimed to have property assets worth 30% or more of their total asset value. In part, the lack of a proactive approach to premises management was caused by the mismatch between business planning horizons and those of property:

> The business and political decisions tended to have shorter lead times than property decisions. The majority of organisations' planning period was between 1 and 5 years. These two factors made it difficult to incorporate property into the strategic planning process.[14]

However, as real estate managers became more organised and more vociferous, and as the building implications of technological advances became more widely understood, so property began to appear on the corporate agenda. Although it remained the Cinderella of corporate management decision-making, attention increasingly focused on property and its role in overall operational performance. Generally, a much closer relationship between business strategy and premises strategy began to emerge. This was reflected in an increasing concern to maximise the utility of property. Frank Duffy and Alex Henney observed that:

> Historically companies have focused their attention on the 'sharp end' of their businesses, such as marketing, controlling costs, and finance. Few have paid much attention to using and operating buildings efficiently and to providing an agreeable environment within which people will work efficiently.[15]

The authors argued that such attitudes were changing because:

> Some modern office buildings in the City are now very large and expensive, and for that reason alone demand more care. In addition buildings

have become technologically far more complex. . . . If such buildings are not looked after carefully . . . then they will not perform properly.[16]

But progress was slow. Surveys consistently found evidence of real estate remaining low on the corporate agenda. In 1992 Debenham Tewson and Chinnocks found that "only around 10% of the Times Top 1000 have a property director responsible for real estate".[17] These findings were consistent with those of Graham Bannock & Partners,[18] Ranko Bon[19] and Arthur Andersen.[20] Property continued to be the poor relation of other corporate resources. Virginia Gibson highlighted the low priority given to property. Referring to the results of a survey, she found that the "vast majority" of organisations considered property to be only moderately important or not relevant to their future success, compared to other resources that were considered critical or very important.[21]

Varcoe made a persuasive case for the strategic role of property in corporate planning. He argued that premises have a fundamental effect on the people who occupy them, which led him to conclude that the primary role of property has to be:

> that of supporting at all levels, from the individual to the strategic, the goals and aims of the businesses and organizations which occupy them. They must be as ready for change as any other business component, if not more so, being able to adapt to or absorb its effects with speed and efficiency.[22]

In a similar way, Alexander argued that the mission of real estate management should be "to improve the quality of the operating environment, to add value to the business and to minimise the exposure of an organisation to risk".[23] What was emerging was a growing sense of the need to professionalise the real estate management role.

Professionalisation of real estate management

McLennan traced the historical development of REM, recording that the Institute of Administrative Managers (IAM) began to recognise the increasingly specialised role of FM as far back as 1965 in their Office Design Group. This was formalised in 1990 with the creation of the Institute of Facilities Management (IFM). The younger Association of Facilities Management (1985), merged with IFM in 1994 to create the British Institute of Facilities Management (BIFM). At that time there were some 6,300 paying members. The BIFM formally defined REM as: "the integration of multi-disciplinary activities within the built environment and the management of their impact upon people and the workplace".[24]

McLennan argued that the "reasons for the development of facilities management as an occupation are varied, simultaneous and complex", citing the rise of knowledge and managerial workers; the organisational demand for labour flexibility; the development of information and communication technology;

and the increasing resource costs of space together with the labour to support the services of that space as key reasons. He concluded:

> the management of the space supporting the organisation became an important part of an organisation's service operation that required a more informed individual to integrate, co-ordinate and manage the facilities and their associated support services.[25]

So, when REM emerged in the UK office economy in the 1980s, how did it slot into the comfortable and established structure of built environment professions? The short answer is that it did not. It can be further stated that it has struggled to do so ever since. The key reason for this seems to be that REM is, in some ways, a hybrid profession. It requires skills that traditionally 'belong' in business administration, construction, design, engineering, HR, technology and surveying.

Further explanation was provided by Nutt, who argued that one of the reasons why REM has struggled to establish its own, exclusive professional jurisdiction is that much of the technical knowledge base of REM resides in the existing professional occupations of design, engineering, surveying, and construction. "Each of these occupations has a corresponding professional body that is claiming facilities management as a part of their jurisdiction", including the Chartered Institute of Building Services Engineers (CIBSE), the Chartered Institute of Building (CIOB), the Royal Institution of Chartered Surveyors (RICS) and the Royal Institute of British Architects (RIBA).[26] According to McLennan, the "jurisdictional disputes suggest that facilities management is seen as a growing occupation, an important activity for clients, and of importance to society and therefore a viable professional occupation in its own right".

While the existing Chartered Institutes extend their knowledge base to include REM, it remains difficult for REM to become a fully separate profession. The formation of the RICS FM faculty and the CIOB's FM specialisation are two examples. Unlike its forebears, the BIFM competes for international membership, with, for example, the International Facilities Management Association based in the USA.

REM is, on the one hand, an example of the continued specialisation of knowledge work and, on the other, an example of the managerialisation of the professions in general. The occupations dealing with knowledge work and service are more fluid than the earlier, traditional 'industrial' professions of engineering, surveying and architecture, for example.

So it was that REM emerged and evolved – professionalised – while continuing to sit tentatively, and sometimes uncomfortably, alongside other, established built environment professions. REM was increasingly recognised by the property supply industry as a force to be reckoned with: it became organised, vociferous and demanding. More rigorous approaches to the management of service suppliers and contracts, and better information and analysis of the characteristics of occupied portfolios increased the efficiency and effectiveness of real estate within the context of organisational performance. The dot.com bubble highlighted the increasingly volatile business environment within which most businesses operated, and in which office space had to be planned, procured and occupied. As such, perceptions of the role of the office began to change; it was increasingly being seen less as a passive, inert by-product of doing business, and more as a corporate resource for corporate management that needed to be planned and provided in the same efficient and effective way as other organisational resources.

Despite this undoubted progress, the results were not uniform. One of the reasons for the continuing inefficiency in the management of real estate was the separation of *property* management and *facilities* management. The former typically dealt with acquisition and disposal of properties together with capital projects, while the latter, often seen as the poor relation, has traditionally dealt with the day-to-day management of service contracts and external service providers. This *Upstairs-Downstairs* cultural feature was exacerbated by the different skill sets in each area: property is typically managed by someone with the letters RICS after their name, while the person responsible for facilities is more likely to have a business management qualification. These are generalisations, but they help to explain some of the on-going inefficiencies in the management of real estate.

The same separation of roles also occurs in the supply process. Property companies employ 'managing agents' to collect rent and administer leases, and these agents in turn will enter into contract with service providers. The provider of the space and the user of the space are in fact separated by layers of decision-making, which in turn means that feedback is inefficient. Writing well in advance of trends in the UK, a team from Cornell University led by Franklin Becker summarised the need clearly:

> In no cases was it possible, or desirable, to try to understand the success or failure of a particular facility approach in terms of its constituent elements. Good facility design requires a systems or ecological view from the beginning. This means considering the nature of technology, management style and organizational practices, use policies, facility design and work practices as interdependent components. The system is not likely to prosper, or even survive, unless all these factors are working in harmony.[27]

Two highly influential research reports were published in 1993, significantly both in the USA, which helped to move the REM agenda on to the next phase. Michael Joroff and colleagues encapsulated the task for real estate managers in their report *Strategic Management of the Fifth Resource: Corporate Real Estate.* They said in the preface that:

> In the 1980s, corporate managers, consultants, and researchers transformed the role of four corporate resources – capital, people, technology, and

information. As a result these four resources now support their core businesses far more effectively than before.[28]

Their vision was to see to real estate as the *fifth resource*, lying at the heart of business planning, and making a direct positive contribution to the performance of the business. To this end, they argued that the under-management of corporate real estate was no longer acceptable. In short, they set out the challenge for REM executives as being:

> to learn the needs of the corporation as a whole and the needs of business units in particular and then to devise a strategy to satisfy them even when the answer may not involve traditional forms of real estate.[29]

In this sense the REM had to integrate disparate workplace disciplines (for example, the impact of space planning on productivity), but it also had itself to integrate into other parts of the corporate plan (for example, by contributing to corporate management objectives).

In the same year as the Joroff study came a report from Arthur Andersen.[30] *Real Estate in the Corporation: the Bottom Line from Senior Management* set out to examine how corporate real estate could be managed to support businesses that were experiencing rapid change. The key conclusions emerging from the research included the need for efficient and innovative work environments, and for flexibility. It concluded that meeting the needs of business units was the most important real estate management activity, and that while cost per square foot was the primary measure of real estate activity, business return on assets was the most meaningful indicator of corporate performance. Leifer encapsulated the nature of the change well:

> The role of corporate real estate has evolved from coordinating real estate transactions and maintaining properties to designing a real estate strategy that ties directly into the business plan of the corporation.[31]

Consultants Arthur Andersen preached a similar line. Recognising that non-involvement in strategic planning and lack of access to senior executives, together with a poor understanding of real estate by senior management, were impediments to change, the study drew a number of important implications. Not least, it concluded that REM activities should evolve to respond to emerging needs of business units, and should develop new skills and management capabilities.[32]

The notion that real estate should be treated as a corporate resource was fleshed out in a number of publications; notable among these was McLennan, Nutt and Kincaid.[33] In another, Apgar argued that "Business real estate is not merely an operating necessity, it's a strategic resource".[34]

As the REM emerged and matured, it passed through distinctive phases: cost reduction and value enhancement. While the first earnt the new profession a role at the corporate decision-making table, the latter gave it far greater significance.

9.2 Phase I: cost reduction agenda

The leaden weight of property on companies' balance sheets was thrown into sharp relief as the white-collar recession of the early 1990s took hold. Businesses sought a reduced cost base and, crucially, more flexibility. Suddenly corporate management's attention was drawn to the need to improve efficiency and increase productivity, so decision-making and planning focused on those issues that would help to equip companies with the ability to adapt and succeed in increasingly competitive and uncertain markets:

> the primary organizational need is for greater adaptability, to anticipate differential rates of change, to accommodate expansion and contraction and to enable rapid response to business opportunities.[35]

Inevitably, the spotlight fell on property and, initially, corporate focus was simply on cost and wherever possible businesses sought to reduce their overall commitment to property. Mahlon Apgar argued that rising occupancy costs impaired a company's earnings, share value and overall performance and that reducing these costs could "increase a company's profitability because every dollar saved drops straight to the bottom line".[36] Apgar went on to suggest that efforts to reduce occupancy costs had never been timelier:

> During the 1980s, the typical service business saw its ratio of occupancy costs to revenues more than double, its real rents increase by 50%, and its space use per employee grow by 80%.[37]

Such stark figures showed how the expanding office economy was driving an expanding real estate industry and ultimately increased demand for office space. They also hinted at the scope for making cost savings, and many organisations instigated severe cost reduction programmes.

As the recession took hold, the pressure was on to reduce costs wherever possible, and the new work agenda of agile working and the digital workplace were unceremoniously pushed to one side as the focus of many real estate management teams shifted dramatically from a strategy of IT-enabled new workstyles to a single-minded focus on reducing occupancy costs. This was a reasonable but narrowly focused reaction that had lasting effects because, while the short-term focus was on cost cutting, the long-term under-current of drivers that had been pushing the restructuring of the office economy continued their work.

New wave management theory

The real estate managers and their colleagues in other corporate resource areas came under pressure from Boards whose broader agenda was simple: reduce the cost base and improve the competitive position. The chill winds of change gathered pace, and jobs for Company Man were severely culled. The term 'core competency' became familiar as corporations sought to shed layers of 'non-core' activities and with them, layers of people. Group structures, divisional structures and hierarchies were subject to the corporate scythe as the notion of corporatism (large, rigid, inflexible) was suddenly viewed with suspicion.

And, the whole ethos was given intellectual credence by a plethora of management gurus and their books, all in various ways proclaiming to have found an angle on the emerging economic landscape: the management guru reached 'popular' status and, in some areas, possibly cult status. New wave management theory and the names of Drucker, Hammer, Handy, Porter, Peters and Senge became very familiar to anyone with even a passing interest in what was happening to business culture. Each had their particular theses and approaches, but through the dense undergrowth of books and articles that emerged, the consistent messages were about downsizing, outsourcing, delayering, re-engineering, mobility and virtual work: in short, methods for becoming leaner, more efficient, more flexible and more competitive, all means by which to dismantle corporate structures.

Many commentators and advisors pursued the new wave management agenda with an almost missionary zeal, interweaving the cost-cutting agenda with theories about the nature of companies in the emerging office economy. We were told, for example, that as the efficiency measures took hold, they would lead to horizontal networks replacing vertical hierarchies, and highly skilled, group-based knowledge workers replacing clerical, departmentalised information processors. Robert Heller picked up the theme, arguing that since "horizontal business processes are all multi-functional and multi-disciplinary, the manager is being led in similar directions. Cross-functional, synergistic and inter-departmental working is unavoidable now: so are task specific teams".[38] And *Business Week* argued that horizontal corporations should organise around processes rather than tasks; flatten hierarchies; use teams to manage everything; let customers drive performance, and reward team performance.[39]

And the leviathans of the office economy seemed to be fulfilling the predictions of the gurus. Outsourcing spread rapidly as companies sought to shrink to a newly defined core of activities, relieving themselves of the overhead burdens associated with support activities such as catering, cleaning, building maintenance, security, technology and even accounting and personnel. Demergers multiplied as the fashion for breaking up conglomerates took hold: BAT, BTR, Courtaulds, Forte, Hanson, ICI, Racal and Thorn EMI all got in on the act. Others, such as AT&T, Eastman Kodak and General Electric, delayered; while yet others underwent full-scale business process re-engineering, including Asea Brown Boveri, Rank Xerox and Shell. By the early 1990s, ABB was able to report that it had 215,000 employees worldwide, with a central staff of just 200; around 5,000 autonomous business units and only three layers of management.[40]

It was a given in much of the new wave management theory that the new corporate structures being propounded would be enabled by technology, but the nature of its influence was less clear. Projections about the future impact of technology on patterns of work during the 1990s tended to polarise between the Panglossian optimism of the technology apostles like Peter Drucker, who believed that the changes taking place then were more profound than the Industrial Revolution; and the Armageddon merchants of doom like Norbert Weiner who forecast that computer technology would destroy enough jobs to make the depression of the 1930s look like a picnic.[41] One contemporary forecast suggested that within 30 years, as little as 2% of all the world's current labour force would be needed to produce all the goods necessary for total demand.[42] A quarter of a century on from this forecast, the reality is somewhat different.

As the 1990s progressed and economic conditions began to ease, the limitations of cost cutting began to emerge; a growing body of evidence suggested that the slash-and-burn management fads had largely failed. For example, one study found that fewer than half the American firms that had downsized between 1990 and 1997 had seen any improvements in quality, profitability or productivity.[43] At the same time, it emerged that North American and western European countries that had invested heavily in new technology had failed to reap the anticipated benefits in terms of increased output, or productivity. Through the 1990s, manufacturing productivity in America rose at more than 3% per annum on average, but in services, annual gains averaged less than 1%. All this despite the fact that technology, as a percentage of American firms' total investment in equipment, rose from around 5% in 1960 to over 45% in 1997.[44]

Then in a dramatic *volte-face*, one of the leading exponents of downsizing – Stephen Roache from US investment bank Morgan Stanley – declared, "If you compete by building, you have a future. If you compete by cutting, you don't".[45] This was a key statement because it helped explain the next phase in the evolution of the workplace agenda. The essence of the statement was that, in the context of a rapidly growing knowledge economy, corporations should recognise the inherent value of the intellectual capital of their workforces. In other words, the slashing of wage bills, which pushed up earnings and share value in the short term, actually led to a reduction in shareholder value over the longer term. Downsizing quickly took on a new persona: dumbsizing.

But despite these tempering events, the fact of fundamental change in corporate life was becoming increasingly apparent. There was to be no return to the unwieldy corporate structures of earlier times: the imperative to maintain competitiveness would ensure that. As corporations continued to grapple with cost efficiencies, while also seeking to nurture the inherent value of their workforces, they began to explore more innovative ways to manage buildings and workers, together with the oil that enabled both – technology.

The Holy Grail of spaceless growth

The recession-driven new wave management zeal coalesced with a growing uptake of office technology to allow different approaches to workstyles: there was the iron fist of economics *forcing* real estate managers to do things differently, supported by the liberating impact of technology *allowing* things to be done differently. These two contrasting and sometimes opposing forces drove workplace management in a particular direction: higher density.

The pressure to reduce costs led to highly prescriptive design solutions. A relentless attack on occupancy standards ensued, involving workplace design aimed at intensifying the use of expensive space, known less formally as *max packing*. There are two principal means of achieving more intensive use of space. First, space allocations per desk (density) are reduced. For employees in an open plan, there is simply less space around their workstations; while for others there are fewer enclosed offices, and those which survive become little more than cells. Over the course of the 1990s, average occupancy density fell by one-third, in round terms, from c150 sq ft (13.9 sq m) per desk to c100 sq ft (9.3 sq m) per desk.

Of course, increased density has its limits, imposed by building regulations relating to fire escapes, sanitary provision, and so on. This is where the second step to intensification comes in: space is used more dynamically through the introduction of desk sharing, or 'hot desking'. It was well known that traditional office layouts are underutilised most of the time due to people being out of the office, and it was a short step to the introduction of flexible working styles and desk sharing as a means of improving their use of space.

In combination, higher densities and more dynamic use of space allowed buildings to support more people in the same amount of space, often reducing an organisation's appetite for space by around 20–30%. Indeed, it led to achieving the Holy Grail of *spaceless growth*, whereby headcount could grow without the traditional search for additional real estate.[46]

Before moving on, it should be recognised that spaceless growth was not an entirely benign phenomenon. 'Workplace transformation' became a commonly used phrase to describe space rationalisation projects in which the transformation was presented as a great leap forward for the worker. Such environments were often described as *liberating* workers, *freeing them from the shackles of the office*; *providing them with work-life balance* and *enabling them to make choices*. But, clearly, the same claim could not be true for all workers. Certain workplace transformation projects attracted a good deal of publicity for their "funky" layouts and "quirky" features. However, what such publicity singularly failed to do was recognise the fact that what might be "liberating" for one type of worker (personality, function, station) might be threatening to another with a different personality. While templates of worker types might be employed to define shades of grey in terms of what office facilities each is "entitled" to, the overall impact is a fundamental change in the relationship between worker, employer and, literally, their common ground – the workplace. Concerns were being raised about the narrow cost-driven focus nearly a decade ago.

> The argument for making more efficient use of expensive real estate is undeniable [it] should enable organisations to adapt to rapidly changing

conditions; to churn and restack, and to reduce the impact on the bottom line. The question is whether the emphasis of the workplace design agenda has perhaps leant too far in one direction. The fact that it has done so without any real, objective evidence makes the question both relevant and important.[47]

Rather like with the management gurus, there was a gradual dawning that cost-cutting had its limitations, and before long a more value-driven approach to real estate management began to emerge.

9.3 Phase II: getting value from real estate

As the 1990s cost reduction phase of REM moved towards its logical conclusion, there was a slow realisation that the workplace could add value rather than just leak cost. The digital office introduced the notion of agile working and underscored the fact that if, as a manager, you could positively influence the output, or productivity, of the workers, then you could add value to overall business performance. Crudely put, a 5% saving on the office cleaning bill paled into insignificance compared to a 5% increase in sales, or fee income or output. The search was on for a value-adding role in real estate.

Planning for uncertainty

As we saw in Chapter 8, the 1990s was the decade of the 'digital office'. PCs, laptops, mobile phones and the internet swept all before them. 'Tech stocks' consequently soared, higher and higher. Tech, we were told, would do little less than change the way we lived. It would revolutionise our patterns of work and consumerism; redefine the relationships between business and consumer and between business and supplier; and open up undreamt-of opportunities through access to unlimited information, enmeshing individuals, companies, communities and societies in a virtual, global network. This new economy would release us from the shackles of locational friction that the old one imposed on our businesses, and create a foot-loose economy where innovation replaced profit as the main goal of business. Wanting to forget the lean times of the early 1990s, and buoyed by a more recent stock market boom, City analysts jumped on to an already dangerously over-loaded bandwagon. The stock market valuations of technology, media and telecoms businesses have since become legendary. For example,

America Online, the largest internet company, has risen by 34,000 per cent since 1992 and is now worth more than 273 times expected earnings for the year to June. That makes it bigger than Ford or Disney . . . in Europe's top 10, ahead of Nestle, Shell or UBS.[48]

The stampede for tech stock was so great that there was a hidden crash in the stock value of 'old economy' companies. In the latter half of 1999 and the first quarter of 2000, when the dot.com hype was at its maximal phase of growth, £150 billion was wiped off the value of Britain's leading industrial companies. This compared to the £50 billion wiped off by Black Monday in October 1987.[49] And amidst all the hype there was, again, the promise that technology would reduce our dependency upon real estate as businesses and consumers conducted their transactional activity across the internet. Retailers and 'e-tailers' were seen to be locked in a battle for customers with only one possible outcome. Banks, insurance companies, travel firms, supermarkets – all manner of providers of goods and services – were widely predicted to become internet-based and, in the process, shed great swathes of unwanted retail outlets. And under such a scenario, who would need all those armies of 'paper pushers' at head office?

Yet by 2001 the rubble of the dot.com boom lay all around. Established companies joined the list of casualties, alongside new dot.coms such as Amazon.com; Boo.Com, Buy.Com, Etoys.com and Letsbuyit.com. AOL's all-share take-over of Time Warner marked the summit of the dot.com boom. Afterwards, its shares plummeted; advertising collapsed and huge debts were racked up; shares fell from a high of around $56 in mid-2001, to around $12 by late 2002. Similar fates befell Adelphia Communications, Cisco Systems, Global Crossing, NTL, Qwest Communications, Tyco International and Xerox Corporation. In an extraordinary chain of events, many companies lost 90% of their share value in the space of two years.

While the dot.com bubble was an extraordinary event, it demonstrated clearly the rapidly changing business context within which office space would have to be planned, procured and occupied in the future. And it demonstrated the heightened sense of uncertainty that now pervaded business planning compared to earlier times. In this sense, the way in which offices are planned and managed began to change dramatically. As we saw earlier, offices could no longer be regarded simply as inert, physical environments within which office processes took place. Rather, they became part of the business infrastructure of the organisations that occupied them, and were increasingly planned and managed as one facet of corporate resource planning.

Corporate resource management

One of the key lessons to emerge from the increasingly turbulent business environment was that the office, or real estate, should not be managed in isolation. The increasingly agile nature of work, the growing dominance of knowledge work and the need to invest in people all meant that real estate should be seen not as an expensive overhead to be cut, but as a corporate resource to be managed alongside other corporate resources areas. Figure 9.3 shows one of the earlier representations of such a model; in this case the model describes the role of 'business infrastructure management'.[50]

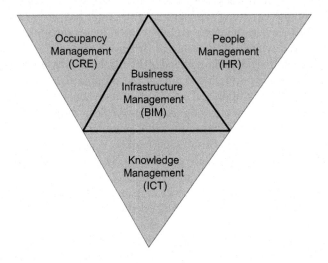

Figure 9.3 Business infrastructure management
Source: Reproduced from Harris (2006).

First, the traditional property or FM function is referred to as occupancy management to recognise its real role: to manage the occupation of space for operational effectiveness. Secondly, the BIM function acts as an intelligent client unit, gathering, interpreting and providing for the needs of the organisation.

Wes McGregor argued that to be effective, the work environment will need to result from the collaborative efforts of all aspects of business, "where IT and HR along with facilities management are combined to achieve an effective integrated support service for business operations".[51] McGregor went on to argue that this new agenda would call for much higher skill levels than what are traditionally evident in facilities managers who would

> need to become 'multi-lingual' in terms of their ability to interpret the needs of their customers (individual, group and corporate) and translate facilities related issues into a language that enables them to communicate effectively with managers of the business.[52]

Traditionally, these aspects of organisational planning have worked in silos with virtually no joined up approaches. But organisational planning is demanding more flexible approaches to the deployment of resources, and there is an increasing need for a more co-ordinated approach to resource planning.

For example, agile work strategies cannot be successfully implemented without there being a clearly integrated approach to all three components. If a business wishes to introduce agile working by providing a work hub, and allowing a high proportion of its workforce to be mobile or home-based, then real estate, HR and technology all have roles to play in supporting such a strategy. It is more efficient and effective to have one plan with these different aspects integrated into a seamless approach, rather than to have three plans with contradictory policies and procedures. In this way, REM is becoming an integral and integrating part of core business planning. At a time when management teams are recognising the direct link between business performance and the quality of the workplace, those responsible for delivering a "high performance" workplace are in a position to take on a front-of-house role.[53]

Economic and technological change are driving changes in business which, in turn, are driving changes in workplaces and workstyles. At each level there are major implications for general management. The pace of change is likely to quicken; and new digital technologies, mobile devices and ubiquitous connectivity are forcing organisations to ask what their workplaces are really for. Tomorrow's corporate workplace will integrate technology, place and people in a more seamless manner. As a recent report for the RICS noted:

> the delivery of a consistent, high-quality, cost-effective built environment and associated services as 'table stakes' in today's economy. Whether facilities are managed in-house or by a contracted service provider, they are a strategic business resource, and must be managed as such.[54]

As the role of the office workplace has evolved; so too has the workplace management function. The term 'real estate management' is increasingly anachronistic as management of the workplace becomes a corporate management function rather than a technical backroom function. Where once the 'property manager' focused on lease management, acquisition and capital programmes, today's workplace manager has a greater focus on managing people and their experience of the workplace.

Workplace management will increasingly blend the fragmented skills of design, construction, real estate and facilities into a coherent management function, allied to its colleagues in HR, procurement and technology, among others. In this way, workplace management will become an integral and integrating part of core business planning and an essential support to general management. It will underline a shift in emphasis in office thinking, "from castle to condominium".[55] The point is that the office should no longer be seen as an impregnable fortress that lasts millennia, and should instead be seen as a venue for guests to enjoy.

From managing desks to welcoming guests

During the 1990s, there was a growing appreciation that the workplace could contribute to corporate performance rather than act simply as a passive, expensive backdrop to business. It had to be managed to support the technology-enabled, agile, digital workplace. As described in Chapter 8, while in the past workers went to the office to access the paraphernalia for work – phone, personal computer, filing and stationery – increasingly they were doing so for different reasons. The office workplace was becoming less and less a static backdrop to process-based, largely routine and solitary work.

Instead, the role of the office was becoming one of enabling people to interact and collaborate; providing a palette of settings in which to work. In other words, the office was becoming less a place to go to work on a set of prescribed tasks, and more somewhere to visit and interact with colleagues. It was an increasingly actively curated environment, managed more like a hotel than a traditional office, with a high level of service and experience for 'guests'.[56]

The 'guests' demanded a high level of service, amenity and experience. This shift of emphasis implied a shift in priorities for the real estate manager from *building* performance to *workplace* performance, from managing desks to welcoming guests. Old-style, janitorial facilities management provision yielded to workplace management that provided concierge-type services for employees with a more diverse range of priorities:

- a workplace to support corporate agility and enable change;
- space as a hub and connector to reinforce corporate culture and values;
- a workplace that brings people together to forge a sense of shared objectives;
- design for continuous adaptability and diverse usage patterns;
- creating a memorable experience to attract talent and nurture skills;
- providing amenities and services (food, wellbeing, events, etc.), and
- activity-based spaces for collaboration, concentration, creativity and confidentiality.

Of course, not all office workplaces were managed in this way. As ever, there remained a broad spectrum of approaches to the use and management of office space. But the overall direction of travel in the early years of the twenty-first century was towards a REM model in which the focus had shifted from the physical parameters and performance of the space (which by now could be largely taken as a given) towards a model that focused on people and business performance.

This shift in focus was given added impetus by the rapid expansion of the flexible space market (Chapter 7). Not only did the flexible space operators demonstrate that it was possible to provide high-quality space with service to SMEs, as the market evolved, they increasingly did so with corporate occupiers. Once global brands rather than just SMEs were opting for flexible space, real estate managers, and their firms more generally, recognised that the design and management of real estate had evolved: it was now a strategic corporate resource; to be managed and deployed like other corporate resources – people, capital and equipment.

Notes

1 Marriott (1967) *Op cit*, p39
2 *Ibid*, p41

3 Duffy (1983) *Op cit*
4 *Ibid*, pp4–8
5 G. Davis, F. Becker, F. Duffy, & W. Sims (1985) *ORBIT 2,* Harbinger Group, Norwalk
6 K. Alexander (1992) Facilities Management in the New Organisation, *Facilities,* Vol 10, No 1, pp6–9, MCB University Press
7 J.H. Renzas (1992) Corporate Restructuring: Evaluating the Real Estate Component, *Area Development,* January, pp33, 50, 54, 57, S/H Publications Inc., Easton, Philadelphia
8 S. Zeckhauser & R. Silverman (1981) *Corporate Real Estate Asset Management in the United States,* Harvard College, Cambridge, MA
9 P. Veale (1987) *Managing Corporate Real Estate Assets: A Survey of US Real Estate Executives,* The Laboratory of Architecture and Planning, Massachusetts Institute of Technology, Cambridge, MA
10 Department of Health and Social Security (1982) *Underused and Surplus Property in the National Health Service,* HMSO, London
11 The Audit Commission (1988) *Local Authority Property – A Management Overview,* HMSO, London
12 National Audit Office (1989) *Home Office: Control and Management of the Metropolitan Police Estate,* HC 455 HMSO, London; National Audit Office (1988a) *Estate Management in the National Health Service,* HC 405 HMSO, London; National Audit Office (1988b) *Property Services Agency: Management of the Civil Estate,* HC 507 HMSO, London
13 M. Avis, V. Gibson, & J. Watts (1989) *Managing Operational Property Assets,* Department of Land Management, University of Reading, Reading
14 Avis *et al* (1989) *Op cit,* pII
15 F. Duffy & A. Henney (1989) *The Changing City,* Bulstrode Press, London, p72
16 *Ibid*, p73
17 Debenham Tewson and Chinnocks (1992) *The Role of Property: Managing Cost and Releasing Value,* DTC, London, pIV
18 Graham Bannock & Partners (1994) *Property in the Boardroom,* Hillier Parker, London
19 R. Bon (1995) Corporate Real Estate Management Practices in Europe and the United States: 1993 and 1994 Surveys, *Facilities,* Vol 13, No 7, pp5–13, MCB University Press
20 Arthur Andersen (1993) *Real Estate in the Corporation: The Bottom Line for Senior Management,* Arthur Andersen & Co, South Carolina
21 V. Gibson (1995) Is Property on the Strategic Agenda? *Property Review,* May, pp104–109, Eclipse Group, London, p107
22 B.J. Varcoe (1992) Premises of Value, *Facilities,* Vol 10, No 3, pp14–20, MCB University Press, p14
23 Alexander (1992) *Op cit,* pp6–9
24 P. McLennan (2003) Facility Management: A Profession or a Knowledge Occupation? In: S.J. North (Editor), *FM Professional Development,* Essential FM Reports 27, LexisNexis, London, pp2–3
25 *Ibid*
26 B. Nutt (1999) Four Competing Futures for Facility Management, *Facilities,* Vol 18, No 3–4, pp124–132
27 F. Becker, W. Sims, & B. Davis (1991) *Managing Space Efficiently,* Cornell University, International Facility Management Program, p103
28 M. Joroff, M. Louargand, S. Lambert, & F. Becker (1993) *Strategic Management of the Fifth Resource: Corporate Real Estate,* Industrial Development Research Foundation, USA, p10
29 *Ibid*, p21
30 Arthur Andersen (1993) *Op cit*
31 W. Leifer (1998) A Primer on the Evolving Role of Corporate Real Estate, *Corporate Real Estate Executive,* Vol 13, No 5, June, pp39, 43
32 Arthur Andersen (1993) *Op cit,* p8

33 P. McLennan, B. Nutt, & D. Kincaid (1999) *Futures in Property and Facility Management,* Conference Proceedings FM Exchange, University College, London

34 M. Apgar (2009) What Every Leader Should Know about Real Estate, *Harvard Business Review,* November, pp100–107

35 Alexander (1992) *Op cit,* p6

36 M. Apgar (1993) Uncovering Your Hidden Occupancy Costs, *Harvard Business Review,* May–June, pp124–136

37 *Ibid,* p124

38 R. Heller (1994) The Managers' Dilemma, *Management Today,* January, pp42–48

39 Cited in: The Horizontal Corporation, *Business Week,* 20 December 1993, pp44–49

40 T. Peters (1992) ABB's Prophet of Smallness Points Way to Profit in the Nineties, *The Times,* 11 November, p25

41 Cited in: A World Without Jobs? *The Economist,* 11th February 1995, pp23–26

42 P. Toynbee (1996) Is Work Good for You? *RSA Journal,* March, pp20–28

43 Cited in: Instant Coffee Management Theory, *The Economist,* 25 January 1997, p83

44 Cited in: Assembling the New Economy, *The Economist,* 13 September 1997, pp105–111

45 J. Carlin (1996) Guru of Downsizing Admits He Got it All Wrong, Reported in the *Independent on Sunday,* 12 May, p20

46 R. Harris (2009) *Space, but Not as We Know It,* Unpublished Paper, www.ramidus.co.uk

47 R. Harris (2012) *Reflections on the Modern Office,* Unpublished Paper, www.ramidus.co.uk

48 R. Taylor & J. Labate (1999) Age of the Day Trader, *Financial Times,* 9 February, p23

49 J. Waples (2000) The Hidden Crash, *Sunday Times,* 5 March, pB5

50 R. Harris (2006) Workplace and FM: A New Agenda Part One, *Essential FM Report 59,* July/August, pp2–3, Tottel Publishing

51 W. McGregor (2000) The Future of Workspace Management. In: B. Nutt & P. McLennan (Editors), *Facility Management: Risks and Opportunities,* Blackwell Science, London, p85

52 *Ibid*

53 R. Harris & H. Cooke (2014) Is Corporate Real Estate at a Crossroads? *Journal of Corporate Real Estate,* Vol 16, No 4, pp275–289

54 J. Ware, R. Harris, M. Bowan, & P. Carder (2017) *Raising the Bar: From Operational Excellence to Strategic Impact in FM,* Royal Institution of Chartered Surveyors, London

55 R. Harris (1996) Less a Castle, More a Condominium, *Facilities Management Journal,* Vol 4, 6 June, p13

56 Brunia *et al* (2016) *Op cit,* pp30–47

10 Divining

From castles to condominiums

In 1967 two American futurologists, Herman Kahn and Anthony Weiner of the Hudson Institute, published *The Year 2000: A Framework for Speculation on the Next Thirty Three Years*.[1] They applied mathematical modelling, simulation techniques and qualitative assessments to a broad spectrum of technological and social trends, in the hope of eliminating the subjectivity inherent in gathering the views of experts. The book presented scenarios in technological, social, economic and political dimensions, in an effort to provide a framework to anticipate future events and prepare policy responses. The work yielded a list of 100 innovations likely to occur in the final third of the twentieth century.

The passing of time has proven some of their forecasts overly optimistic. Personal flying platforms and interplanetary travel have failed to materialise, and we do not have programmed dreams, chemical methods of improving the memory, cities lit by artificial moons, deep-sea mining or much influence over the weather. However, the book was more successful in anticipating some of the major changes that would change the world of work. Among the work-related speculations were the following.

- General use of automation and cybernation in management and production.
- Extensive . . . centralisation of . . . information in high-speed data processors.
- Extensive use of robots and machines "slaved" to humans.
- Automated universal (real time) credit, audit and banking systems.
- Inexpensive high-capacity, worldwide . . . communication (perhaps using satellites. . .).
- Personal "pagers" (perhaps even two-way pocket phones) and other electronic equipment for communication computing, and data processing program.
- Inexpensive (less than one cent a page), rapid high quality black and white reproduction; followed by color . . . perhaps for home as well as office use.[2]

The scale of Kahn and Weiner's undertaking demonstrates that, even with a 'scientific' approach, the future is highly unpredictable. This should act as a health warning, several decades later, to the trend towards technological determinism in popular discourse: 'because something can happen, it will'. There is today much prescription that is in fact little more than speculation, based as it often is on confirmation bias from social media posts rather than on an evidence base. The book also raises an interesting dimension, which is that the impact of technological change, and the possibilities opened up by it, are often tempered by our social and cultural abilities or aptitude to adapt.

In this final chapter, we turn from our historical account of the office economy to consider the future. In doing so, it is imperative to distinguish short-term and current events from long-term change. For example, as this chapter was being written, the Covid-19 pandemic was being played out, along with a public discussion suggesting little short of an existential crisis for the office. This chapter speculates on the future of the office economy in the context of long-term change; not the immediacy of today's headlines.

10.1 Agglomeration and weightlessness

Technological determinism has led to many false prognoses, and this is no less the case than with the 'future of cities' narrative. The dissolution of the city has been discussed extensively since the dawn of information technology. A good example is the 'death of distance' discussion which accompanied the rise of the internet: if information is equally available everywhere, at all times, then why be in the city centre when we can be telecommuting from a bucolic idyll of our choosing? Frances Cairncross was forthright: "The death of distance will mean that any activity that relies on a screen or a telephone can be carried out anywhere in the world".[3] The fact that this has not happened on a widespread basis (apart from in some transactional environments such as contact centres) suggests that more subtle forces are at work.

In its widest sense, technology enables some things to happen that could not happen before; it opens up new opportunities; it even creates new industries and activities. But it rarely follows a linear route from cause to effect: plentiful forecasts of technology decimating jobs in the economy failed to take account of its ability to *transform* jobs and *create* new jobs. So, because technology allows for urban dissolution, it does not follow that society will follow that trajectory. One problem is that 'death of distance' operates everywhere, equally: and because cities are competitive, one city's gain will be another's loss.

One of the logical constructs of the 'death of distance' narrative is that it overcomes many of the issues inherent to urban economics that were discussed in Chapter 3 with, for example, CBDs commanding high rents and labour costs. Of course, this is attractive: why pay high rents downtown, when you can move to a regional town and pay half? But the high costs are borne by businesses, and have been for over a century, by and large because they are seen to be offset by the advantages and externalities of being agglomerated in the CBD. Further, the evidence on cost is weak: actual long-term data show that occupancy costs in London today are lower in real terms than they were before the oil crisis of 1973.

As we have seen in this book, the economy has evolved over a century from being primarily manufacturing-based to one that is services-based and, within the latter, information and knowledge are determining characteristics. While the advancement of technology was relatively modest for most of the twentieth century, towards the end it sped up and led to a transformational stage in the economy, which we are living through in the first decades of the twenty-first century. But the critical question here is whether such transformation will involve the weakening of city structures.

Urban form

Agglomeration is the invisible glue that holds the office economy together in dense clusters. Whether in accounting, advertising, banking, consulting, finance, legal and media; or in coding, developing, graphic designing, programming and software engineering; or in architecture, construction, design, engineering, planning and surveying, the office economy is bound together in enormously complex webs of relationships, involving partnering, collaborating, networking and socialising, and which drive innovation and higher output. At their most basic, they form vast job markets; on a more complex level, they form sustainable and competitive business ecosystems. These strengths are not replicable in a fully dispersed world.

It is something of a paradox that, since the dawning of the digital office, the pulling power of some city centres has strengthened rather than weakened. Despite the growing ability of individuals to work remotely and for organisations to move to the suburbs or set up a distributed model, centripetal forces have grown.

At the same time, an agglomeration is not fixed in time and space. There are countless historic examples of clustering activity that have emerged, evolved and extinguished, for all manner of reasons. London's transformation from a physical trading cluster, based on its docks, to a services trading cluster is a case in point. This book has shown how, over the past two centuries, London has evolved: it has adapted to new challenges; it has changed shape; activities have come and gone. Today's Shoreditch is yesterday's Hatton Garden. The office economy has been at the heart of this change, as we saw in Chapter 6, with the transformation of London's business geography since the 1980s.

Urbanist Michael Glaeser is positive about the future of cities: "a few decades of high technology can't trump millions of years of evolution", going on to suggest that connecting in cyberspace "will never be the same as sharing a meal or a smile or a kiss. Our species learns primarily from the aural, visual and olfactory clues given off by our fellow humans".[4] More seriously, Glaeser points to the paradox in which "the declining cost of connecting over long distances has only increased the returns to clustering close together", while humanity's ability to rise to new challenges "needs all the strength it can muster, and that strength resides in the connecting corridors of dense urban areas".

But this positivism comes with a caveat: cities will not continue to operate as they have in the past. And neither will London. In terms of the scope of this book, the office economy will change rapidly as buildings and the spaces between them come to be used differently. Just like the monolithic corporations of the 1970s, monolithic office blocks have had their day. The binary choice of CBD for work and suburbs for living must yield to a far richer, more nuanced urban experience of living and working. But to achieve this, cities will need to innovate and change: their buildings will need to appeal more to society, to people, and less to corporate interest.

Inevitably, some cities will be less successful: fortunes will vary and some will decline, just as we saw cities such as Amsterdam, Antwerp, Bruges, Florence and Milan lose their power and influence centuries ago, and more recently Detroit. London's population fell from 8.6 million in 1941 to 6.3 million in 1991. It was a shrinking city; it was failing. It is now pushing nine million people, having experienced nearly three decades of rejuvenation. It has re-confirmed its place in the top flight of global cities (Chapter 6). This growth and status must not be taken for granted: London competes on a global stage; and the office economy is an essential ingredient of its competitive advantage.

Economy

In the midst of the Industrial Revolution, Benjamin Disraeli observed that change is inevitable, and in true Victorian style he added, "In a progressive country change is constant".[5] He was speaking at a time when western society was undergoing immense change and on the threshold of major advances such as the introduction of electricity, the motorcar and the skyscraper, to mention but three world-changing innovations.

Yet his words might have been spoken today. Whether in retailing, logistics, corporate offices, industrial processes or life science labs, the nature of business is changing at a faster rate than ever before. One of the defining features of business today is the speed and ubiquity of change. Businesses must be capable of continuously adapting to changing market conditions, which means that they must be fleet of foot. This is achieved through flatter, leaner and more agile organisational structures and business processes. The need for adaptability and responsiveness is common to both large and small firms.

The role of the modern office economy was described in detail in the early chapters of this book. We saw how it originated in the coffee houses and counting houses of eighteenth century London; before evolving in the nineteenth century with rapidly expanding global trade in goods and growing wealth in the domestic population. Clerical factories grew rapidly. Organisations became larger and more complex to meet the demands of the rapidly evolving trading environment. We also saw how the office economy emerged as a distinct facet of the city – no longer an ancillary function to industry, but an activity in its own right. In Chapter 8, we met 'Company Man', the iconic image of the rise of post-war corporatism. This was the time when sprawling international organisations brought together thousands of sedentary, clerical workers into single headquarter buildings (Mumford's 'human filing cases' for the 'circumspect care of paper'), where command-and-control management oversaw unchanging and predictable 'corporate islands' which planned ahead with a comparatively high degree of certainty, providing jobs for life.

By the 1990s, there were more people going to work in offices than there were going to factories; but then a perfect storm of technology and recession suddenly saw global corporations being 're-engineered'. Layers of management, non-core activities and complex divisional structures fell with each swing of the management consultants' scythe. A lifetime career based on graduated promotion within a single corporation became as rare as hen's teeth. The digital economy had changed the office economy utterly, completely.

The concept of the corporate island was found wanting when workers could come and go more or less at their choosing because *they*, not the corporate body, held the intellectual capital. Knowledge workers have far more freedom than their clerical factory or corporate office forebears. The technological revolution that has seen the rise of the knowledge economy has acted as a spur to the growth of small businesses by stripping away many of the barriers to entry that setting up a new business once involved.

The journey thus described, as summarised in Figure 1.4, saw London's coffee houses and counting houses evolve into clerical factories, then corporate offices and then digital offices. There are now signs that the digital office is yielding to the fifth age, the network office. The emergence of the network office economy is characterised by connectivity, linkages, relationships, agility and knowledge.

Knowledge is intangible; it is difficult to 'value', it is aspatial. But it is also part of an even broader phenomenon, sometimes referred to as the 'weightless economy'. It is widely anticipated that most future economic growth will derive from 'things' that have no weight: computer software is perhaps the most obvious example; branding is another of increasing importance. And accounting for the economy, or quantifying it, becomes more problematic, as eloquently described by Danny Quah:

> International trade becomes not a matter of shipping wine and textiles from one country to the next, but of bouncing bits off satellites. With economic value having no clear points of physical entry and exit, international trade statistics become that much murkier and ambiguous. Keeping track of trade is no longer just counting the bottles and bales that pile up on the loading docks in a port.[6]

Haskell and Westlake described intangible investment in *Capitalism without Capital*, noting along the way that intangible investment overtook tangible investment at the time of the Global Financial Crisis in 2008.

> Investment used to be mostly physical or tangible, that is, in machinery, vehicles and buildings and, in the case of government, in infrastructure.

> Now, much investment is intangible, that is, in knowledge-related prod-
> ucts like software, R&D, design artistic originals, market research, training
> and new business processes.[7]

For large parts of the twentieth century, owner occupation and long leaseholds
of offices were common, as buildings provided: financial meaning in the form
of assets and security; operational meaning in the form of bespoke environments, and symbolic meaning in the form of architecture. As first the service
economy and then the knowledge economy spread, so the new work processes
and the work environments dismantled these meanings. With certain caveats,
today, a corporate lawyer can vacate a building and be replaced by a bank, a
tech company or an advertising agency.

The meanings of the space, in the terms described, have largely disappeared.
Many companies in the office economy have become 'weightless': they do not
benefit from controlling physical assets; their operational needs are generic, and
they attach little symbolic meaning to the exterior of their building. Yet, they
are looking out to a property supply process that weighs a great deal indeed.

The economy is demanding a new approach. The supply process of the
twentieth century does not match the demand profile of the twenty-first. The
supply process must – as it has in the distant past – re-connect with the customers that live and work in the environments that it creates rather than solely with
the short-term priorities of those that provide the capital. In the long-term, the
latter's interests will be better served.

Technology

Czech painter, writer and poet Joseph Čapek (1887–1945) was murdered at
Bergen-Belsen. He invented the term *robot*, using the original Czech word
robota, which specifically refers to repetitive or drudge work, to describe what
the new machines would be used for. Curiously, the term is often attributed to
Joseph's science fiction-writing brother Karel, who introduced the word into
literature in his play entitled *R.U.R.* (Rossum's Universal Robots) in 1920. In
the century since Čapek's invention, machines have been replacing workers in
countless tasks.

The fear that machines, or technology, would replace humans in the workplace is not a new one. We saw in Chapter 2 how the introduction of adding
machines in the nineteenth century led to disputes between bookkeepers and
their employers over concerns about their job security. More recently, automation and office technology have laid waste to swathes of jobs in the office economy: consider the back offices that no longer exist, the typing pools, the drafting
rooms, the publishing departments, the secretaries and accounts departments.

The number of robots is rising quickly: by 2019, there were 2.6 million
industrial robots in operation worldwide.[8] But while robots and automation
are replacing workers, what is less clear is the net balance between creation
and destruction. It is often speculated that the disruption has created more jobs

Figure 10.1 Three ages of the office economy: 22 Bishopsgate, Tower 42 (reflected) and Threadneedle Street

Source: © Erik Brown, 2020.

than have been lost and that it has contributed to much greater prosperity. For example, overall, technological change that replaces routine work is estimated to have created more than 23 million jobs across Europe from 1999 to 2016, or almost half of the total increase in employment over the same period. Recent evidence for European countries suggests that although technology may be replacing workers in some jobs, overall it raises the demand for labour.[9]

Technology has brought higher productivity to many sectors by reducing the demand for workers for routine tasks, and it has opened doors to new activities once imagined only in the world of science fiction. Yet, as we move towards the third decade of the twenty-first century, a new wave of fear and uncertainty is stirred by the threat of robotics (with its potential to decimate jobs) and artificial intelligence (with its ethical and moral implications). The World Economic Forum (WEF) argued in 2016 that "the Fourth Industrial Revolution, combined with other socio-economic and demographic changes, will transform labour markets in the next five years, leading to a net loss of over 5 million jobs in 15 major developed and emerging economies".[10] The WEF suggested that, in fields such as artificial intelligence and machine-learning, robotics, nanotechnology, 3-D printing, genetics and biotechnology, innovation "will cause widespread disruption not only to business models but also to labour markets . . . with enormous change predicted in the skill sets needed to thrive in the new landscape".

What is clear is that many children currently in primary school will have jobs as adults that do not exist today. As tasks and jobs are replaced by machines, it is no coincidence that the skills required in the emerging economy are those that are more difficult to replace with machines, such as critical thinking and socio-behavioural skills; or managing and recognising emotions that enhance teamwork. Workers with these skills are more adaptable in labour markets. Our buildings in the office economy must come to reflect this new reality and provide spaces appropriate to the new work.

The firm

A little-known Nobel economist, Ronald Coase, sought to explain why we have firms, rather than countless individuals selling goods and services to one another. His basic thesis was that there are costs attached to doing business which are more effectively and efficiently dealt with by a single decision-making structure. Coase described his task in *The Nature of the Firm* as being to "discover why a firm emerges at all in a specialised exchange economy", and he went on to suggest that the main reason "why it is profitable to establish a firm would seem to be that there is a cost of using the price mechanism":

> We may sum up . . . by saying that the operation of a market costs something and by forming an organisation and allowing some authority (an 'entrepreneur') to direct resources, certain marketing costs are saved.[11]

The knowledge economy questions Coase's underlying thesis because it challenges the notion of his 'marketing costs' or 'friction': when we are not dealing with goods and trade, but with knowledge, where is the friction? The critical nature of connectivity, changing structures, the priorities of knowledge workers and the reduced importance of the 'corporate island' in favour of a more complex web of supply chain relationships and contingent workers, are all altering the nature of the firm, with huge implications for the real estate industry.

The key question here is whether the dawn of the network economy will, in fact, successfully challenge the historically strong centripetal forces of the office economy. Whether this turns out to be the case or not, the point made earlier about the need to overcome the binary choice of CBD for work and suburbs for living remains salient.

As we saw earlier, the previous dominance of large corporate organisations has ceded in the knowledge economy to a more variegated landscape of large and small businesses. The 'pre-digital' London office economy was dominated by corporate behemoths: global enterprises that changed slowly and predictably. In the 1970s, London was exporting 'back office' jobs to the suburbs, not only because wages were cheaper (a differential that has since disappeared) but because the predictable, process-based work could happen anywhere. In today's digital economy, where innovation is lifeblood, collaboration critical and specialist input often required, a sterile, a mono-use business park 60 miles from the CBD is not necessarily the place to be.

The knowledge economy allows individuals to trade their intellectual capital using cheap and ubiquitous technology. While in the past, new businesses faced "difficult choices about when to invest in large and lumpy assets such as property and computer systems", today's technology enables them "to go global without being big themselves".[12]

> Today they can expand very fast by buying in services as and when they need them. They can incorporate online for a few hundred dollars, raise money from crowdsourcing . . . hire programmers from Upwork, rent computer processing power from Amazon, find manufacturers on Alibaba, arrange payments at Square, and immediately set about conquering the world.[13]

The fundamental role of technology in driving new business formation has been accompanied by a more human trait – the exercise of choice. The knowledge economy allows more people to choose how, when and for whom they work. "Most people in the UK who start up a business do so because they view it as an opportunity rather than a necessity".[14] Greater numbers of people are choosing to control their own destiny by exploiting technology.

Seen this way, we now have a mixed economy of large corporations and small firms – *undertaking pretty much the same work*. Unlike in the recent past, they have symbiotic relationships; their level playing field is knowledge. One is not disadvantaged by the scale and spending power of the other; while the other gains external benefits from specialist input and contingent labour. New cities are now being created where national governments are procuring the services of masterplanners, architects, engineers, planners, technologists, financiers and countless others from around the globe, based on their knowledge and experience, not scale or distance. And the same is true in all corners of the economy.

The dynamics of the small firm economy are illustrated by research on the City of London. Here, while around two-thirds of the total office stock is in units over 100,000 sq ft (9,290 sq m), more than half (52%) of all occupiers are in units of less than 5,000 sq ft (464 sq m), and 72% occupy less than 10,000 sq ft (929 sq m). By contrast, only 2% of occupied units are larger than 100,000 sq ft (9,290 sq m). The data also show that over time, the share of total stock occupied in small units has increased at the expense of large units.[15] More widely, the number of UK businesses employing over 250 actually *shrunk* by 6% between 2000 and 2014. Against this, the number of businesses employing 50–249 workers grew by 17% and those with 1–49 workers grew by 15%.[16]

These dynamics have important ramifications for a potential decentralised office economy, not least that the number of large occupiers is in fact quite limited, especially in the European rather than the North American context. Conversely, for small businesses whose operating models depend upon collaboration and complex supply chain relationships, there is a strong centralising force.

Smaller businesses are a particular feature of the emerging network office economy. The real estate supply process will need to adapt to this reality with products and services more akin to consumer markets than corporate markets. Small firms are driving demand for flexible space; many are no longer satisfied with secondary space on secondary streets. They aspire to the quality of space that their corporate cousins occupy. And the flexible space market is providing that – along with intensively managed, commodity space that customers can turn on and off as their business cycles demand.

There is an implied assumption in many economic texts – and indeed, in real estate discourse – that small firms are either start-ups or companies working towards growth. This is a misreading of the office economy in the twenty-first century, and especially in the emerging network economy. Many small companies are simply that – small companies. They do not aspire to become larger corporates; they comprise highly skilled people, with significant intellectual capital, who are doing what they might have been doing in a corporate environment, but choose instead to 'do their own thing'. This is likely to become even more common, and the supply industry will have to become more sensitive to this growing source of demand.

10.2 The supply process

The Polish Renaissance mathematician Nicolaus Copernicus pointed out the momentous fact that the sun, rather than our own inconsequential piece of astral real estate, sat at the centre of the universe. A similar 'Copernican Revolution' is required in real estate, whereby the supply industry comes to terms with the fact that in the network economy, we must place the occupier, or customer, at the centre of the supply process.

Structure

The structure of the real estate industry is a product of the nineteenth century. It reflected the social structures of the time as much as functional structures. And ever since, new skills and activities have accreted, coral reef-like in every direction, resulting in a highly fragmented, complex, inefficient and slow industry.

Chapters 5 and 6 amply demonstrated that the real estate industry is a supply-led process. But it was not always so: the first speculative office building in London appeared only in the 1830s. Even during the late nineteenth and early twentieth centuries, speculative development was modest. It was not until the 1950s that the current model of ownership, funding and development was widely practiced. Even with the rise of the flexible space market in the digital economy, the underlying structure remained in place – just.

For much of the post-war period, commercial buildings were built primarily to meet the needs of the organisations that owned them, rather than those that occupied them. Pension funds and insurance companies have been the clients of the developers who deliver the product, not the businesses that lease and occupy their buildings. Many of the ugliest and most dysfunctional contributions to the urban landscape over the past 50 years, built to crushingly tight budgets, are the direct outcome of this separation of interests between the supply process and occupation.

The immediate post-war supply industry responded to the critically difficult financial and economic imperatives of the time. The scale of re-building required across the country's built environment and economy was immense. The private sector had to make a pivotal contribution. And it did. The achievements have to be recognised; and some of the consequences given sympathetic historic treatment.

However, the fact is that the supply industry failed 'to change with the times'. The industry structure, cultural, financial and legal, ossified. In the office sector, it became a 'mono-product industry' with the institutional specification for buildings and an adversarial contractual relationship between supplier and occupier, wrapped in feudal language.

The dawn of the digital office in the 1980s was a wake-up call. The structure was found wanting with the arrival of the digital office economy, when the customers' priorities shifted but the industry continued to supply the same product. The arrival (and success) of the flexible space operators was a second wake-up call. In the context of a weightless economy and all the drivers of change described earlier, the operators were suddenly offering everything the customers wanted: speed, flexibility and certainty, all in a service wrap with a focus on people rather than bricks and mortar. But the sclerotic supply process and the deep-rooted professional jealousies therein continue to exert themselves.

In an echo of Marion Bowley's comments nearly three decades before (Chapter 7), in the aftermath of the early 1990s property market collapse, management consultants Ernst & Young argued that the industry was "poorly managed because it is both flawed and fragmented".[17] They went on to suggest it was fragmented because "occupiers, developers and investors have different and non-complementary agendas, objectives, market leverage, and decision-making standards".[18] The consultants argued that most developers tended to focus on projects as investment opportunities rather than their attractiveness to occupiers. Little has changed since the consultants drew their conclusions.

A first step in the direction of reform would be the complete abandonment of the feudal language of 'landlords and tenants', together with the structurally adversarial nature of 'landlord *versus* tenant' negotiations. Neither is appropriate, and in contemporary terms, they carry the wrong connotations: occupier businesses today are not tied to landlords in the manner of tenant farmers. Owners/providers should be signing modern business contracts with their customers, not full repairing and insuring leases with alienation clauses. Consider the commercial aircraft industry, with its funders, manufacturers and airlines; this is a sector with similar capital intensity, but where partnering replaces adversarial relationships.

Recent years have seen the rise of the flexible space market (see 'Commodity space' later in this chapter). But the 'industry' must go much further to recognise the fundamental changes affecting not only the office, but of the wider built environment. Today's economy requires innovative approaches to real estate supply. The old model which relied on capital growth will have to yield to one where income and service define value rather than yield.

Actors

At the core of the real estate industry is a triumvirate of interests: the developer who spots opportunities and manages risk, the investor who buys the end product and the broker who arranges the deals. In this sense the industry is deal-driven, riding the ups and downs of changing economic fortunes. Around these key players there floats a whole host of professional advisors – architects, surveyors, engineers, cost consultants, planners and so on. The advisors range across an enormous tapestry of activities. They advise on offices, shops, warehouses and industrial premises (with very few individuals working in more than one sector); they deal with schools, homes, hospitals, leisure premises, restaurants and universities; they deal with farmland, oil rigs and quarries; and they acquire, cost, design, dispose, fund, litigate, manage, masterplan, measure, plan, research, survey and value. The breadth of activities is staggering. And therein lies the problem.

Weightless firms in the network economy have no more interest in this labyrinth of specialisms than the average car owner cares about what happens under the bonnet. They simply want to conduct their business, in an experiential workplace when they need to do so.

The valuation community will have to come to terms with the fact that the capital value of a building is based on the value that it adds to occupiers (and those beyond), not on the perceived covenant strength of the occupiers and their onerous leases. This requires a fundamental shift in thinking, from a capital growth model to an income based model. And it requires a cultural shift in the key skill sets required to make the transformation. Changes such as these will require enormous cultural changes within the traditional supply process.

While owners will continue to offer 'vanilla space' on long, less flexible leases on particularly large-scale buildings where the occupiers also benefit from security of tenure, a growing proportion of the occupational market will move towards a service-based model, where flexibility and intensive management are the hallmarks of success. The role of 'Property Management' will evolve, or be replaced by, one more akin to curation than contract management. It will involve the recruitment of non-property skills, mainly from the hospitality sector, with people who are given the same career structures as those with property skills. Office hoteliers will replace traditional owners, offering commoditised space and services to guests.

The aspect of service is one that presents particular challenges for the traditional owning community. Put simply, traditional owners are accustomed to managing 'assets', not customers who demand intensive and personalised management function. To move from one to the other requires a change in culture and the adoption of new skills, which have more in common with the hospitality industry than with the conventional property industry.

Just as hotels offer swimming pools, gymnasia, health care, personal services, shops and restaurants, so could 'office hotels', with the additional benefits including serviced bedrooms and apartments. All with a technology and service wrap. Figure 10.2 shows different levels of service provision that could be provided for different customers.[19]

The cast of actors will then be more akin to the hotel industry. Here, owners, brands, operators and managers now combine in different permutations, using operating models to suit locational and cultural factors as well as each actor's respective goals and business model. For example, rather like in the office sector, owners will often lease a building to a brand who will in turn subcontract the operational aspects on a franchise basis or in a management agreement.

Figure 10.2 Levels of service in the office economy

Source: Image by kind permission of Investment Property Forum.

Commodity space

This change of emphasis will also be driven by different patterns of occupation. Traditional, monolithic, multi-let buildings will increasingly become 'office hotels', or multi-use buildings, containing a range of different activities, crucially with ground-level public access (Figure 10.3). Street frontages will become more permeable, and there will be greater provision of serviced space for occupiers looking for licences rather than leases, and there might even be serviced apartments for workers to occupy when they stay in town.[20]

With its short-term licences replacing long-term leases, all-in unitary costs replacing upward-only rent reviews and opaque service charging, and easy-in, easy-out terms replacing dilapidations and alienation clauses, the flexible space market is in the process of commoditising real estate. Perhaps the next stage of evolution will see more 24/7 use of buildings, with overlapping uses, a richer blend of private and public space and more interaction between occupiers.

The vast majority of the market – relatively smaller occupiers – will move towards flexible space that will, at some point, be as easy to hire as an Airbnb stay: an office will be ordered, created and delivered via a credit card, in an 'Easy Office' version of 'Easy Hotel'. We will then cease to refer to flexible space as a 'sub-market', because it will be the norm. The same processes will transfer themselves to other property sectors, including shops, laboratories and light industrial premises.

The flexible space market has challenged traditional property provision and found it wanting. While the challenge has been made before, the most recent wave of change has created the conditions for it to succeed this time. There will be no return to the *status quo ante*. Accepting that there will always be demand for long-term space from large corporates looking for core estate security, the

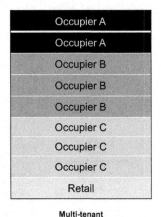
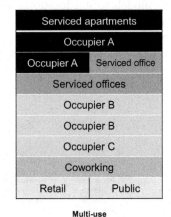

Figure 10.3 Traditional multi-tenant and future multi-use building

Source: Adapted from Ramidus Consulting (2018).[21]

rest of the market could evolve to flexible space in one form or another. This suggestion is in line with Simonetti and Braseth's prediction that, by 2030, most companies' real estate requirements will be outsourced and consumed on demand, while only 20% will be taken via traditional long-term obligations.[22]

The 'old world' real estate language of 'asset management' must change; real estate is not an asset (in the financial sense of the term) for an occupier (unless an owner occupier). In the network office economy real estate is fast becoming a resource, or commodity, to be drawn upon when needed, to support corporate and operational objectives. Like other corporate resources, including capital, people, technology and information, it is now being managed as a flexible, efficient and productive resource. Owners and managers will need to adapt to this new reality.

10.3 From market to network

This book has traced the evolution of the modern office over five distinct 'ages'. The precise start and end of each age is less important than the overall picture (Figure 10.4). The office has moved from market, to factory, to corporate, to digital. The thesis of this book is that it is about to move to network. The table characterises the work and place themes of each age. Curiously, the key feature of both the first and final phases is knowledge.

As stated earlier, this chapter seeks to distinguish short-term events from long-term trends. But as the book was being finalised, there was a singular event that has the potential to influence long-term trends. It could even have a determining role in the emergence of the network office.

2020 existential crisis

Suddenly, in March 2020, the office economy closed, in a manner that no one had expected or planned. Most disaster recovery plans were based on a single incident premise – a terrorist event, a weather event or a building event; and involved moving staff home and to temporary locations. But in March 2020, a world health crisis, in the form of Covid-19, suddenly led to the closure of all offices, with no sense of how and when they might re-open. Footfall in London's business district collapsed by 90% and more. The streets were empty, cafes were empty, trains were empty.

Yet, despite everything, the office economy did not grind to a halt. In a feat of historic ingenuity, companies and organisations found ways of managing 'distributed work'. The ubiquity of home-based and mobile technology enabled a continuity of work that would have been impossible as recently as a decade ago. Indeed, such has the scale of the crisis been, and so successful has our means of coping been, that it has led to the inevitable: "if we can manage as we are, then why go back to the old ways?"

The mainstream media normally treats the situation in this binary, 'old way-new way' manner: contrasting and discussing working from the home (WFH)

Focus	Approx. era	Work and place themes	Technology	Brands
Market	1700–1830	Knowledge. Coffee Houses. Counting Houses. Exchange. News. Social.	Adding machines. Copying. Ink pen. Ledgers. Paper. Stenography.	Electric Telegraph Company. Gestetner.
Clerical Factory	1830–1920	Production. Fixed. Manual. Mechanical. Production line. Repetitive.	Carbon paper. Electricity. Filing. Lifts. Paper clip. Telegraph. Telephone. Typewriter.	Burroughs. Edison. Marconi. Remington. Waterman.
Corporate	1920s–1970s	Process. Scale. Departments. Layers. Command and control. Predictable. Processes. System furniture.	Action Office. Air conditioning. Biro. Calculator. Fax. Fluorescent lamp. Golf Ball typewriter. Mainframe. Photocopier. System furniture.	3M. BIC. General Electric. Herman Miller. IBM. Xerox.
Digital	1980s–2020	Power. Agile. Changing. Connectivity. Data. Flexible. Power. Processing. Speed. PC left the building.	Cloud. Email. Internet. Mobile. Laptop. PC. PDF. Social media.	Amazon. Apple. Blackberry. Facebook. Google. Microsoft. Motorola. Twitter.
Network	2020–onwards	Knowledge. Commodity. Distributed. Exchange. Hubs. On demand. Service. Social. Weightless.	Artificial intelligence. Augmented reality. Automation. Robotics. Smart cities/ buildings. Virtual reality.	Watch this space!

Figure 10.4 The five ages of the modern office economy

with working in the office (WIO). But of course, the real situation is somewhat more complex, and the long-term response will be similarly nuanced. At this stage – late 2020 – the long-term impact is not known; but it would simply be foolish to assume all outcomes will be benign, spread as evenly as jam.

First, some areas of caution. At a societal level, WFH must not be allowed to reinforce socio-economic inequality. Cities provide large and liquid job markets, maximising opportunities for the greatest number of people. There is a danger that WFH becomes the reserve of one quite narrow demographic slice and a privilege unavailable to many whose domestic circumstances make such a lifestyle problematic. Mental health is now a major social priority, and for those living on their own, loneliness can be a factor; while anxiety and stress have increased for many. For others, being confined to the same property and partners/family each day is not a preference. These issues cannot be ignored when discussing the merits of WFH.

At a corporate level, mechanisms will need to be put in place to ensure that the social capital of organisations is not eroded. A long-term separation of people and workplace (whether on a part-time or full-time basis) might lead to ossified structures or even reduced creativity. Management will need to adapt and evolve with new tools and processes to ensure that productivity and corporate standards are being attained across disparate locations and circumstances. Employers will also need to agree with workers how to deal with legal responsibilities (insurance, duty of care, health and safety) and perhaps costs. And this is before we ask questions about sustainability, such as the benefits of lighting and heating, say, 1,000 homes versus 1,000 desks; or hundreds of people driving to their local community office rather than boarding a single train.

At the worker level, the WFH/WIO balance must support individuals in their early career, for whom training, mentoring, informal learning, socialising, networking and experiencing are vital career building blocks. There are also ethical issues at this level: whether, for example, individuals choose or are told to work from home; what level of monitoring takes place; what 'rules' are put in place, and where the employer's duty of care begins and ends. The latter covers health and safety, ergonomics and so on. Many younger people rent property and share with others, and there is often not the space or furniture to work comfortably and safely for lengthy hours each day.

Bearing these cautionary comments in mind, it has indeed been proven beyond doubt that, if it has to, office work can be undertaken on a distributed model. Management can no longer insist that requests for flexible working must be rejected out of hand. People are being productive, and various platforms have allowed meetings to continue. For those who commute an hour or more a day, the time and cost saving is significant. There have been many reports of improved productivity and effectiveness; work-life balance can be enhanced. There are, no doubt, significant advantages for many.

Whether this results in swathes of vacant office space is another issue. The past two decades have involved occupiers accepting ever more dense occupation (from c150 sq ft [13.9 sq m] per person and a desk each) towards 80 sq ft per desk, with the latter shared at a ratio of 8:10. These trends are likely to be reversed with a lower-density environment, fewer formal workstations and a richer palette of work settings.

The pandemic has magnified and accentuated a trend that has been growing for two decades: agile working. For the early adopters, the main barriers were technological: connectivity, compatibility and ubiquity (or lack of it). More recently, the barriers have been cultural, as managers have sought to preserve line-of-sight management techniques. Both these main barriers have now been breached, and the office economy will move quickly towards more pervasive agile working. But this raises a fundamental question: what will be the role of the office?

What role for the office?

The answer to this question is critical to how the office will evolve as we move inexorably towards 2050. In the 'clerical factory ', armies of obedient white-collar workers turned up at their designated desk each day, fulfilled the tasks allotted to them and travelled home in the evening; unquestioning, undemanding. This was the *employer's office*. In the 'digital office', more discerning, agile workers selected their workstyle to more closely match their personal circumstances; they had more choice over their employer and their place of work. They chose their work pattern, and they demanded service. This was the *employee's office*.

The tensions between employer and employee (corporate versus individual) have always been present (as we saw in Chapter 2, with Charles Lamb and the East India office in the early nineteenth century); but in the digital office they were exposed in a manner like never before. The fact is that an organisation and an individual will have different perspectives on the workplace at any point in time. The physical attributes of the workplace, the organisational systems and processes, the workplace culture and the demands of the work itself will all be subject to arbitration.

A central part of workplace design and management is to mediate this tension between the different perspectives. One of the key problems, however, is the underlying sense that a perfect workplace is waiting to be discovered, and that if we can only correct this or that issue, then we will move 'forward', make progress. This is, of course, a chimera. No matter how well planned and responsive the work environment is, if individuals are not comfortable, or aligned, with the organisation, their effectiveness will suffer. Conversely, individuals who are highly aligned to an organisation, and deeply motivated by their work, might put up with all manner of workplace discomforts and shortcomings while at the same time being highly effective.

A new furniture system or a new layout can be corporately imposed – and normally are – without recourse to the organisational ecology. At which point there is push-back and organisational friction. There was a period in the early 2000s when it was considered 'hip' to have pool tables and bean bags for staff to 'enjoy' themselves (forgetting, of course, that half the workforce are introverts who do not wish to display themselves) along with the faintly threatening (or, perhaps, absurd) notion that you must be happy in the workplace.

Seeking a balance between corporate and individual perspectives is a never-ending task for management – never-ending because the business, the work, the people are constantly changing and adapting. For example, as technology has evolved, the opportunities for remote working have expanded, assisting many in their desire to live a life away from the city while enjoying the benefits of its employment opportunities. Long-distance commuting has increased as a result; and it was only a matter of time before the cost, time and stress involved in commuting came to be questioned, given that most work could be completed without a daily visit to the office. The widespread nature of this phenomenon (brought into absolute clarity by the pandemic of 2020) has led many to question the role of the office.

Yet all too often, the office is referred to in simplistically narrow terms, as one-half of a binary discussion about the office and home. There are dangers with this approach. Not least, we need to recall Studs Terkel's plea that work should be *a search for daily meaning as well as daily bread*. If we reduce the role of the workplace to a transaction between employer and employee, then it simply joins the other items on the corporate potential savings list. An example of this is wage cuts for those not commuting based on cost of living. Alternatively, a knowledge workers' desire for a four-day week presents for the company an opportunity to cut 20% of the workforce. Similarly, if office work becomes simply a transactional relationship in which the worker gets paid for ticking task boxes, then those boxes might be ticked as easily in Laos as they can be in London, at much less cost.

So, what is the role of the office in the twenty-first century – the network office?

From an employer perspective, perhaps first and foremost, the office is the physical embodiment of the legal document that is the firm. 'Brass plate' firms somehow are less interesting than firms that have physical presence. They are, after all, just brass plates. But a firm that has physical presence, no matter what size, carries with it a history; shared experiences, achievements, highs and lows, networks of friends and colleagues, relationships. It is the setting, and the expression of the culture and ideas of those who work toward a common purpose.

Secondly, it provides an environment in which to 'direct operations', to provide a hub from which everyone employed by a company can be managed, engaged, trained, mentored, directed (not, to be clear, five days a week). The office allows the employer to engage everyone with the business plan and develop a sense of common purpose (crudely, enhancing profit and pay). Around this can be built a 'company culture', a sense of competitive superiority, common systems and processes, regulatory compliance and so on.

Thirdly, it provides an environment in which the enterprise can develop, expand, innovate, create new markets. Moving to an entirely distributed model risks inadvertently cutting off one avenue of innovation: serendipitous meetings and conversations, whether around water coolers, in break-out rooms or at colleagues' desks.

From an employee perspective, the workplace provides a locus in which to share purpose, direction, objectives, knowledge, experience, war stories and personal troubles. Much of this sharing satisfies our social attributes as humans. We socialise and in so doing become culturally richer. We share in a common purpose. Even when we sometimes feel distant from the organisation due to personality clashes or career disappointments – the office remains the Maypole.

In recent years there has been much focus on the 'multi-generational' workplace. And it is important to recognise that young, mid-career and twilight career workers have different motivations, drawing upon the workplace for different purposes. For some, the office provides structure in the form of career development, time management and social networking. For others, it might just provide corporate or social support in an otherwise entirely mobile workstyle.

The draw to the office remains strong for many employees not least because of a push from the domestic environment. For many workers, the quality of public realm, access to shops, leisure and services, and office support in the form of nutrition and wellbeing are often far superior in the office to that which is available in the home environment or locale. And many workers, particularly the young and those in shared accommodation, lack the facilities to work in a safe and healthy manner. In this way, the office acts as an alternative, or a balance. Often the journey between office and home is seen as a punctuation mark in the day; a shift from one role to another.

Experiential workplace

It is now a given that work is not something that happens solely at an assigned desk in a dedicated, single-purpose work environment called an office building. Work in the office economy now takes place across a range of work settings, ranging from the home, to different settings within the office, to cafes, public transport, client premises and, indeed, anywhere that the knowledge worker deems appropriate to do what needs to be done.

From both the employer and employee perspective, the office workplace today is no longer a passive backdrop to work activities; it is part of a broader work experience. And whether the workplace is provided by the employer or an operator, the management of the space assumes critical importance. The office is an increasingly actively curated environment, managed more like a hotel than a traditional office, with a rich palette of work and social settings, and a high level of service and experience for 'guests'. There is a clear recognition that offices provide a place to bring people together.[23] Meeting face-to-face, and tacit knowledge via co-location of colleagues, are still key aspects for businesses, to encourage innovation and mentoring. The workplace is also recognised as a social hub for colleagues, particularly new joiners who need to become connected to the organisation.

The emerging workplace will provide an interesting blend of business and domestic design attributes; a pleasant, welcoming atmosphere in which to collaborate, innovate, socialise and learn. A richer palette of work settings and services, which might be tailored to individual requirements and available 'on demand', will be provided in a highly connected environment, with a far more sophisticated, or smarter, management regime. Buildings will be greener and healthier, and they will have the ability to create experiences rather than simply provide static backdrops. Priorities are shifting from performance to experience.

Smart building technology will allow many aspects of the building's hard and soft infrastructure – lighting, heating, ventilation, meeting facilities, purchase of services – to respond to individual preferences. A further physical impact as the office becomes a network of on-demand places will be a high degree of customisation. Like a stage-set, it will be reconfigured to respond to shifting demands. Already today, smartphones have the capability to tell us the quality of air and light in the spaces we occupy and give us the option to adjust that through the building sensors that exist in the system.

The office workplace will be shaped more by the people that choose to occupy it on a particular day for a specific task, interaction or activity than has been the case in the past. Rather than the 'one-size-fits-all' uniformity of the clerical factory era, a less rigid, multi-function, multi-setting workplace is emerging.[24] Digital technology is a key enabler to this trend, but it is happening in response to the workforce trends described earlier, and the resulting demand to create more stimulating, experiential environments that provide choice, support and variety through the working day.

The network office economy

Figure 10.5 proposes a 'model' for what the office network economy might look like in organisational terms. Diagram (a) sets the scene with the 'Traditional, Concentrated' model. This is where individuals, typically, travel from home to a downtown office five days a week. The CBD forms the single focus for everyone, and there is little variation in workstyles. Diagram (b) then moves to the 'Hub and Spoke' model in which the employer establishes regional offices and encourages staff to travel more locally to one of those each day. Digital communications and virtual meetings are maintained directly with the 'mothership'.

Diagram (c) proposes the 'Split Mode: WFH/WIO' model, in which workers divide their time, by mutual agreement, between home and the head office, typically two days at home and three in the office. Diagram (d) offers 'Split Mode: Community Office/HQ'. Here, workers elect to use

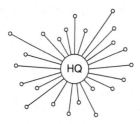

A. Traditonal, Concentrated - HQ only

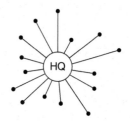

B. Split Mode - WFH & HQ

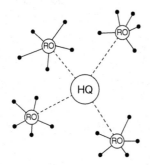

C. Hub & Spoke - Regional Office + WFH

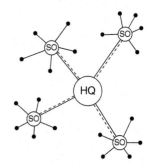

D. Hub & Serviced - HQ + Serviced Office + WFH

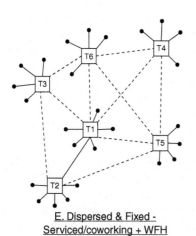

E. Dispersed & Fixed - Serviced/coworking + WFH

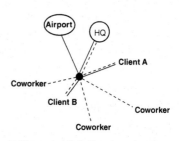

F. Dispersed & Agile - Anywhere

Tx	Town	○	Home (non working)
WFH	Working from Home	●	Home (working)
RO	Regional Office		
SO	Serviced Office	╱	Virtual Comms
HQ	Central Business District	╱	Travel-to-work

Modelling The Network Office Economy

Figure 10.5 The network office

community offices (serviced/coworking facilities) in their residential location on a part-time basis, travelling to the office on an 'as needed' basis, connecting virtually on other days.

Diagram (e) proposes the 'Dispersed and Fixed' model in which a corporate organisation dispenses altogether with its CBD office and establishes a network of dispersed offices in preferred locations. In this model there is minimal physical connectivity between locations, and a near total reliance on digital communications.

Finally, diagram (f) proposes the 'Dispersed and Agile' model. This also does not have the CBD as the focus for activities. Instead, it is centred on the individual, showing how workers might be based from home, and from there they sometimes travel to the CBD or to a clients' premises, or the airport, but on other occasions connect virtually with co-workers.

These six models are not mutually exclusive; they merely describe alternative modus operandi, and it is perfectly expected that different models will work *within* organisations as well as between organisations. The chosen blend of models, for the organisation and the individual, will be the one that best matches a compromise of their interests.

10.4 The office economy in 2050: towards un-real estate

> The greatest danger in times of turbulence is not the turbulence – it is to act with yesterday's logic.
>
> **—Peter Drucker**

The modern office economy has evolved over nearly four centuries, but the degree of change over the past four decades has been transformative compared to the gradual change of the earlier, much longer period. Some of the underlying tenets of the office economy have been severely challenged, and many of yesterday's truths, assumptions and norms are becoming tomorrow's archaic practices.

Cities will continue to exert strong pull powers on the office economy in terms of, for example, agglomeration economics. But the role of offices and office buildings in cities is evolving, and cities will need to provide a richer working and living experience. Office space is fast becoming a weightless commodity for firms whose value lies in their intellectual power rather than in their assets. Relationships between large firms and small firms are evolving, and the latter will grow in number more quickly than the former.

Large, monolithic, single-use buildings are losing their attraction in an era when depth and breadth of experience replace hours worked as the measure of success. The profile of demand is far more complex than the traditional large occupier base of relatively stable corporate bodies. Buildings will increasingly take on the profile of hotels in their operation and management.

The workplace experience and the manner in which it is curated will assume greater importance. Obviously, design will continue to provide innovative solutions to emerging problems; but the real focus in the coming years will be a management issue. How workplaces are managed will assume critical importance, not only in an era of heightened health awareness, but also in an era when space is a commodity that customers turn on and off as their businesses require.

As we move from the digital economy to the network economy, the supply industry will have to change, and change radically. The asset-driven approach will become a historic novelty. Real estate has to become an agent of change, not a barrier. And this means that old attitudes and perceptions must change. Where once a building was seen as a castle, it must now be seen as a condominium. It exists to meet a customer need, and everything around its design, delivery and management must be focused on that need.

Welcome to 2050, the network office and the era of un-real estate.

Notes

1 H. Kahn & A.J. Weiner (1967) *The Year 2000: A Framework for Speculation on the Next Thirty Three Years,* Macmillan, New York.
2 Kahn and Weiner (1967) *Op cit,* pp51-55
3 Cairncross (1995) *Op cit,* p39
4 M. Glaeser (2011) *Triumph of the City,* Macmillan, New York, p248
5 Benjamin Disraeli Speech in Edinburgh, 29th October 1867, cited in: *The Times,* 30th October
6 D. Quah (2019) The Invisible Hand and the Weightless Economy. In: R. Fouquet (Editor), *Handbook on Green Growth,* Edward Elgar, Cheltenham, p468
7 J. Haskel & S. Westlake (2018) *Capitalism without Capital: The Rise of the Intangible Economy,* Princeton University Press, Princeton, NJ, p239
8 International Federation of Robotics, Frankfurt, https://ifr.org/. Cited in: World Bank Group (2019) *The Changing Nature of Work,* The World Bank, Washington, DC
9 G. Terry, A. Salomons, & U. Xierahn (2016) *Racing with or against the Machine? Evidence from Europe,* ZEW Discussion Paper 16–053, Centre for European Economic Research, Mannheim
10 World Economic Forum (2016) *The Future of Jobs,* WEF, Geneva
11 R. Coase (1937) The Nature of the Firm, *Economica,* Vol 4, No 16, pp386–405
12 The Economist (2015) Reinventing the Company, *The Economist,* 24th October
13 *Ibid*
14 Lord Young (2013) *Growing Your Business: A Report on Growing Micro Businesses,* HMG, London
15 R. Harris (2016) New Organisations and New Workplaces, *Journal of Corporate Real Estate,* Vol 18, No 1, pp4–16
16 Office for National Statistics (2014) *Business Population Estimates for United Kingdom and Regions 2014,* Department for Business Innovation & Skills ONS, Sheffield
17 Ernst & Young (1993) *The Property Cycle: The Management Issue,* Ernst & Young, London, p5
18 *Ibid*
19 Ramidus Consulting (2020) *Op cit,* p21
20 *Ibid,* p24

21 Harris (2016) *Op cit*, pp4–16
22 Simonetti & Braseth (2017) *Op cit*
23 R. Appel-Meulenbroek, P. Groenen, & I. Janssen (2011) An End-user's Perspective on Activity-based Office Concepts, *Journal of Corporate Real Estate,* Vol 13, No 2, pp122–135
24 S. Brunia & A. Hartjes-Gosselink (2009) Personalisation in Non-territorial Offices: A Study of a Human, Need *Journal of Corporate Real Estate,* Vol 11, No 3, pp169–181

Index

Note: Page numbers in *italics* indicate a figure on the corresponding page.

Printed in the United States
By Bookmasters